Advances in Environmental Sustainability, Energy and Earth Science

Pankaj Pathak • Sadia Ilyas •
Rajiv Ranjan Srivastava • Javid Dar •
Subashree Kothandaraman
Editors

Advances in Environmental Sustainability, Energy and Earth Science

Proceedings of the 1st International Conference AESEE-2024

Springer

Editors
Pankaj Pathak
Department of Environmental Science and Engineering
SRM University AP
Amaravati, Andhra Pradesh, India

Sadia Ilyas
Department of Civil, Environmental and Natural Resources Engineering
Luleå University of Technology
Luleå, Sweden

Rajiv Ranjan Srivastava
Center for Advanced Chemistry, Institute of Research and Development
Duy Tan University
Da Nang, Vietnam

Javid Dar
Department of Environmental Science and Engineering
SRM University AP
Amaravati, Andhra Pradesh, India

Subashree Kothandaraman
Department of Environmental Science and Engineering
SRM University AP
Amaravati, Andhra Pradesh, India

ISBN 978-3-031-73819-7 ISBN 978-3-031-73820-3 (eBook)
https://doi.org/10.1007/978-3-031-73820-3

This Springer imprint is published by the registered company Springer Nature Switzerland AG
The registered company address is: Gewerbestrasse 11, 6330 Cham, Switzerland

Contents

Part I
Circular Economy and Resource Recycling

Automated Segregation of Plastic Wastes: Pioneering a Greener Frontier in Waste Management

Pankaj Kumar, Hephzi Jones, Pankaj Pathak, and Priyanka Singh

Abstract Segregation of waste is a key component of the waste management system. To attain high-quality recycled plastics, reduce pollution, and for a cleaner environment, sorting waste is vital. Manual sorting of waste can be inefficient due to human error and difficulty in separating different plastic types. It also causes several health issues for the workers. In medium- to high-volume factories, automated technology plays a critical role in the sorting of plastic materials. Accordingly, in this chapter, we have focused on different types of automated sorting techniques like direct and indirect sorting of plastics from municipal solid waste. The use of robots in sorting materials, such as plastic, paper, metals, glass, etc., from municipal solid waste can reduce manufacturing costs and energy consumption while facilitating the creation of secondary raw materials. This chapter is set forth by critically evaluating the different types of segregation technologies and their efficiency. By combining artificial intelligence (AI), deep learning, and robotics, we're creating a more automated and eco-friendly way to manage waste, ensuring a healthier future for generations to come.

Keywords Waste management · Waste Plastics · Automated Segregation · Artificial Intelligence · Environmental pollution

---

P. Kumar · H. Jones
Resource Management Lab, Department of Environmental Science and Engineering, SRM University, Amaravati, Andhra Pradesh, India

P. Pathak (✉)
Resource Management Lab, Department of Environmental Science and Engineering, SRM University, Amaravati, Andhra Pradesh, India

Waste to Wealth Center, SRM University, Amaravati, Andhra Pradesh, India

P. Singh
Department of Computer Science and Engineering, SRM University, Amaravati, Andhra Pradesh, India

P. Pathak et al. (eds.), *Advances in Environmental Sustainability, Energy and Earth Science*, https://doi.org/10.1007/978-3-031-73820-3_1

1 Introduction

The persistent increase in solid waste is giving rise to significant global concerns over sustainable waste management, leading to the implementation of various solid waste management (SWM) technologies [1]. Segregation, recycling, and solid waste disposal are major issues faced across the globe [2]. It is reported that globally around 2.01 billion tonnes of municipal solid waste are generated annually. Of which, not even 33% of waste is maintained properly [3], as can be seen from Fig. 1. About 62 million tonnes of waste is generated annually in India. Metropolitan urban communities produce 10 million tonnes of waste annually. With a population of 1.35 billion, India has a per capita waste generation ranging from 0.12 to 5.1 kg per person and a daily average of 0.45 kg/capita [4].

Solid wastes consist of recyclable, inert, biodegradable, and nonbiodegradable materials, as shown in Fig. 2. In today's world, handling and managing nonbiodegradable inert wastes are becoming more difficult [5].

Plastic is found everywhere, and the infrastructure of modern-day society needs this material significantly. This is a versatile material of choice for various applications across diverse fields, including food packaging, 3D printing, and cell phones,

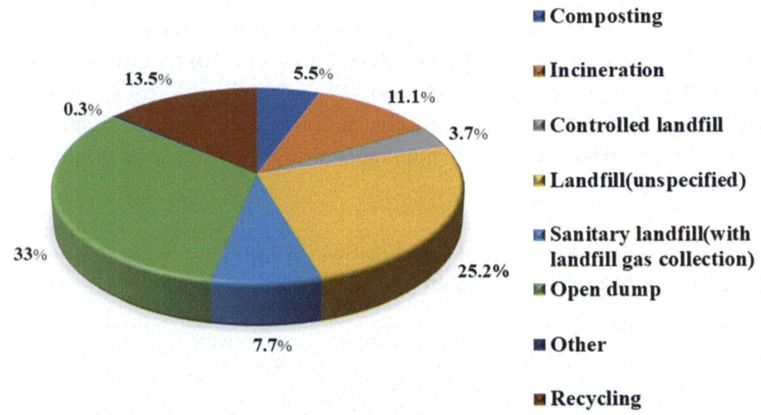

Fig. 1 Different technologies to treat the municipal solid waste

Fig. 2 Composition of municipal solid waste

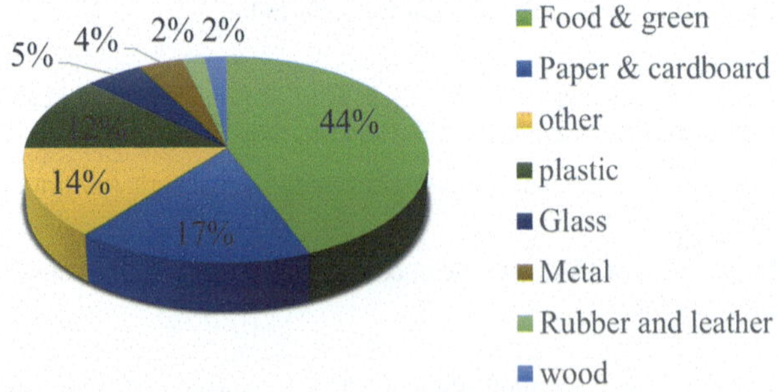

and formability is made possible by its low cost, bio-inertness, and adaptability in design. Plastic manufacture is closely associated with petrochemistry and fossil feedstocks, which is equivalent to 8–9% of the world's oil and gas usage [6]. Globally, 400.3 Mt. of plastic is produced [7]. Plastic has many beneficial applications, but there are drawbacks as well. Nowadays, plastic pollution may be found throughout the world, including in the air, water, soil, drinking water, and oceans. Most of these urban areas have landfills that are overflowing and cannot accommodate any more waste.

"Waste management hierarchy" has been accepted by the majority of nations as a step towards developing municipal solid waste (MSW) management techniques. Waste management frameworks based on the Internet of Things (IoT) reduce the risk associated with working with waste workers. A sustainable society requires recycling. After they are used, consumers discard their plastic bottles. Reusing and recycling also contribute to a decrease in the quantity of waste dumped in landfills [8]. Among the numerous environmental and health hazards associated with improper waste disposal include congenital disabilities, increased cancer incidence, groundwater contamination, land degradation, and child mortality. Waste management procedures were less sophisticated in the past; a few workers would gather waste from the streets and place it in specific locations. Waste was dumped at these specified locations after the vehicles were filled. Yet, as artificial intelligence (AI) becomes more prevalent, the waste management sector is changing dramatically to become more profitable and sustainable [9]. AI is an emerging technology that is becoming more and more common in many different sectors, especially waste management. Urban waste treatment plant design and operation can be revolutionized by integrating robotics and AI, which will result in more sustainable waste management techniques and improved operational efficiency for SWM [10].

However, AI-based technologies are already being adopted by several developed nations, such as Austria, Germany, New Zealand, the USA, the UK, Japan, Singapore, Switzerland, South Korea, and Canada, to optimize resource utilization, efficiency, and recycling opportunities throughout the SWM cycle. AI technologies are becoming more and more important in waste management, especially for the sorting and treatment of solid waste [11]. Therefore, AI is essential for creating sustainable waste management models that take into account social, economic, and environmental concerns, especially when moving towards a "zero waste circular economy" [12].

Researchers have worked on a variety of frameworks to address waste management concerns, such as an automatic waste segregator that is enabled by the IoT and is a crucial component of smart bins that are used to segregate and store waste. Waste collection bins with Wi-Fi connectivity are used, along with an IoT-based Automatic Waste Segregator for effective recycling—GSM-enabled automatic waste segregator and monitoring system [13]. Susha (2021) has designed a wheeled robot vehicle with a robot arm that uses metal, ultrasonic, and moisture sensors in addition to an infrared (IR) sensor to detect waste. A significant limitation of this study is that, despite the use of metal sensors, waste is ultimately only sorted into the categories of dry and wet waste [14]. Researchers have also worked on applications such as the Household Waste Management System Using IoT and Machine Learning; an AWS that uses an equal resonant impedance detecting component to identify metallic objects and capacitive sensors to identify wet and dry waste; an automatic waste segregator that uses a mechanical arm; an automatic waste segregator that uses an Arduino Mega 2560 load up microcontroller-based stage Arduino Mega MCU V3 module, which is interfaced with ESP8266 12E LoL in Node MCU V3 module, inductive vicinity sensor, laser module, light dependent resistor (LDR), liquid sensor module, and ultrasonic sensor; an AWS that uses Arduino UNO; a system that allows for the continuous remote checking of strong burn through receptacles using ZigBee-PRO and GPRS for supporting robust waste management procedures, robust waste monitoring, and the board framework that uses radio frequency identification (RFID) in conjunction with smart frameworks [15].

Two distinct approaches are being used in automated waste sorting and segregation research: sorting by sensor technology and AI via machine learning. Combining several sorting technologies is possible to increase sorting efficiency or product yield. Innovative technologies for sorting, like digital watermark-assisted sorting and tracer-based sorting, are being created and showcased. Several studies on waste management have been conducted, including ones that aim to summarize developments in automated garbage sorting and sort and classify plastic waste streams into multiple classifications [16].

2 Different Technologies for Automated Segregation

Automated segregation of waste represents a transformative approach to modern waste management, leveraging advanced technologies to streamline sorting processes and enhance recycling efficiency. Through the integration of robotics, sensors, AI, and other cutting-edge tools, waste facilities can achieve higher levels of accuracy and throughput in separating recyclable materials from mixed waste streams. These systems are capable of identifying and sorting various materials, including plastics, paper, metals, and organic waste, with remarkable speed and precision as illustrated in Fig. 4. By reducing reliance on manual labor and minimizing human error, automated segregation not only increases the overall efficiency of recycling operations but also improves the quality of recovered materials [15]. Automated segregation of waste relies on various technologies, each offering unique benefits and applications. Here are some key technologies used in automated waste segregation.

2.1 Detection of Waste

To identify waste, a waste detection system must be developed. This can be accomplished by utilizing various sensor types or object detection algorithms, as shown in Table 1.

Table 1 Commercially available sorters for segregation of plastics [16–18]

Types of plastics identified	Name of the equipment	Method of sorting	Accuracy	Black plastic sorted/ colors sorted	Application
PET, PE, PVC, PLA, PET-G, HDPE, PS, and PP	MSS CIRRUS plastic max	High-resolution NIR/color sensor	Up to 98%	Yes/–	All resins are sorted.
Color sorting, flakes, and pellets	L-VIS	Color camera sorting	98%	Yes/ yes	Automatically sorts shredded, granulated materials and flakes, optical for small particle applications.
All resins	MSS Aladdin	NIR and vision spectroscopy	92–98%	Yes/ yes	HDPE and PET.
PP, ABS, PVC, POM, PMMA, PE, and PS	Unisort P4000	NIR and vision spectroscopy	99%	–/Yes	Sorts PS, PP, PE, PVC, ABS, POM, and others.
HDPE, PET, PVC, EPS, PP, mixed LDPE, sorting film (HDPE/LDPE), ABS	Eco pack-model EP optical plastic sorting machine	NIR, VIS, deep learning	–	–/Yes	PET/PE recycling, plastic film
Removes PVC and BFR-containing plastics	Red wave QXR	XRF	80%	No/no	Used for PET and WEEE stream purification.
Sorting of flakes by color	Red wave C	Vision spectroscopy with CCD camera	Up to 99%	No/yes	Separation of plastics by color.
LDPE, films, PVC, PET, LLDPE, and PS	Film max	NIR, color, and metal sensors	Up to 98%	–/–	Nonmetals

2.1.1 Sensor-based Detection

Based on the wavelength, emission intensity, absorption, or scattering resulting from molecular energy transition, spectroscopic methods identify the chemical compositions, elemental contents, and molecular structures of waste materials. IoT has been a major factor in the development of automated waste segregation systems that use a variety of sensors, actuators, etc. Multiple sensors are employed to detect waste, and these sensors aid in the classification of waste according to distinct material qualities, as illustrated in Fig. 3. Metallic sensors are used to detect and separate metal from waste, facilitating the recycling of metal batteries, cans, and other items. Metal detectors can be made of electromagnetic and inductive sensors.

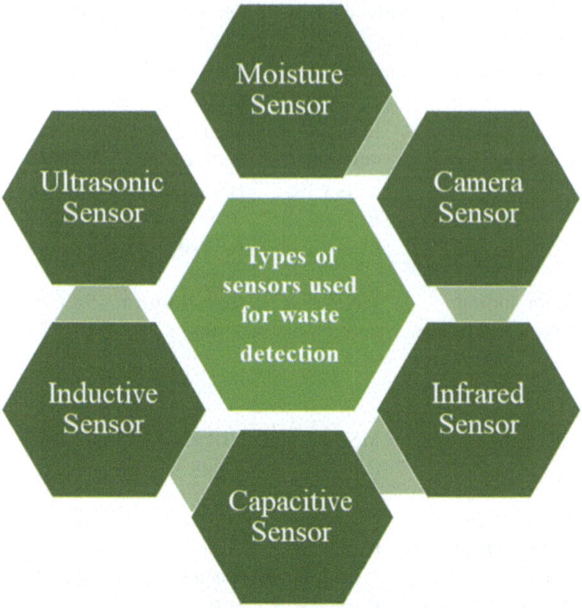

Fig. 3 Types of sensors used for the detection of waste

Capacitive proximity sensors are utilized for noncontact detection. They can identify things that are metallic and nonmetallic, such as liquid, paper, plastic, etc. Ultrasonic sensors are object detectors that assist in identifying whether or not an object is detected, such as a level indicator or an obstacle. The object detection method is a more accurate and efficient way to separate waste into different categories, even though the sensors used help with waste segregation based on factors like material, color, odor, moisture, etc. [19].

Eddy Current Separators

Eddy current separator employs a strong magnetic field to extract nonferrous metals from waste after all ferrous metals have been extracted. Ferrous metals that heat up inside the eddy current field cannot be sorted using eddy current separators. This may cause the belt of the eddy current separator unit to become damaged. A thin layer of mixed garbage is transported by a conveyor belt that has an eddy current separator attached to it. There is an eddy current rotor at the end of the conveyor belt. Metals just fall off the belt because of gravity, whereas nonferrous metals are hurled forward off the belt into a container [15].

Inductive Sensors

Nonferrous metal sorting techniques and magnetic separation may not always be able to recover metallic components, such as composite materials and stainless steel. The solution to this issue is the induction sorting system. The material's electrical conductivity serves as the primary measurement criterion. Under a belt, tiny sensors that are only slightly wider than a thumb's width release electromagnetic waves. These waves are altered by electrical conductors moving through the field, and a computer recognizes the signal difference for every particle. This enables precise particle sorting with conductors as thick as 1 mm [20].

Near Infrared Spectroscopy

The wavelength of near infrared spectroscopy (NIR) is between 750 and 2500 nm. NIR spectroscopy is frequently used to identify metal ions with attached molecular groups. However, it is rarely used to identify C, N, O, and S, as well as chemical bonds like C=O, due to the weak signal and large spectrum peaks that constantly overlap each other. NIR has therefore always been appropriate for organic materials and is frequently used in the categorization of waste plastic [21].

NIR plastics classification procedures were implemented in the beginning of 2010. Researchers used PCA to help classify HDPE/LDPE and PLS-DA to classify PLA/PET. Since then, the classification has been expanded to include all major polymer types (PVC, PP, PE, PET, and PS) with suitable chemometric instruments, as shown in Fig. 4, and it has achieved extremely high accuracy except PS/ABS and LDPE/HDPE. Wastes that are made of plastic and those that are not separated using NIR; on the other hand, surface contamination has minimal effect on the percentage of reflectance, while the sample's thickness and the labels' coating on transparent plastic have an impact. Furthermore, NIR is inappropriate for dark materials since black plastic absorbs light [22].

Raman Spectroscopy

The information on molecule vibration and rotation is obtained using the scattering spectra, which are produced by Raman spectra and have a different frequency than the input light (laser). Raman spectroscopy has proven to be a useful tool in recent years due to its ability to produce strong, distinct peaks in the spectrum and its efficacy in quantitative analysis. Researchers examined Raman's effectiveness in categorizing PP, PS, ABS, PVC, and PET by combining PCA and SVM. But Raman finds PS/ABS confusing. Raman may be less reliable than NIR due to its sensitivity to contaminants and background fluorescence. Additionally, as NIR has trouble identifying dark-colored plastic, researchers looked into Raman's ability to do so [22]. With no data processing involved, researchers classified PP, PS, and ABS using Raman with a 95% accuracy rate. With an accuracy of 96.20%, Bae et al. classified PP, PS, and PET in 2019 using Raman spectra and a PCA-neural network [23].

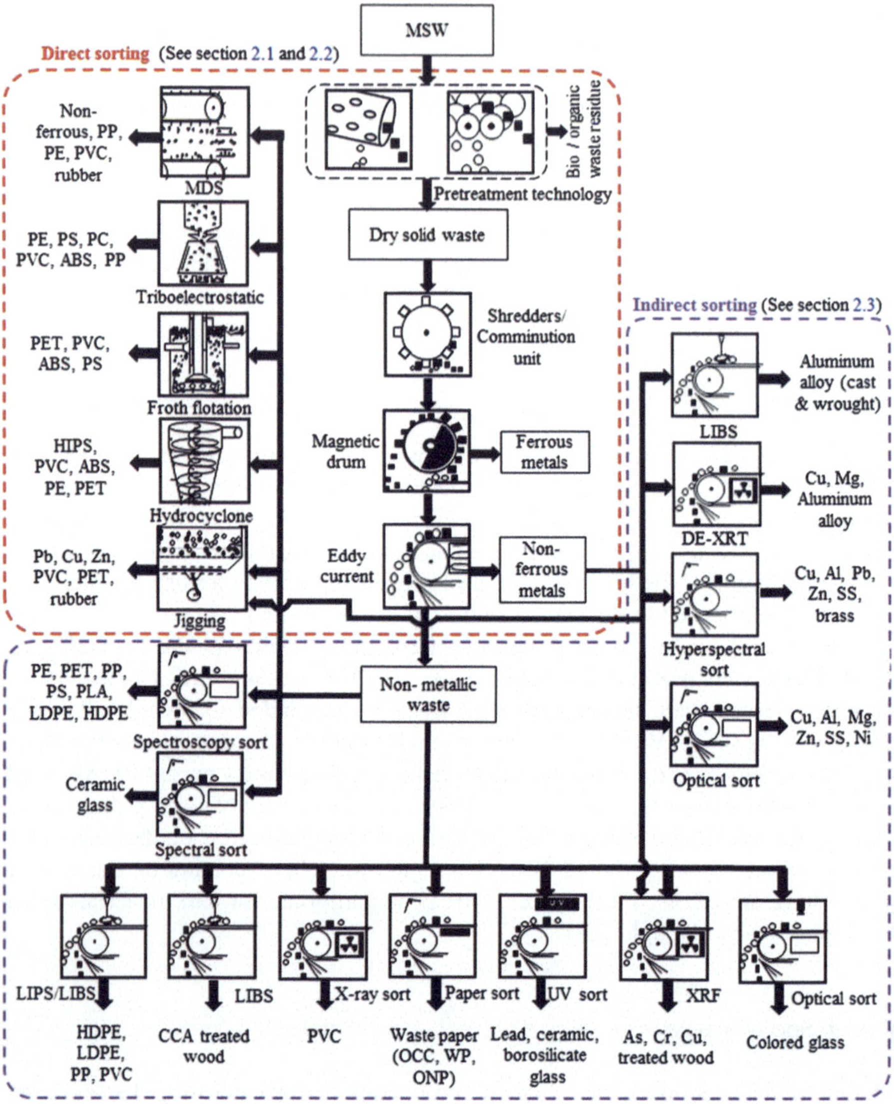

Fig. 4 Automated sorting of recyclables from municipal solid waste (adopted with permission from Sathish et al. 2017)

Laser-Induced Breakdown Spectroscopy

An ultrashort pulse laser used in laser-induced breakdown spectroscopy (LIBS) creates plasma on the matter surface, and a sample's composition can be determined via the plasma emission spectrum. Almost every component of an ultrashort pulse laser can produce a unique, characteristic spectral line due to its exceptionally high energy density. As a result, everything can theoretically be categorized, but chemical structures cannot be identified. In recent years, LIBS has been successfully applied to the classification of polymers. ANN combined with LIBS produced an accuracy that surpassed 97.8% for LDPE/HDPE, which is challenging to differentiate using NIR and Raman. Furthermore, LIBS and PLS-DA achieved mAP at a surprising 99.55% after a variable significance adjustment and a detection speed that exceeded 10 fps. Applications for LIBS are wider than those for NIR and Raman [22].

X-Ray Fluorescence Spectroscopy

Plastics encountered in the recycling of durable items are analyzed using X-ray fluorescence spectroscopy (XRF). XRF is capable of identifying elements present in plastics, such as Cl, Br, Sb, and Ti.

Positive aspects of XRF include its comparatively simple operation, quick sample preparation and analysis, low need for recalibration, a wide spectrum of elements that may be detected, and wide concentration range for analysis. For plastics recycling settings in the real world, XRF is therefore appropriate. Several XRF instrument uses in plastics recycling are demonstrated by examples from both qualitative and quantitative investigations [24].

2.1.2 Object Detection Based on Computer Vision

Waste sorting with computer vision (CV) has been conceptualized and studied for almost two decades. With applications in waste management, recycling, and environmental monitoring, waste detection has grown in importance as a field of study. AI and deep learning (DL) algorithms have been applied to create automated waste detection systems that can recognize various trash kinds and categorize them for additional processing, as shown in Fig. 5. Machine learning (ML) is a branch of AI that uses statistical models and algorithms to help computer systems get better over time at a particular task. A subclass of ML, DL uses multiple-layer artificial neural networks to learn representations of data. DL is especially well-suited for challenging tasks like speech and image recognition [19].

Using AI applications is one way to speed up the implementation of sustainable SWM. Object detection is a method that combines deep learning, machine learning, and image processing to classify things in images. The development of a general-purpose method for high-accuracy and high-efficiency detection uses generic object detection. Convolutional neural networks (CNNs) are a popular form of deep learning algorithm used for object detection. Popular CNN-based models for object detection include R-CNN, Faster RCNN, SSD, Fast RCNN, YOLO models, and FCOS are represented in Table 2. The newest, fastest, and most accurate model for object recognition is called YOLOv7. The application of deep learning algorithms with optical sensors, such as RGB and depth cameras, to the detection and classification of objects [28].

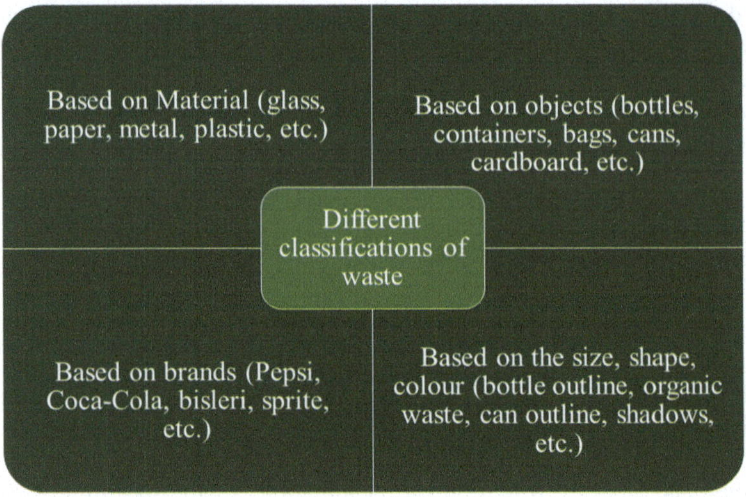

Fig. 5 Different classifications of waste

In addition to providing a thorough explanation of the many object detection frameworks, including R-CNN, Fast R-CNN, Faster R-CNN, YOLO, and SSD, the researchers also examine the progression of object detection from conventional computer vision methods to deep learning-based approaches and conclude that deep learning-based techniques have outperformed conventional computer vision methods, but that these models' robustness and generalizability still need to be improved [19].

3 Suitability of Automated Segregation of Plastic Waste

Automated segregation of waste is an emerging technology that holds great promise in improving waste management practices and promoting sustainability. This process involves the use of advanced sorting systems, often employing a combination of sensors, robotics, and AI, to segregate different types of waste materials automatically [29]. The suitability of automated waste segregation can be evaluated based on several factors. Automated systems can process large volumes of waste quickly and accurately, reducing the need for manual sorting, which is labor-intensive, time-consuming, and prone to human error. Advanced sensors and machine learning algorithms can accurately identify and categorize different materials, ensuring proper segregation [30]. Manual sorting of waste can expose workers to potential hazards, such as sharp objects, toxic substances, or infectious materials. Automated systems minimize human exposure to these risks, improving overall safety in waste management operations [31]. As populations grow and waste generation increases, automated segregation systems can be scaled up to handle larger quantities of waste efficiently. This scalability is essential for meeting the growing demand for effective waste management solutions [32].

Table 2 Detection of waste materials using various sensors and algorithms [19, 25–27]

Sorted materials	Type of algorithm	Accuracy	Limitations
Metal, glass, plastic bottles, PET plastic	ResNet-50 (CNN mode of 50 layers)	IoT-based sorting machine: 98.53% Detection model classification accuracy: 89.77%	You should consider the set of images to be providing both an empty bottle together with a bottle cap to make a powerful classifier.
Glass, metal, plastic, paper	VGG-16, Alex net, KNN, SVM, RF	VGG-16: 93.0% AlexNet: 91.0% SVM: 80.0% RF: 85.0% KNN: 88.0%	The less accurate and most challenging material to categorize is glass.
Opaque plastic containers, cardboard boxes, drink cans, opaque and clear plastic bottles	Mask-RCNN	Max validation accuracy: 89.6% Max test accuracy: 55.6%	Due to shadows and reflections of objects classified inaccurately.
Cardboard, plastic, glass, paper, metal, and trash	Faster R-CNN	91%	Inconsistent prediction failed to identify things with multiple objects.
Glass, fabric, metal, plastic, and paper	YOLOv4	92.23%	The majority of waste collection utilized are single photos insufficient to meet needs in real life.
Glass, paper, cardboard, plastic, metal, trash (indigestible and digestible waste)	CNN	95.3125%	Glasses come in a variety of thicknesses and colors, difficult to categorize.
Plastic, metal, and paper	CNN, SVM	CNN: 83% SVM: 94.8%	The limitation was small number of photos in the training set.

While the initial investment in automated segregation systems can be substantial, the long-term operational costs and potential revenue from recovered materials can offset these expenses. Automated systems can also reduce labor costs associated with manual sorting [33]. Automated segregation facilitates the separation of recyclable materials, such as paper, plastics, metals, and glass, from nonrecyclable waste. This separation enables more effective recycling and resource recovery processes, contributing to a circular economy and reducing the environmental impact of waste [5]. However, the suitability of automated waste segregation also depends on certain factors, such as the availability of advanced technology, infrastructure, and skilled personnel to operate and maintain these systems. Additionally, the effectiveness of these systems may vary depending on the complexity and composition of the waste stream, as well as the level of public participation in proper waste sorting at the source. Overall, automated segregation of waste presents a promising solution for efficient and sustainable waste management, particularly in urban areas and regions with high waste generation rates. Continuous research, development, and implementation of these technologies, along with public awareness and participation, can contribute to a cleaner and more sustainable future.

4 Conclusions

In the face of escalating plastic pollution and the urgent need for sustainable waste management solutions, automated segregation emerges as a beacon of hope, offering a greener and more effective approach to tackling the complex challenges posed by plastic waste. Through the integration of advanced technologies such as robotics, sensors, and AI, automated segregation streamlines sorting processes, enhances recycling efficiency, and minimizes contamination in recycling streams. This transformative approach not only optimizes resource utilization and reduces reliance on virgin materials but also contributes to the conservation of natural resources, mitigation of environmental pollution, and mitigation of greenhouse gas emissions associated with plastic production and disposal. Moreover, automated segregation fosters economic resilience within the waste management and recycling sectors, catalyzing innovation, and investment in sustainable technologies and practices. As governments, businesses, and communities worldwide increasingly prioritize sustainability and circularity in their development agendas, automated segregation represents a pivotal tool in the transition toward a more resilient, equitable, and environmentally responsible future. By harnessing the power of automation to revolutionize plastic waste management practices, we can forge a path towards a greener and more sustainable world, where waste is minimized, resources are conserved, and the health of our planet and communities is safeguarded for generations to come. Automated segregation of plastic waste stands as not just a solution to waste management problems but as a catalyst for positive change, driving progress towards a circular economy founded on principles of sustainability, innovation, and environmental stewardship.

Acknowledgments This work acknowledges the support from SRM University—AP through the SRM University—AP Research Grant.

References

1. Pujara Y, Pathak P, Sharma A, Govani J (2019) Review on Indian Municipal Solid Waste Management practices for reduction of environmental impacts to achieve sustainable development goals. J Environ Manag 248. https://doi.org/10.1016/j.jenvman.2019.07.009
2. Rafeeq M, Alam S. Automation of plastic, metal and glass waste materials segregation using arduino in scrap industry

3. Maalouf A, Mavropoulos A (2023) Re-assessing global municipal solid waste generation. Waste Manag Res 41(4):936–947. https://doi.org/10.1177/0734242X221074116
4. Mapari R, Narkhede S, Navale A, Babrah J (2020) Automatic waste segregator and monitoring system. Int J Adv Comp Res 10(49):172–181. https://doi.org/10.19101/ijacr.2020.1048053
5. Mohammed MA et al (2023) Automated waste-sorting and recycling classification using artificial neural network and features fusion: a digital-enabled circular economy vision for smart cities. Multimed Tools Appl 82(25):39617–39632. https://doi.org/10.1007/s11042-021-11537-0
6. Nielsen TD, Hasselbalch J, Holmberg K, Stripple J (2020) Politics and the plastic crisis: a review throughout the plastic life cycle. Wiley Interdisciplinary Rev: Energy Environ 9(1). https://doi.org/10.1002/wene.360
7. Geyer R, Jambeck JR, Law KL (2017) Production, use, and fate of all plastics ever made. Available: https://www.science.org
8. Raju SK (2020) Automatic plastic waste segregation and sorting using deep learning model. Int J Sci Technol Res 9:2. Available: www.ijstr.org
9. Brancoli P, Bolton K, Eriksson M (2020) Environmental impacts of waste management and valorisation pathways for surplus bread in Sweden. Waste Manag 117:136–145. https://doi.org/10.1016/j.wasman.2020.07.043
10. Goutam Mukherjee A et al (2021) A review on modern and smart technologies for efficient waste disposal and management. J Environ Manag 297:113347. https://doi.org/10.1016/j.jenvman.2021.113347
11. Abunama T, Othman F, Ansari M, El-Shafie A (2019) Leachate generation rate modeling using artificial intelligence algorithms aided by input optimization method for an MSW landfill. Environ Sci Pollut Res 26(4):3368–3381. https://doi.org/10.1007/s11356-018-3749-5
12. Fang B et al (2023) Artificial intelligence for waste management in smart cities: a review. Environ Chem Lett 21(4) Springer Science and Business Media Deutschland GmbH:1959–1989. https://doi.org/10.1007/s10311-023-01604-3
13. Arebey M, Hannan MA, Basri H, Abdullah H (2009) Solid waste monitoring and management using RFID, GIS and GSM. In: SCOReD2009–Proceedings of 2009. IEEE Student Conference on Research and Development, pp 37–40. https://doi.org/10.1109/SCORED.2009.5443382
14. Susha B, Madhukara Shanbog R, Hussain S, Professor A, Student U (2021) Automatic segregation of waste using robotic ARM. [Online]. Available: www.ijcrt.org
15. Balakrishnan K, Krishna SK, Subha T (2016) Automatic waste segregator and monitoring system. [Online]. Available: www.stmjournals.com
16. Alexander Dubanowitz Advisor J, Themelis NJ, Dubanowitz AJ, (2000) Design of a materials recovery facility (MRF) for processing the recyclable materials of New York City's municipal solid waste design of a materials recovery facility (MRF) for processing the recyclable materials of New York City's municipal solid waste executive summary
17. Damgacioglu H, Hornilla M, Bafail O, Celik N (2020) Recovering value from single stream material recovery facilities—an outbound contamination analysis in Florida. Waste Manag. 102:804–814. https://doi.org/10.1016/j.wasman.2019.11.020
18. Lubongo C, Alexandridis P (2022) Assessment of performance and challenges in use of commercial automated sorting technology for plastic waste. Recycling 7(2). https://doi.org/10.3390/recycling7020011
19. Satav AG, Kubade S, Amrutkar C, Arya G, Pawar A (2023) A state-of-the-art review on robotics in waste sorting: scope and challenges. Int J Interact Des Manuf 17(6) Springer-Verlag Italia s.r.l.:2789–2806. https://doi.org/10.1007/s12008-023-01320-w
20. Chandramohan A, Mendonca J, Shankar R, Baheti NU, Krishnan NK, Vidyalaya R (2016) Automated waste segregator. https://doi.org/10.1109/TIIEC.2014.9
21. Pasquini C (2018) Near infrared spectroscopy: A mature analytical technique with new perspectives—A review. Anal Chim Acta 1026:8–36. https://doi.org/10.1016/j.aca.2018.04.004

22. Zhao Y, Li J (2022) Sensor-based technologies in effective solid waste sorting: successful applications, sensor combination, and future directions. Environ Sci Technol 56(24) American Chemical Society:17531–17544. https://doi.org/10.1021/acs.est.2c05874
23. Bae JS, Oh SK, Pedrycz W, Fu Z (2019) Design of fuzzy radial basis function neural network classifier based on information data preprocessing for recycling black plastic wastes: comparative studies of ATR FT-IR and Raman spectroscopy. Appl Intell 49(3):929–949. https://doi.org/10.1007/s10489-018-1300-5
24. Riise BL, Biddle MB, Fisher MM. X-ray fluorescence spectroscopy in plastics recycling
25. Zhang Q et al (2022) A multi-label waste detection model based on transfer learning. Resour Conserv Recycl 181:106235. https://doi.org/10.1016/j.resconrec.2022.106235
26. Rahman MW, Islam R, Hasan A, Bithi NI, Hasan MM, Rahman MM (2022) Intelligent waste management system using deep learning with IoT. J King Saud Univ–Comp Inform Sci 34(5): 2072–2087. https://doi.org/10.1016/j.jksuci.2020.08.016
27. Liu C et al (2022) A domestic trash detection model based on improved YOLOX. Sensors 22(18). https://doi.org/10.3390/s22186974
28. John A, Meva D (2020) A comparative study of various object detection algorithms and performance analysis. Int J Comp Sci Engg Open Access Res Paper 8(10). https://doi.org/10.26438/ijcse/v8i10.158163
29. Pai RR, Bangera PJ, Shaikh MM, Prajwal P, Shetty RP (2022) Design and development of an automatic dry waste segregator for household and institutional wastes. In: Materials today: proceedings. Elsevier Ltd, pp 1048–1054. https://doi.org/10.1016/j.matpr.2021.10.487
30. Choi J, Lim B, Yoo Y (2023) Advancing plastic waste classification and recycling efficiency: integrating image sensors and deep learning algorithms. Appl Sci (Switzerland) 13(18). https://doi.org/10.3390/app131810224
31. Jahan R, Jhansi Rani MG, Manasa ULN, Susmitha G, Girija MK, Nikitha P (2020) Automated Waste Segregation System using IoT. Int Res J Eng Technol. Available: www.irjet.net
32. Murugan N, Sivathanu A, Vaidyanathan K, Tiwari A, Varma A (2023) Automated home waste segregation and management system. Int J Elect Comp Engg 13(4):3903–3912. https://doi.org/10.11591/ijece.v13i4.pp3903-3912
33. Abhinav Nishanth K, Gaikwad GA, Students U, Professor A, (2021) Issue 8 www.jetir.org (ISSN-2349-5162) JETIRFA06009 Journal of Emerging Technologies and Innovative Research (JETIR) www.jetir. [Online]. Available: www.jetir.org
34. Townsend GT. Benefits of construction and demolition debris recycling in the United States. 2017. Available from https://cdrecycling.org/. Accessed on 6th May, 2024

Investigations on Water Pollutant Load Reduction Using Inverse Fluidized Bed Bioreactor

Dharmesh H. Sur and Abhishek Gupta

Abstract A fluidized bed reactor is employed for three-phase interactions for the treatment of wastewaters using the inversed bed. Specific parameters like gas holdup, bed heights, gas velocities, and spargers are investigated in an experimental setup. The bubble movements governing the flow patterns across the entire process contributed to fabricated wastewater that mimicked the textile effluent (initial value of 1110 ppm) for its aerobic digestion. The purpose of the current study is to reduce chemical oxygen demand (COD) by employing four gas spargers specifically designed as per open percentage area. This inverse phase bioreactor (solid media particle density lower than water) was investigated for its influence by the static bed heights (H) of 4.85 cm and 2.43 cm, which was accompanied by the low gas velocities of the range of 0.343 cm/s and 0.4068 cm/s. $\varepsilon g = 0.0062$ Ug 0.995 was the deduced correlation for the gas holdup. The accuracy in the measurements of COD was supported by the statistical values of regression coefficient and standard errors. Attainment of up to 96.9% COD reduction is achieved for synthetic wastewater.

Keywords Fluidized bed · Flow method · Three phases · Spargers · Gas holdup correlation

1 Introduction

The current scenario of employing a fluidized bed bioreactor in an aerobic mode is effective in its applications, especially while treating diversely characterized wastewater from industries [1–7]. The optimization strategies have also been set up [8]. The bioreactor operation involves three phases—media particles (solid phase), wastewater (liquid phase), and air (gas phase). The microorganisms retain themselves during the bioreactor operation and grow into solid media particles to digest

D. H. Sur (✉) · A. Gupta
Department of Chemical Engineering, Marwadi University, Rajkot, India
e-mail: dharmesh.sur@marwadieducation.edu.in

P. Pathak et al. (eds.), *Advances in Environmental Sustainability, Energy and Earth Science*, https://doi.org/10.1007/978-3-031-73820-3_2

organic waste [9, 10]. This happens through the layer of biofilms formed over the solid media particles [11–13]. This three-phase fluidized bed bioreactor uses only the gas as a fluidizing agent. The low gas velocity with high mass transfer rates is adopted to favor less energy consumption in an aerobic mode. The coefficient of mass transfer is almost twice the value in a three-phase fluidized bed if it is compared with traditional beds because of higher values of gas holdup as well as residence time [14]. When a three-phase fluidized bed is employed with media particles (solid), which has density lower than that of water, it is termed a three-phase inverse fluidized bed (TPIFB). Batch mode is adopted in this experimental study, which implies that only gas velocity with zero liquid velocity is employed [15].

Several parameters control the performance of TPIFB based on the behavior of the three phases taking part in it. The bed expands downwards. Gas spargers distribute the gas uniformly in the bioreactor [15, 16]. Various gas spargers are fabricated to check the effects of operational parameters. Each of the gas spargers has characteristic openings on its surface, which is referred to as an "open percentage area," where air passes through wastewater and hence dominates the gas holdup of the TPIFB. This is one of the major factors affecting the typical mass transfer. Knowledge of the performance of TPIFB needs attention and enrichment in a batch mode [2, 7, 9, 13, 15]. Exactly, this research aims to investigate the COD reductions in the fabricated TPIFB using designed spargers.

2 Experimental Procedure

Designing bioreactor is very challenging because of the specificity and its deployment for intended use [17]. The bioreactor column (acrylic: 9.7 cm diameter and 100 cm height) was designed to treat the fabricated wastewater (liquid phase) with sparging air (density of 1.20 kg/m^3 and viscosity of 0.018 Pa.s) through it in the presence of solid media particles (polypropylene: 4 mm average diameter, 722.6 kg/m^3 density). For the composition of the fabricated wastewater (an initial COD of 1110 ppm), as given in Table 1, the performance of various gas spargers was examined in a batch mode for COD reduction. The bacterial cultures of *Escherichia*

Table 1 Composition of synthetic wastewater

Components	Concentration (mg/L)
Dextrose	1
Urea	1
Yeast extract	0.4
Ferric chloride	0.0864
Calcium chloride	0.6
Cobalt chloride	0.036
Di-potassium hydrogen phosphate	1
Potassium dihydrogen phosphate	1
Magnesium sulphate	0.12

coli (NCIM 2805) and *Pseudomonas* spp. (NCIM 2026) were acquired from the National Chemical Laboratory, Pune, India. To maintain the TPIFB in a way that did not cause contamination, an air filter was placed before the point of air entry from the compressor. Not only that but solid media particles were also washed with an 8% hydrochloric acid solution to ensure their contamination-free status.

The media particles and the air entered from the top and bottom, respectively, into TPIFB. The bioreactor column also has liquid entry as well as sampling point. The number of media particles was estimated based on the fraction of static bed height to column diameter (H/D) of 0.5 and 0.25. Such mentioned values are found to be equal to 4.85 cm and 2.43 cm bed heights for the media particles, respectively. The experiments were supervised at discrete combinations of parameters incorporating gas velocities (0.6 cm/s, 0.4068 cm/s, and 0.343 cm/s), bed heights (30 cm, 60 cm, and 90 cm), and the static bed heights (H = 4.85 cm and 2.43 cm). When the bioreactor operation began, gas bubbles started moving in an individual way, making patterns, each representing flow style. The gas flow required for bringing about the inversion (downward movement of the particles), as illustrated in Fig. 1. Manometer measured the pressure difference required to ensure the steady state as realized by constant bed expansion, at which COD reductions were measured. Very

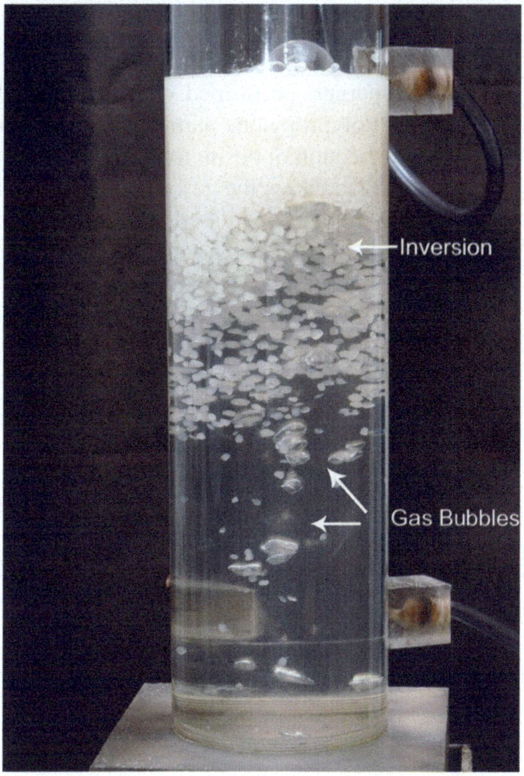

Fig. 1 Onset of fluidization

low gas velocity during the operation was the major advantage of the present work. The correlation between gas velocity (U_g) and gas holdup (ε_g) was also derived.

3 Results and Discussion

The organic load was digested aerobically by the air that passed through the gas spargers. Depending on the open percentage area of the gas spargers, bubbles of different sizes were formed. The larger and smaller bubbles moved upward together and made a flow pattern. The solid phase comprising media particles makes a layer on the top part of the column, where they lie in layers. The total height of this layer was considered as static bed height (H) in a bioreactor. The H/D ratio contributed to TPIFB performance along with the role of gas velocities [18]. The bed height measured the total volume of contents of particles and wastewater together in a TPIFB column. The main parameters affecting bioreactor operations were flow patterns, gas holdup, and different spargers, which were discussed separately in the following sections in detail.

3.1 Flow Patterns

Each of the spargers performed with two possible gas velocities at 0.343 cm/s and 0.4068 cm/s. Polypropylene particles acted as solid media and took their entry from the top (in a direction opposite to the sparged gas) of the TPIFB, wherein fabricated effluent was filled. Now, the gas velocity gradually increased to zero liquid velocity. When the gas velocity increased, the packed portion started moving progressively, as seen in Fig. 2a. Bed of particles underwent few structural changes. The bottom section of the bed started to fluidize. The velocity at which bubbles started fluidizing was regarded as minimum gas fluidization velocity (0.214 cm/s), and above that, there existed a partially fluidized bed where the bottom portion of the bed showed inversion (Fig. 2b). The unsettled segment of the bed gradually moved when the entire bed fluidized. Particles were in motion by virtue of their buoyancy force when minimum superficial gas velocity was maintained. Homogeneous bed expansion was observed at this stage, and the expansion of the bed continued. Bubbles of different sizes were formed upon an increase in gas velocity and continuously coalesced (coalesced bubble regime). Once the gas velocity maintained is constant, bubbles of uniform size (dispersed bubble regime) with spherical or ellipsoidal shapes form. The time period to reach the steady state was approximately 17 min. After the system reached a steady state, the readings were noted down for reduction in COD at the regular intervals of 2 h. The refinements to the readings were for 10 min at a point of change in COD.

Fig. 2 (**a**) Bioreactor column in operation; (**b**) Bed in Inversion

3.2 Bed Height

Experiments were accomplished on three bed heights—30 cm, 60 cm, and 90 cm. There was no remarkable difference in the reduction of COD in relation to bed heights. Trials amongst all bed heights gave only a 2.5% maximum difference in COD reduction at the most. The pattern of formation of bubbles and inversion of fluidized beds remained similar for all these bed heights. However, the time taken for bubbles to reach the solid media particles at the top changed, but that was negligible from the viewpoint of overall COD decrease. Further, none of the bed heights affected the minimum fluidization velocity (= 0.214 cm/s) for all experiments. Therefore, the bed height was taken as 30 cm for further experiments.

3.3 Gas Holdup

The volumetric expansion was used to measure the average gas holdup. Volumetric expansion refers to the state as observed by the increase in bed height. The initial height of the bed was subtracted from the final height, and the value received this way was divided by the final height of the TPIFB column to get the gas holdup. For stagnant liquid effluent, gas holdup increased with the rise in gas velocity. The values of gas velocity corresponding to gas holdup are given in Table 2.

Table 3 illustrates a few correlations (ε_g and U_g) deduced from the literature, indicating almost a linear relation as $\varepsilon_g = k\ U_g{}^n$ at low gas flow rates [19]. The

Table 2 Average values of gas holdup

	H (cm)	Flow rate in LPM	Flow rate in (cm/s)	Sparger 1	Sparger 2	Sparger 3	Sparger 4
				Open area in percentage			
H/D				17.99%	10.42%	8.08%	1.53%
0.5	4.85	1.52	0.4068	0.00934	0.0108	0.00868	0.0103
		1.28	0.343	0.00822	0.0084	0.0077	0.00948
0.25	2.43	1.52	0.4068	0.0095	0.0109	0.0089	0.01
		1.28	0.343	0.0084	0.0096	0.0079	0.0098

Table 3 Correlations for the gas holdup

Sr. No.	Ug (cm/s)	ε_g	Correlation	References
1	0.343–0.4068	0.0078–0.0109	$\varepsilon_g = 0.0062\ U_g^{0.995}$	Current work
2	1–8	0.05–0.3	$\varepsilon_g = 0.71\ U_g^{0.535}$	[19]
3	0.1–1.8	0.003–0.045	$\varepsilon_g = 1.45\ U_g^{0.8}$	[20]
4	20–800	0.2–0.9	$\varepsilon_g = 0.417\ U_g^{0.44}$	[21]

present work confirmed the same. 0.995 was the value of "n" achieved in the current work. It showed the well-built bubble regime. The interrelationship for the gas holdup is proposed as follows:

$$\varepsilon_g = 0.0062\ U_g^{0.995}$$

3.4 Gas Spargers

The flow patterns observed for sparger 1 (Fig. 3a) was a mix of bubble flow and slug flow. Bubble flow persisted until 25 h. Slugs were found at the end of the batch after 25 h. Therefore, until 25 h, COD removal mainly contributed to bubble flow. After 25 h, there was a gradual decrease in the rate of reduction of digestion of organic matter. This was due to the higher gas holdup at a velocity of 0.4068 cm/s. There was a continuous merging of smaller bubbles into bigger ones, and those bigger bubbles later contributed to the formation of slugs. These slugs were responsible for poor mass transfer as very little surface area was available for gas–liquid mass transfer. Slugs disturbed an upper-packed portion of the bed too. The experimental trials consistently finished with such patterns and lasted up to 34 h. Maximum 93% COD removal was achieved at 0.4068 cm/s gas velocity in this sparger with 18% openings by its design.

On the other side, homogeneous bubbly flow existed (Sparger 2: open surface area of 10.42%) and was responsible for the reduction in COD. Sparger 2 (Fig. 3b) showed a steep reduction in COD between 12 h and 25 h because microorganisms

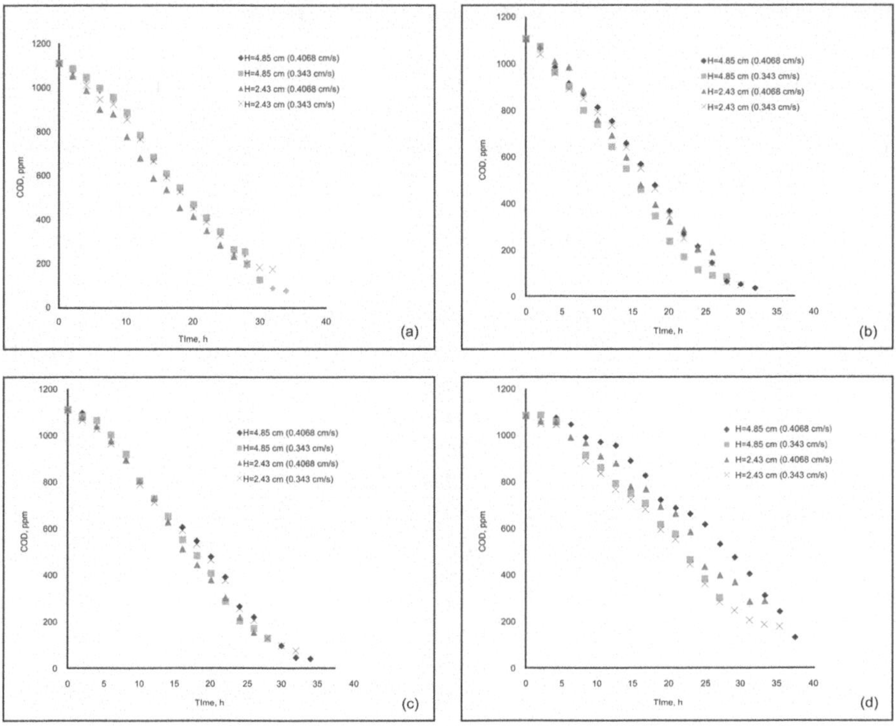

Fig. 3 Performance of each sparger for reduction in COD: (**a**) sparger 1, (**b**) sparger 2, (**c**) sparger 3, and (**d**) sparger 4

were adapted to the bioreactor environment during this period and started digestion of COD from sample wastewater. The gas holdup existed during this period of time. It favored the growth of *Pseudomonas* and *E. coli* on the surface of media particles (polypropylene). This resulted in faster degradation of the organic contents of the sample. The maximum removal in COD achieved was 96.9% at H (static bed height) = 4.85 cm and 0.4068 cm/s gas velocity. As the H increased, more gas velocity was required to move a top-packed portion of the bed. Sparger 3 (Fig. 3c), with an open surface area of 8.077%, exhibited a similar pattern of flow and COD reduction as that of Sparger 2 with the removal of 96.6% COD at H of 4.85 cm and 0.4068 cm/s gas velocity.

Sparger 2 and sparger 3 exhibited similar trends in flow patterns. They were compared, as shown by the values of reduction in COD in Tables 4 and 5. For both spargers 2 and 3, there was a uniform distribution of bubbles and the existence of consistent flow. Therefore, it favored participation of all the solid media particles in the inversion process equally by turning them upside down at regular interval. However, a slight amount of turbidity was detected for spargers 2 and 3. It was contributed by overgrown biofilm due to the digestion of organic waste on media particles. Therefore, media particles needed replacement after each batch operation.

Table 4 Summary of all results at 24 h and 26 h for lateral comparisons

Sparger	1		2		3		4	
Open percent area	17.99%		10.42%		8.08%		1.53%	
Gas velocity (cm/s)	0.4068	0.343	0.4068	0.343	0.4068	0.343	0.4068	0.343
H = 4.85 cm, 24 h	63.4	63.08	76	63.4	63.09	74.58	39	64.91
H = 4.85 cm, 26 h	69	68.73	80.9	69	68.74	82.14	43.17	72.55
H = 2.43 cm, 24 h	68.33	63.58	73.95	76.6	73.11	66.08	46.16	58.74
H = 2.43 cm, 26 h	74.44	69.16	81.54	81.38	80.77	77.17	59.77	67.82

Table 5 Summary of maximum possible performance of each sparger

Sparger	Time (h)	H (cm)	Gas velocity (cm/s)	COD	% Removal
1	28	2.43	0.4068	204.9	81.54
	31	4.85	0.343	128.85	88.39
	32.4	2.43	0.343	176.85	84.07
	33	4.85	0.4068	77.7	93
2	26	2.43	0.4068	190.55	82.83
	30	2.43	0.343	47.61	95.71
	32	4.85	0.4068	34.41	96.9
	32	4.85	0.343	93.47	91.9
3	28.4	4.85	0.343	123.44	88.88
	30	2.43	0.4068	150.96	86.4
	33.45	2.43	0.343	71.32	93.57
	34	4.85	0.4068	37.74	96.6
4	26	4.85	0.343	304.69	72.55
	32	2.43	0.4068	288.47	74.01
	34	2.43	0.343	180.44	83.74
	37.5	4.85	0.4068	131.76	88.12

Further, it was important to understand that nonuniform flow or slugs restricted all the particles from participating in the process of inversion.

In case of Sparger 4 (Fig. 3d), the gas entered very slowly because of less opening (1.53%) of gas sparger. Therefore, it took up to 4 h for the initial adoption of microbes into the biological environment of the bioreactor.

The COD digestion was considerable between 15 h and 25 h. There existed very small bubbles, and they moved up and accumulated. Merging of smaller bubbles into larger bubbles started after 25 h during this batch. They converted to slugs and continued then after. The least value of removal in COD was 72.55% at 26 h, H = 0.485 cm, and 0.343 cm/s gas velocity. This was in COD reduction amongst all spargers. So, keeping this as a reference value, Table 4 was illustrated for comparison with all other spargers.

A summary of the overall performance of the bioreactor was given for comparison in Table 5. All the work with the proposed designs of spargers supported the objective of working at much-reduced velocities of the gas phase in a batch mode of TPIFB. Many authors worked on the maximum removal of COD in different

Table 6 Comparison of various researches towards COD reduction

S. N.	Effluent type	Mode	Maximum (%)	Process	Author
			COD removal	Characteristics	
1	Synthetic	Inverse three-phase	96.90%	Aerobic	Present work
	Wastewater	Fluidized bed bioreactor			
2	Starch	Three-phase fluidized bed	93.80%	Aerobic	[9]
	Wastewater	Bioreactor			
3	Fertilizer Production	Three-phase fluidized bed Bioreactor	92%	Biochemical	[19]
4	Textile	Submerged filtration	91.50%	Physical-chemical	[22]
5	Refinery	Three-phase fluidized	90%	Aerobic	[23]
	Wastewater	Bed bioreactor			
6	Petroleum	Sequencing batch reactor	90%	Organic, toxic	[24]
	Refinery			Hydraulic shocks	
7	Cotton textile	Catalytic thermal treatment (thermolysis)	90%	Physical-chemical	[25]
8	Pulp and paper	Continuously aerated bench-top bioreactor	89%	Bioremediation	[26]
9	Synthetic	Fluidized-bed reactors		Chemical	
	Wastewater	Up-flow	87–95%		[27]
		Down-flow	55–95%		
10	Pulp and paper	Laccase-polymerized Membrane filtration	60%	Physical-chemical	[28]
11	Textile (azo dyes)	Fluidized-bed reactor	49%	Chemical	[29]

systems with many parameters, as shown in Table 6. In the present work, a maximum of 96.9% COD removal was achieved in the case of sparger 2 (open percentage area of 10.42%), with the use of very low fluidization and gas velocity.

3.5 *Calculation of Standard Error and Statistical Analysis*

A random error existed for the data of COD reduction in each case of sparger. So, standard error was the most important concept in reaching inference [30]. Standard error was found out from the standard deviation of the sampling distribution. The calculated standard errors for present trials are given in Table 7. When the standard error was small, the average deviation between the estimates and the actual population parameter (COD reduction) was small. It meant that the estimate of the reduction in COD was more "precise" during performance.

Table 7 Standard error calculation of each sparger

		H = 4.85 cm		H = 2.43 cm	
Sparger details		Gas velocity (cm/s)			
Sparger no.	Open percent area	0.4068	0.343	0.4068	0.343
1	17.99	79.77	78.02	169.67	160.39
2	10.42	85.23	83.68	169.38	156.84
3	8.077	83.57	84.91	82.38	80.36
4	1.53	71.83	71.51	70.13	77.38

Further, least squares analysis was applied in determining how well the data on COD reduction confirmed the particular model, as shown in Fig. 4. The techniques presented estimations of the deviations or residuals of each datum point from the line representing the model. So it minimized the residuals.

4 Conclusions

Up to 96.9% COD reduction is achieved for the fabricated wastewater due to low gas fluidization velocity and the use of low-density solid media particles. Bed height does not have an influence on minimum gas fluidization velocity. All the types of spargers—1, 2, 3, and 4—affect the flow regimes. Observations exhibit that increasing static bed height causes a rise in solid concentrations, as evidenced mainly from the middle part of the TPIFB. At the same time, the wall region has had no significant changes. Increasing gas velocity enhances the mixing in the bed. The present work proposes a correlation for the gas holdup, as:

$$\varepsilon_g = 0.0062\, U_g^{0.995}$$

Notations

H: Static bed height (cm)

U_g: Superficial gas velocity (cm s^{-1})

ε_g: Gas phase holdup

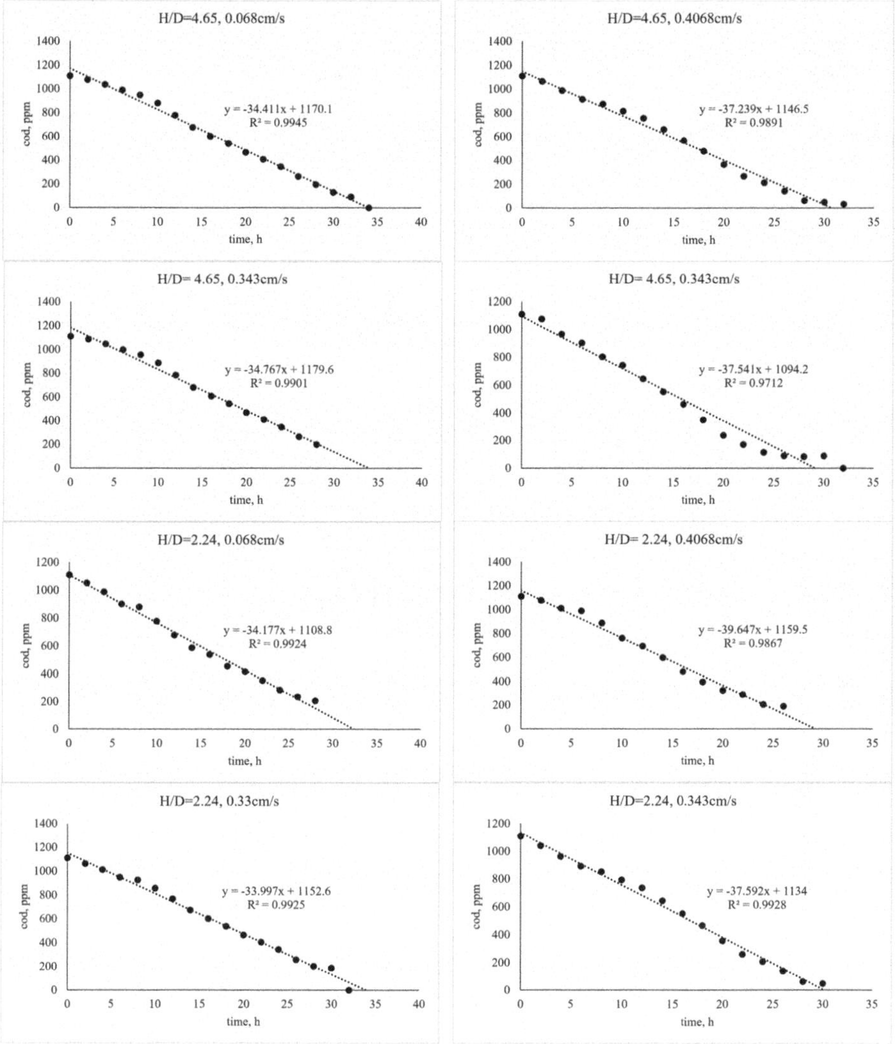

[a] [b]

Fig. 4 Regression with least squares method. (**a**) Sparger 1, (**b**) Sparger 2, (**c**) Sparger 3, and (**d**) Sparger 4

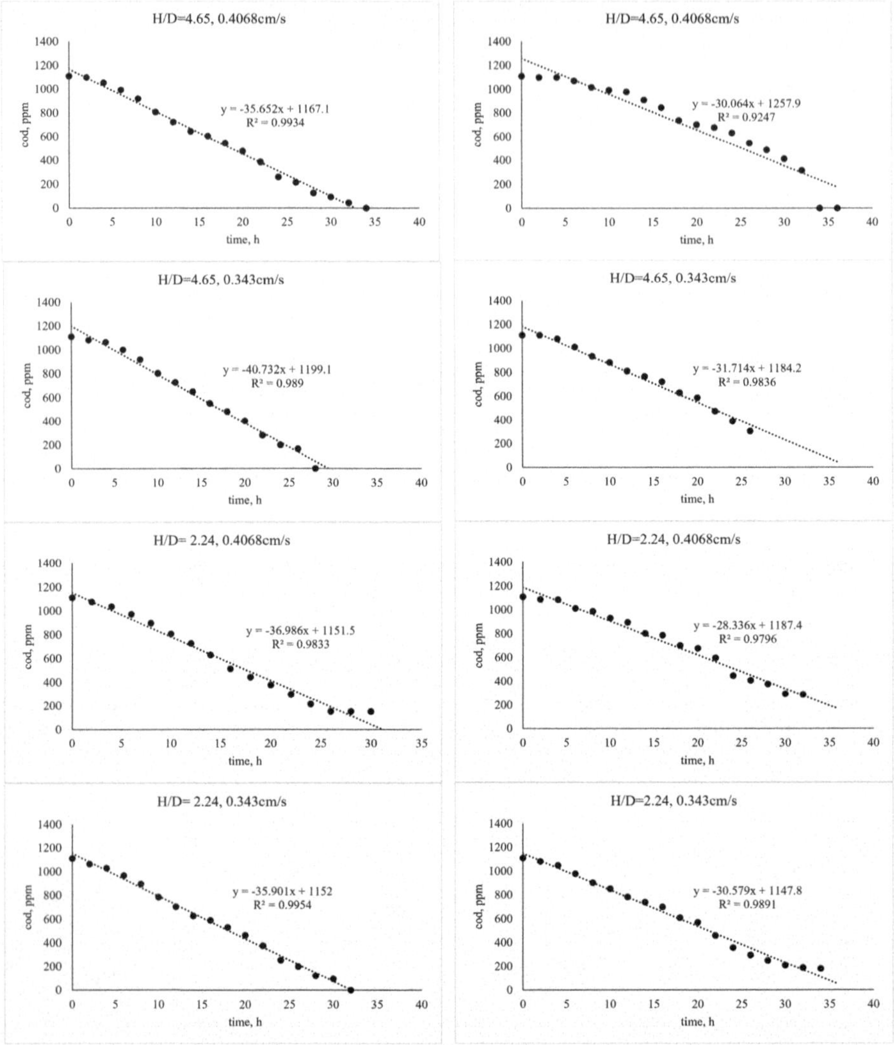

[c] [d]

Fig. 4 (continued)

Acknowledgements The authors are grateful to Prof. (Dr.) R B Jadeja (Pro-VC and Dean of Research) and the Management of Marwadi University, Rajkot (India), for the support and encouragement provided.

Conflict of Interest This manuscript is the authors' original work and has not been published, nor has it been submitted simultaneously elsewhere. All authors have checked the manuscript and have agreed to submit it.

References

1. Challa M, Dash RR (2023) Statistical analysis of treatment of rice mill wastewater using the aerobic inverse fluidized bed biofilm reactor, process Saf. Environ Prot 171:470–481
2. Swain AK, Sahoo A, Jena HM, Bhuyan SC (2022) Effect of superficial gas velocity and ratio of bed volume to reactor volume of inverse fluidized bed biofilm reactor on the removal of ammonia-nitrogen from wastewater. Ind J Chem Technol 28(6):709–716
3. Cheng KY, Acuña CR, Ginige MP, Yan S, Mesquita A, Pentzien M, Zheng G, Zhou L, Kaksonen AH (2021) Inverse fluidised bed bioreactor enabled high-rate selenate reduction for wastewater treatment. Environ Adv 5:1–7
4. Liu Y, Sun X, Sun Z, Zhang C, Zhu J (2021) Experimental and numerical studies on a bubble-induced inverse gas-liquid-solids fluidized bed. Adv Powder Technol 32(12):4496–4508
5. Begum SS, Sikkandar MY, Prakash NB, Bakouri M, Alanazi AB, Alkhatlan NMS (2021) Investigation of flow dynamic characteristics of inverse fluidized bed biofilm reactor for degrading pharmaceutical based biomedical wastewater. Environ Eng Res 26:1–8
6. Sur DH, Mukhopadhyay M (2018) Process parametric study for COD removal of electroplating industry effluent. 3 Biotech 8:1–10
7. Sur DH, Mukhopadhyay M (2017) COD reduction of textile effluent in Three-phase fluidized bed bioreactor using Pseudomonas aureofaciens and Escherichia coli. 3 Biotech 141:1–11
8. Namburath M, Alappat BJ, Ramaswamy ST (2023) A critical review of inverse fluidized bed reactors—start-up optimization strategies and wastewater treatment. Environ Sci Pollut Res 30(50):108370–108392
9. Haribabu K, Sivasubramanian V (2014) Treatment of wastewater in fluidized bed bioreactor using low density biosupport. Energy Procedia 50:214–221
10. Begum SS, Radha KV (2013) Hydrodynamic behavior of inverse fluidized bed biofilm reactor for phenol biodegradation using Pseudomonas fluorescens. Korean J Chem Eng 31:436–445
11. Shieh WK, Keenan JD (1986) Fluidised bed biofilm reactor for wastewater treatment. Water Res 34:725–734
12. Sokol W, Halfani MR (1999) Hydrodynamics of a gas-liquid-solid fluidized bed bioreactor with a low density biomass support. Biochem Eng J 3:185–192
13. Buffiere P, Moletta R (1999) Some hydrodynamic characteristics of turbulent three phase fluidized-bed reactors. Chem Eng Sci 54:1233–1242
14. Sokol W (2001) Operating parameters for a gas-liquid-solid fluidized bed bioreactor with a low density biomass support. Biochem Eng J 8:203–212
15. Comte MP, Bastoul D, Hebrard G, Roustan M (1997) Hydrodynamics of a three-phase fluidized bed-the turbulent turbulent bed. Chem Eng Sci 52:3971–3977
16. Sokol W, Korpal W (2006) Aerobic treatment of wastewaters in the turbulent fluidized bed biofilm reactor. Chem Eng J 118:199–205
17. Sur DH, Mukhopadhyay M (2017a) Process aspects of three-phase inverse fluidized bed bioreactor: a review. J Environ Chem Eng 5:3518–3528
18. Rittman BE, McCarty PL (2001) Environmental biotechnology: principles and applications, 2nd edn. McGraw-Hill International Editions: Biological Sciences Series, New York

19. Mutsago M, Ochieng A, Odiyo JO (2003) Biological treatment of mixed industrial wastewaters in a fluidized bed reactor. J Hazard Mater B96:79–90
20. Chisti MY (1989) Air-Lift Bioreactors. Elsevier, London, pp 32–39
21. Kito M, Shimada M, Sakai T, Sugiyama S, Wen CY (1976) Performance of turbulent bed contactor: gas holdup and interfacial area under liquid stagnant flow. Proceedings of International Fluid Conference on Fluid Technology, pp 411–429
22. Guozhu M, Jianping W, Lei P, Liping D (2003) The denitrification treatment of low C/N ratio nitrate-nitrogen wastewater in a gas–liquid–solid fluidized bed bioreactor. Chem Eng J 94:155–159
23. Cheng Q, Gao C, Liu M, Shuai S, Yu S, Zheng Y, Zho Q (2013) Color removal and COD reduction of biologically treated textile effluent through submerged filtration using hollow fiber nanofiltration membrane. Desalination 314:89–95
24. Sokol W (2003) Treatment of refinery wastewater in a three-phase fluidized bed bio-reactor with a low density bio-mass support. Biochem Eng J 15:1–10
25. Mizzouria NS, Shaaban MG (2013) Individual and combined effects of organic, toxic, and hydraulic shocks on sequencing batch reactor in treating petroleum refinery wastewater. J Hazard Mater 250:333–344
26. Chand S, Kumar P, Mishra IM, Prasad B (2007) Decolorization and COD reduction of dyeing wastewater from a cotton textile mill using thermolysis and coagulation. J Hazard Mater 153: 635–645
27. Rathore VS, Malaviya P (2007) Bioremediation of pulp and paper mill effluent by a novel fungal consortium isolated from polluted soil. Bioresour Technol 98:3647–3651
28. Gungor M, Sahinkaya E (2010) Comparison of sulfidogenic up-flow and down-flow bed reactors for the bio-treatment of acidic metal-containing waste water. Bioresour Technol 101: 9508–9514
29. Fan C, Ko CH (2010) Enhanced chemical oxygen demand removal and flux reduction in pulp and paper wastewater treatment using laccase-polymerized membrane filtration. J Hazard Mater 181:763–770
30. Antonisamy B, Christopher S, Prassana Samuel P (2010) Biostatistics: principles and practice, 1st edn. McGraw-Hill Education (India) Private Limited, New Delhi

Energy Sustainability by Adopting Resource Recycling: A Case Study in Coir Industry

Preethi Ravichandran, Vishal Devarajan, Salai Muthu Kumaran Muthuvadivelan, Kesava Moorthi Veerapandi, Lalith Pankaj Raj Nadimuthu, and Kirubakaran Victor

Abstract Globally, annual coconut production for domestic and commercial purposes is estimated at 62.41 million metric tonnes. The coir is a naturally available elastic fibre, which is abundantly available and unhandled coir waste (i.e., short fibre/peat/dust) has significant potential to maintain the economy and ensure sustainability through an effective waste-to-energy recovery perspective. The primary energy demand in India is predicted to climb up to around 1500 Mtoe. The Indian coir industry is considered a significant energy-intensive under micro, small, and medium enterprises (MSME). The 3 M management conception has transformed into the 4 M management conception, emphasizing energy management as its key focus. The coir industry is an intermittent/interlinked process; supply chain optimization and energy management must join to maintain overall energy efficiency. The manuscript aims to carry out a detailed systematic energy analysis focusing on supply chain and power quality optimization with significant scope for implementing resource recycling measures in the Indian rural coir industry of capacity LTCT 109.9 kW with an average energy consumption of 3500 kWh/billing cycle. Power quality optimization and recommendations were proposed based on the power quality analyser and infrared thermography results to achieve better energy efficiency. Coir industries produce coir dust as bioresidues, which should be treated effectively; otherwise, it will settle over the land, preventing the locking of water bodies and creating environmental degradation. As resource recycling, the above coir industry roughly produces 75–100 kg of coir dust daily, which will be used as feedstock for a gasifier that produces 15 kW_e and 45 kW_{th}. This reduces the grid dependency and is self-sufficient in the power generation of the coir industry. The techno-economic feasibility study is also reported in this manuscript.

P. Ravichandran · V. Devarajan · S. M. K. Muthuvadivelan · K. M. Veerapandi · L. P. R. Nadimuthu (✉) · K. Victor
Centre for Rural Energy, The Gandhigram Rural Institute-Deemed to be University, Gandhigram, Tamil Nadu, India

P. Pathak et al. (eds.), *Advances in Environmental Sustainability, Energy and Earth Science*, https://doi.org/10.1007/978-3-031-73820-3_3

Keywords Energy Analysis · Power Quality · Supply Chain Management · Resource recycling · Circular Economy · Sustainability

1 Introduction

The study delves into the innovative utilization of coconut coir waste for electricity generation, addressing both economic and environmental dimensions. Globally, coconut production reaches a staggering 62.41 million metric tons annually, providing an abundant source of coir—a naturally available elastic fibre [1]. The unhandled coir waste, comprising short fibres, peat, and dust, presents a significant untapped resource with the potential to contribute to the economical sustainability through a waste-to-energy recovery perspective.

In the context of India, where the primary energy demand is predicted to escalate to around 1500 Mtoe, the research focuses on the energy-intensive Indian coir industry, primarily comprising micro, small, and medium enterprises (MSMEs). With a transformative shift from the traditional 3 M to the 4 M management conception, the industry emphasizes energy management as a core focus area. Recognizing the intermittent and interlinked nature of coir industry processes, this study advocates for integrating supply chain optimization and energy management to maintain overall energy efficiency. The research explicitly targets the Indian rural coir industry, operating at a capacity of LTCT 109.9 kW with an average energy consumption of 3500 kWh per billing cycle. The aim is to conduct a comprehensive energy analysis focusing on supply chain and power quality optimization.

The general overview of the manuscript and baseline research details were summarized. The research seeks to enhance energy efficiency by proposing power quality optimization measures based on the results obtained from the class "A" power quality analyser and infrared thermography. Notably, coir industries generate substantial amounts of coir dust as bioresidues, posing environmental challenges if not treated effectively [2].

As a sustainable solution, the study explores the daily production of 75–100 kg of coir dust as a resource for a gasifier, producing 15 kWe and 45 kWth. Here, the average calorific of the coir pith is 2700 kcal/kg [3, 4]. The selected coir industry produces 75 kg/day of coir waste. This is equivalent to 5.6 kcal/sec. Since, industry's operating hours range from 10 to 12 hours per day. Hence, this coir industry has the potential of generating 24 kW of energy, assuming gasification efficiency as 70% [5, 6]. Therefore, the industry has the potential for generating 15–16 kW_e, and three times the thermal energy can be generated. This approach reduces grid dependency and promotes self-sufficiency in power generation for the coir industry.

The study concludes with a techno-economic feasibility analysis, providing insights into the viability of implementing coir waste-based electricity generation. Overall, this research contributes to the evolving landscape of renewable energy solutions like solar photovoltaics [7], emphasizing the circular economy [8], sustainability, and efficient resource utilization in the coir industry [9].

The study aligns with the imperative to diversify energy sources and reduce dependency on conventional grids in the broader context of the nation's energy security status and the evolving energy scenario. It underscores the need for a strategic approach to energy management, explicitly emphasizing the importance of energy optimization through power quality [10]. Comparatively, the study also aligns with the discourse on energy conservation, supply chain management, and power quality, emphasizing the interconnectedness of these elements in achieving overall energy efficiency. The integration of supply chain optimization and energy management resonates with the broader discussion on optimizing energy consumption and reducing waste.

Moreover, the study contributes to the discourse on resource recycling and the circular economy by showcasing a practical application in the coir industry. The utilization of coir waste for electricity generation exemplifies a circular economy approach, wherein waste products are repurposed, reducing environmental impact and promoting sustainable practices. Finally, the study intersects with the concept of achieving carbon neutrality and net-zero emissions. By introducing a sustainable solution for electricity generation from coir waste, the research aligns with the broader global goal of mitigating carbon emissions and transitioning towards cleaner, renewable energy sources [11]. This holistic approach to energy management underscores the importance of integrating environmental responsibility with economic viability.

Using air as the gasification medium, the study examines the gasification properties of coir pith in a circulating fluidized bed gasifier. It investigates how temperature and ER affect a number of factors, including yield, heating value, composition of the gas, carbon conversion, and thermal efficiency. The findings show that higher temperatures favour both gas heating value and hydrogen yield, with a maximum hydrogen yield recorded at 1028.6°C. While gas output and carbon conversion improve with increasing ER, hydrogen content declines. Furthermore, the temperature-dependent increases in cold gas and overall thermal efficiencies peak at 1028.6°C but decrease with increasing ER. The results emphasize how crucial temperature management and ER tuning are to biomass gasification systems to maximize gas yield and efficiency and reduce tar generation. This study offers insightful information for improving biomass gasification systems to generate sustainable energy [12, 13]. It also points to possible increases in the efficiency of diesel substitution, especially when using coir pith through briquetting. The analysis emphasizes how economical biomass-based power generation is compared to traditional techniques, stressing the applicability of agricultural biomass as fuel with slight alteration. It ends by arguing in favour of life cycle cost analysis as a thorough way to evaluate the viability of these power plants. Overall, the manuscript summarizes important discoveries on emissions, engine performance, and the financial sustainability of biomass-based power generation systems [14].

The study looks into the biomass properties of coir pith, which are essential for choosing and assessing gasification systems. The analysis included kinetic parameters, chemical composition, and physical attributes. The average particle size of the coir pith was 0.551 mm, and a considerable mass fraction was between +0.5 and

1.0 mm. It had a porosity of 71.4%, a bulk density of 87.0 kg m^{-3}, and a particle density of 305.0 kg m^{-3}. Higher volatile content (62.0%) was found by chemical analysis in comparison to fixed carbon (26.0%) and ash (12.0%). 48.3% of the elements were oxygen, and 47.1% of the elements were carbon. The thermograms showed discrete thermal breakdown stages with an ignition temperature of 550°C, and the calculated calorific value was 18.24 MJ kg^{-1}. Additionally, kinetic characteristics were evaluated in the study, and a frequency factor (A) of 874–1951 s^{-1} and an apparent activation energy (E) of 32–35 KJ mol^{-1}. These results highlight coir pith's aptitude for gasification procedures and offer insightful information about using it as a renewable energy source [15].

This manuscript highlights the growing interest in utilizing coconut shells biomass for downdraft gasification as a renewable energy resource. While significant progress has been made in understanding the feasibility and potential of his technology, further research is needed to address challenges related to feedstock variability, reactor design optimization, and environmental sustainability. The integration of experimental investigations, computational modelling, and deployment of coconut shell biomass gasification systems was depicted as a step towards commercialization and widespread adoption in the renewable energy sector [9]. This study emphasizes the specific properties of coconut shells as a feedstock pyrolysis. This includes the chemical composition, physical characteristics, and moisture content. Biochar and syngas production from coconut shells via pyrolysis is a subject of growing significance within the border domain of sustainable energy and waste management. The process involves the thermal decomposition of coconut shell biomass, emphasizing crucial factors such as temperature, heating rate, and residence time to influence the yield and quality of the resultant biochar and syngas. The produced biochar, with its distinct physical and chemical attributes, holds promise for applications such as soil enhancement and environmental remediation. Meanwhile, the composition of syngas resulting from coconut shell pyrolysis, particularly the proportion of carbon monoxide, hydrogen, and methane, has been scrutinized for its potential applications in power generation and chemical synthesis. The comparative analysis of findings, especially from this manuscript, highlights advancements and insights, ultimately guiding future research endeavours in this evolving field [16].

2 Literature Review

The literature reveals a growing interest in utilizing coconut and palm kernel shells, abundant agricultural residues, for clean energy production through downdraft fixed-bed gasification. Previous research underscores the significance of particle size and temperature as pivotal parameters affecting gasification efficiency. Investigation into the gasification performance of different biomasses elucidates the complex interplay between these factors and their impact on syngas composition and tar formation. The insights derived from this study contribute to the border discourse on sustainable energy generation, offering valuable implications for designing and operating downdraft fixed-bed gasification systems utilizing coconut and palm kernel shells as renewable biomass feedstocks [17]. The co-digestion (or) co-combustion [18] emphasizes the synergy of blending different feedstocks to enhance gasification performance. The study likely explores the impact of varying mixtures, particle sizes, and operating conditions on the syngas composition, tar formation, and overall gasification efficiency. Insights derived from this research hold promise for advancing the development of efficient and versatile gasification systems, offering valuable implications for renewable energy production and utilizing diverse biomass feedstocks. This investigation aligns with the broader goal of achieving cleaner and more sustainable energy solutions through innovative biomass gasification technology [19].

Gasification is a thermochemical process that partially oxidizes carbon-based compounds with air, oxygen, or water vapour to produce gaseous fuel. Along with CO_2, N_2, O_2, and H_2O, this sub-stoichiometric process produces combustible gases such as CO, H_2, and CH_4. The factors that affect gasification in downdraft reactors with air, with a focus on moisture content (<25%), biomass particle size (<5 cm), and equivalency ratio (0.2–0.4). The literature suggests low heating values (4–6 MJ/Nm^3), cold efficiencies (50–70%), and combustion zone temperatures of approximately 1000 °C. Engines, especially RICEs, have trouble using producer gas because of its poor energy density and volumetric efficiency, which necessitates adjusting the time of spark ignition. However, because nitrogen acts as a knock suppressor due to air oxidation during gasification, greater compression ratios can be used without increasing the susceptibility to knock the promising path towards the production of renewable energy, albeit there are certain operational requirements for effective engine use [20].

The literature study included in the section highlights the evolution of management principles from the 3Ms to the 4Ms, emphasizing the growing significance of energy management as a crucial component of industrial processes. Because it directly affects price mechanisms, it focuses on electrical energy management. Even if there are many contemporary energy-conserving products on the market, small-scale enterprises sometimes lack knowledge and expertise. The study describes an energy audit at a small-scale tile manufacturing business using tools like Power Analysers. According to the audit's conclusions, there are several ways to improve power quality [21]. These include switching to controlled sun drying techniques and LED illumination instead of traditional lighting. It is envisaged that the execution of these recommendations will lead to decreased power usage, enhanced power factor, evasion of fines for low power factor, and ultimately, improvements in power quality [22].

A review of the literature would probably cover a range of studies and research on energy auditing, power quality, and potential energy savings in various settings, such as homes and businesses. Scholars may have examined how contemporary electronics affect power quality and harmonic levels and the efficacy of various strategies for cutting energy use and boosting productivity. One could consult studies on energy-saving strategies put into practice, like replacing old equipment, upgrading lighting, and increasing air circulation. The literature on the travel and tourism sector and the expanding farmhouse trend may also shed light on the reasons for incorporating contemporary technology into these types of facilities and the possible difficulties in managing energy in rural areas. The approach used in the energy audit mentioned in the paragraph could be informed by research on methods for conducting energy audits, methods for evaluating power quality, and suggestions for optimizing energy usage in varied situations. Researchers might identify knowledge gaps their research attempts to fill and provide context for their work by synthesizing data from pertinent literature [23].

The need for energy conservation in rural industries is discussed in the text, with particular attention to the coir sector. It draws attention to how little is known about energy audits and possible studies on conservation in these kinds of environments. In order to evaluate power use and pinpoint areas for savings, the study uses thermal imagers and online power analysers. The analysis identifies imbalanced loading, which reduces power consumption and power factor penalties when load balancing occurs with appropriate sequencing of the critical loads [24]. The conclusion also highlights the need for the coir industry to complete energy audits. It offers doable solutions for increased energy efficiency, such as converting asbestos roofs to transport roofs for natural lighting, implementing LED lighting, properly balancing motor shafts, and comparing specific fuel consumption (SFC) [25].

The above literature on the gasification of coir bioresidue emphasizes the potential of coir bioresidues as high calorific fuel and the optimal and less time way for the conversion of coir waste to fuel in the form of syngas for electricity generation using diesel engines is gasification especially downdraft gasification because it produces tar-free syngas as mentioned in journals [9, 12, 14]. Also, some literature on energy auditing indicates that energy management is essential for energy saving in both residential and industrial sectors. Energy auditing is the critical step in energy management, which is achieved by analysing the power consumption of the industry or residential houses where auditing is done. The above literatures indicate that auditing the power consumption of each load could facilitate the scheduling for the balanced operation of loads and promote the conversion of inefficient loads to efficient ones. The conversion from traditional incandescent or fluorescent bulbs to modern energy-efficient LEDs leads to higher energy savings.

The integration of supply chain management concepts like Demand Predictive Analytics (DPA), Inventory Management System (IMS) [26], and Enterprise Resource Planning (ERP) with Artificial Intelligence (AI) [27] was significantly essential for maintaining the optimal operating point of the machinery in the coir processing industry. Power quality improvement is ensured through systematic supply chain management techniques for effective power quality mitigation and energy utilization [28].

From the overall literature review, the lack of supply chain management and intermittent operation of the coir processing industry is the significant bottleneck of the existing system. Therefore, the proposed solution to the coir industry is providing sustainability through supply chain optimization-based power quality enhancement. The proposed technique is significant in energy efficiency enhancement with the effective resource recycling of the coir industry waste to create a sustainable circular economy.

3 Methodology

3.1 Energy Auditing of Rural Coir Industry

The overall research flow is highlighted in Fig. 1. Energy auditing in the coir industry involves systematically examining energy consumption production process and equipment efficiency with coir manufacturing facilities; this process helps identify energy-saving opportunities to optimize energy utilization and reduce operational cost and environmental impact. Systematic energy auditing went through three phases: pre-, detailed, and post-audit phases. A Class "A" IEC-certified power quality analyser is used for a complete power quality audit, as shown in Fig. 2. In addition, infrared thermographic analysis was carried out through a thermal imager of Flir E50. A thermal imager captures infrared images of equipment and machinery to detect temperature variation, and it helps to identify overheating of components, insulation defects, and energy losses due to inefficiency in the coir manufacturing process. Both instruments play a crucial role in conducting energy audits. In the coir industry, infrared thermal imaging helps to improve overall operational efficiency and energy cost savings.

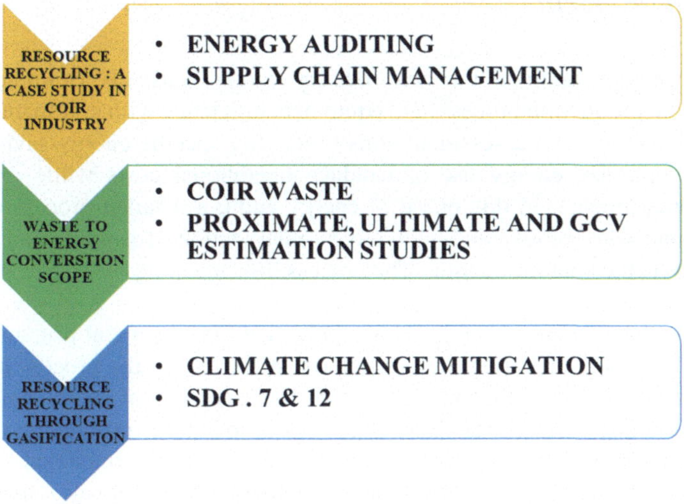

Fig. 1 Research flow

Fig. 2 Energy analysis point of common coupling

3.2 *Supply Chain Management on Energy Efficiency Enhancement*

A detailed energy audit in the coir industry comprehensively examines energy consumption, production process, and equipment efficiency with coir manufacturing facilities. This thorough assessment aims to identify specific energy-saving opportunities to optimize energy use and reduce operational costs while minimizing environmental impact. In this phase of energy audit, a detailed process of energy flow mapping is developed and analysed by intervention of supply chain management based on the core production. The process flow mapping of the coir processing unit is shown in Fig. 3.

From the detailed energy audit with supply chain management principles, it has been found that significant waste is generated at the end of the process in the coir industry, such as coir pith and baby coir. The quantum of waste generated is highlighted in Fig. 4. So, the analysis of coir waste for energy production possibility has been recommended. Bioresidue analysis and tests were conducted to find the energy content in the coir. Based on the integration of logistics management principles into the coir processing system deliberates the quantifiable benefits of energy conservation and operational effectiveness through energy densification of bioresidues and optimal resource management of the coir industry [29].

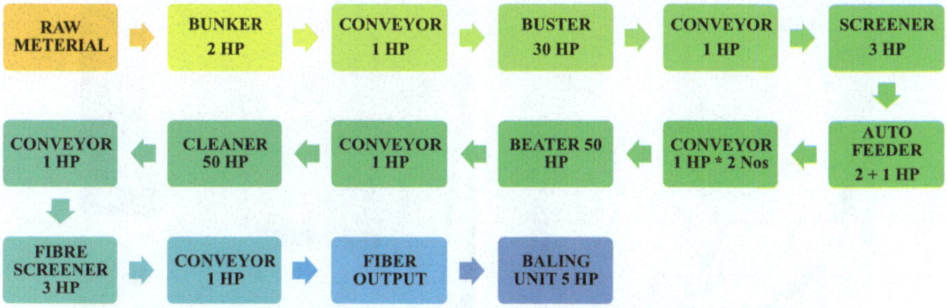

Fig. 3 Energy flow mapping

Fig. 4 Coir industry waste

3.3 Elemental Analysis of Biomass

Proximate analysis is a crucial technique used in various fields, notably in food science, agriculture, and environmental science, to determine the composition of organic materials, as shown in Fig. 5. It provides insights into the fundamental constituents of a substance, aiding in understanding its properties and potential applications. The analysis typically involves evaluating the percentages of critical components such as moisture, ash, volatile matter, and fixed carbon. These components collectively offer valuable information about the material's calorific value, stability, combustibility, and nutritional content. Proximate analysis is a fundamental tool in research, quality control, and product development across diverse industries, facilitating informed decision-making and optimization of processes. Here,

Fig. 5 Coir waste (baby coir) waste sample for investigation

proximate analysis has been carried out as per the ASTM E871. Eqs. 1, 2, 3, 4, and 5 are used to analyse the samples under proximate analysis.

The standard systematic procedure is accurately followed. Formulas used for calculation are given below:

$$\text{Weight of biomass} = (\mathrm{W}_2 - \mathrm{W}_1)\ \text{grams} \quad (1)$$

$$\%\text{of } Moisture\ Content = (W_2 - W_3)/(W_2 - W_1) * 100 \quad (2)$$

$$\%of\ Volatile\ Matter = (W3 - W4)/(W2 - W1) * 100 \quad (3)$$

$$\%of\ Ash\ Content = (W5 - W1)/(W2 - W1) * 100 \quad (4)$$

$$\%of\ Fixed\ Carbon = 100 - (\%of\ Moisture + \%of\ Volatile\ matter + \%of\ Ash) \quad (5)$$

Ultimate analysis is a method used to assess the elemental composition of biomass, typically focusing on carbon, hydrogen, nitrogen, sulphur, and oxygen. It provides insights into biomass properties crucial for energy production, environmental impact assessment, and material utilization. Researchers can evaluate biomass suitability for various applications such as biofuel production, soil improvement, and chemical processing by determining the precise elemental makeup. The ultimate analysis offers essential data for optimizing biomass utilization strategies, enhancing efficiency, and promoting sustainable resource management practices. Equations 6–9 were taken for ultimate analysis, derived from the principles of stoichiometry and elemental mass balance. These formulas are based on the assumption that the biomass consists primarily of carbon, hydrogen, nitrogen, sulphur, and oxygen, with minor amounts of other elements. The ultimate analysis of biomass typically involves determining the weight percentages of carbon (C), hydrogen (H), nitrogen (N), and oxygen (O) present in the sample. Here are the formulas commonly used for calculating various parameters. The complex chemical composition of the coir waste enhances the various constraints like sample variability (i.e., availability, source, and processing), analytical techniques, and effective combustion. All the experimental analyses (proximate and ultimate analysis) were carried out as per the standard test conditions. The proposed recommendation includes the utilization of bioresidues (i.e., coir waste) for the thermochemical energy conversion process. Also, a case consideration-based pre-drying process is recommended through the solar drying process.

$$\begin{aligned}\%Carbon &= 0.97 * fixed\ carbon + 0.7 * (volatile\ matter - 0.1\ ash) \\ &- moisture\ content * (0.6 - 0.1 * moisture\ content)\end{aligned} \tag{6}$$

$$\begin{aligned}\%Hydrogen &= 0.36 * fixed\ carbon \\ &+ 0.086\ (volatile\ matter\ 0.1\ ash) - (0.0035 * moisture\ content^2)\end{aligned} \tag{7}$$

$$\%Nitrogen = 2.10 - (0.020 * volatile\ matter) \tag{8}$$

$$\%Oxygen = 100 - (carbon + hydrogen + nitrogen + ash) \tag{9}$$

Bomb calorimetry is a laboratory technique used to measure a substance's combustion heat. It involves placing a sample inside a sealed container, called a bomb, surrounded by a known volume of water, as shown in Fig. 6. The sample is ignited, and the heat released during combustion raises the temperature of the water. The heat released by the sample can be calculated by measuring the temperature change and knowing the calorimeter's heat capacity. Bomb calorimetry is widely used in chemical and biological research to determine the energy content of fuels, foods, and other organic materials. The following formula is used to calculate the calorific value of coir waste.

Fig. 6 Estimation of calorific value of the sample under investigation

$$H = W\,X\,\Delta t - (E_1 + E_2)/M \tag{10}$$

where W = energy equivalent of calorimeter in Cal/C, 2330 Cal/ °C; H = calorific value of fuel in Cal/g; Δt = temperature rise (3.05° C); M = mass of fuel taken in g; E1 = correction for heat of combustion of firing wires in calories, 20.1 Calories; E2 = correction for heat of combustion of cotton thread in calories, 20.9 Calories.

4 Results and Discussion

4.1 Results and Discussion on Preliminary Energy Audit

Preauditing in the coir industry refers to the preliminary assessment conducted before a detailed energy audit. It involves gathering initial data on energy consumption, production process, and equipment efficiency to identify potential areas of improvement by using data collected from Tamil Nadu and distribution corporation (TANGEDCO). The historical energy utilization pattern and cost details are illustrated in Fig. 7. The baseline dataset is instrumental to carry out the detailed systematic energy audit in the coir industry.

4.2 Results and Discussion on Detailed Energy Audit [Power Quality]

The power analysis results reveal the time versus voltage profile of the coir industry. Here, the overall voltage profile is healthy. During the peak load operation of the industry, no abnormal variation of voltage is logged in the analysis period, as shown in Fig. 8. Therefore, the overall voltage variation ranges from ±5% from the rated voltage level.

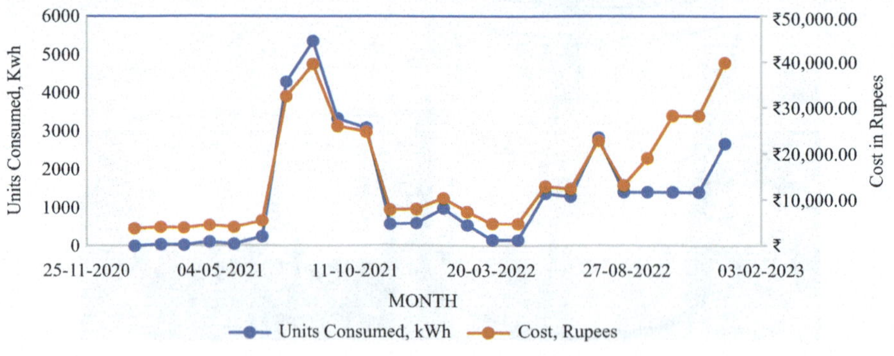

Fig. 7 Historical energy utilization pattern (baseline dataset)

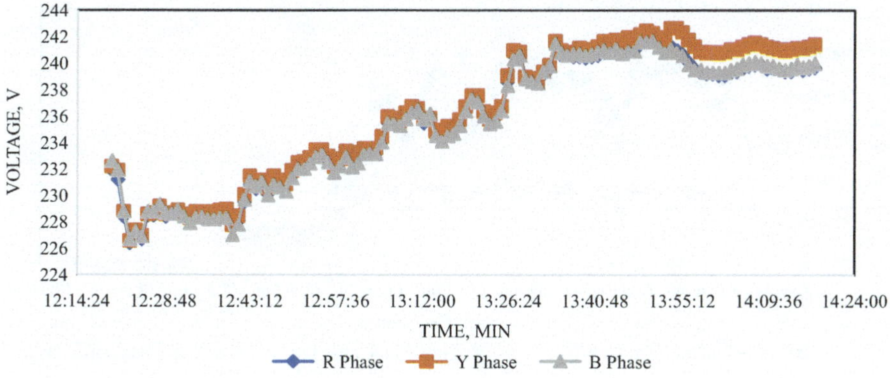

Fig. 8 Time versus voltage

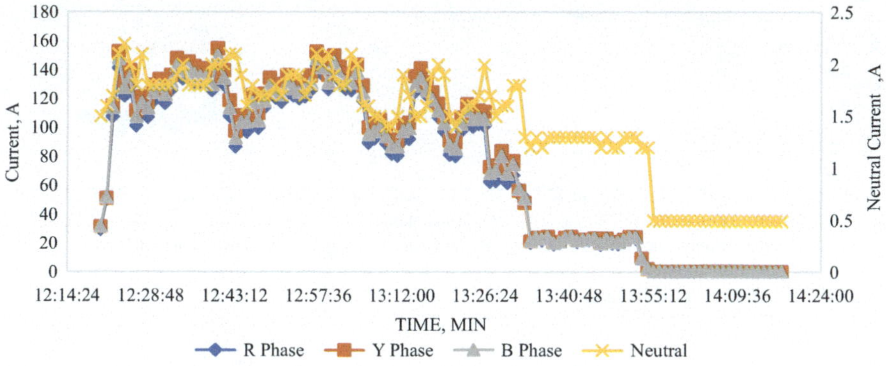

Fig. 9 Time versus current

Figure 9 illustrates the time versus current profile of the coir industry. The primary axis indicates the industry's three-phase current consumption during the analysis. From the current profile, the system behaviour pattern is healthy during peak or load greater than 70%. Therefore, proper load balancing is essential to maintain the limits even during the part load conditions. Also, the secondary axis indicates the quantum of neutral loading in the system. The maximum neutral loading of less than 2A indicates the system's health. The presence of harmonics is under limit.

The power consumption of the industry defines the capacity of the industry, as shown in Fig. 10.

The primary axis indicates the time versus power of the coir processing industry. Here, the active, apparent and reactive power are monitored and reported. The maximum demand of the industry is maintained under the sanctioned demand of the DISCOM. The appropriate reactive power compensation is carried out during the peak load operation. Therefore, it is recommended that the power system be effectively managed during the baseload operation through a sensitive Automatic Power Factor Correction (APFC) capacitor bank system. From the overall power curves, the stability

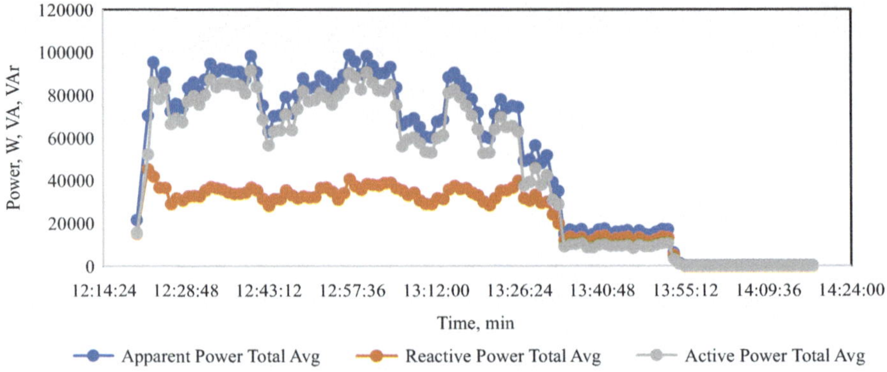

Fig. 10 Time versus power

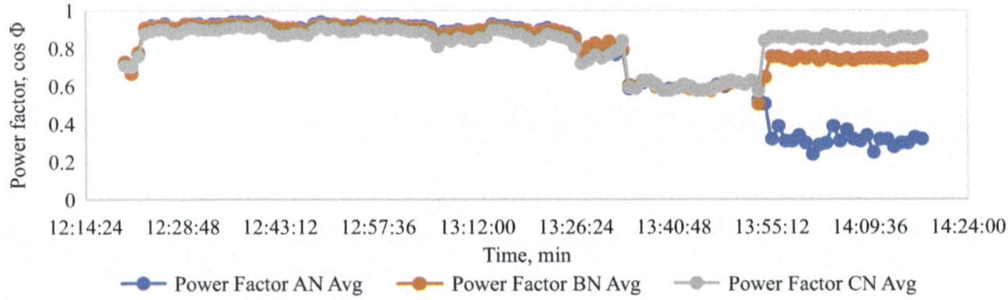

Fig. 11 Time versus power factor

of the system is maintained throughout the analysis phase. Figure 11 illustrates the time versus power factor of the coir industry. Here, the coir unit's overall power factor is maintained appreciably but needs caution during unbalanced conditions. Power factor drop in the industry is identified due to the unbalance of the loads during the part load operation. Therefore, installing a 75 kVAr APFC panel is recommended to effectively manage the power factor at the Point of Common Coupling (PCC). Also, the effect of APFC is significantly less during the part (or) baseload conditions. During the loaded condition, the power factor is maintained at up to 0.93. However, during the base (or) part load condition, the power factor seems to drop to 0.52. It can be increased by applying supply chain management techniques for effective process optimization without compromising energy utilization.

Figure 12 illustrates the overall Voltage Total Harmonic Distortion (V-THDF) percentage of the coir industry. Here, the value ranges from 0.9 to 1.6%, which is maintained under the nominal value prescribed by IEEE 519. Also, the voltage harmonics limit of the coir industry is completely under the prescribed limit as per the Central Electricity Authority (CEA) and Central Electricity Regulatory Commission (CERC) guidelines for the harmonics of LTCT consumers.

Figure 13 represents the time versus percentage Current Total Harmonic Distortion (I-THDF) of the coir industry. During the analysis phase, under peak load operation, the overall percentage of I-THDF values is maintained under the limits

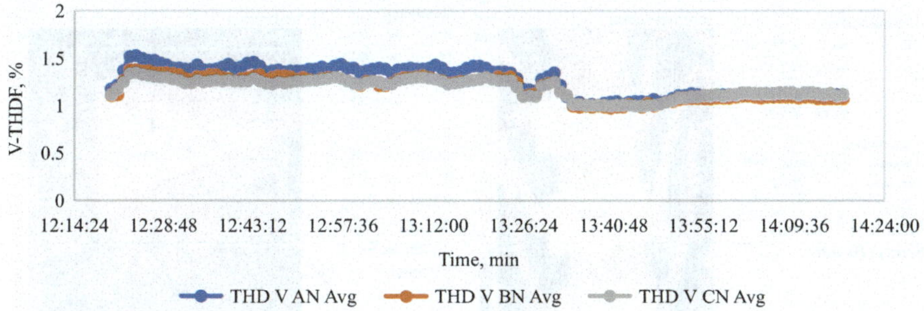

Fig. 12 Time versus % voltage total harmonic distortion factor

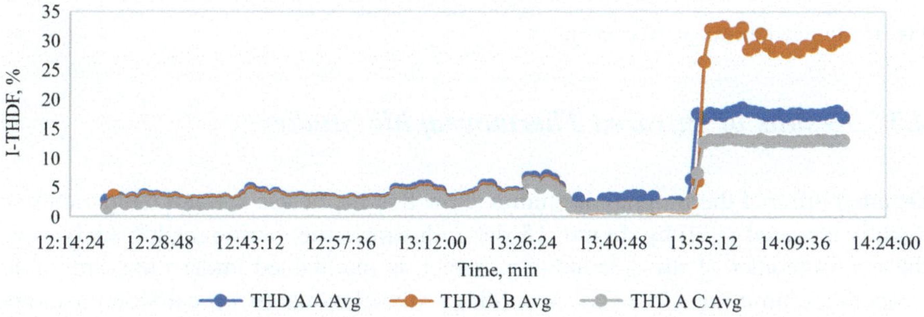

Fig. 13 Time versus % current–voltage total harmonic distortion factor (I-THDF)

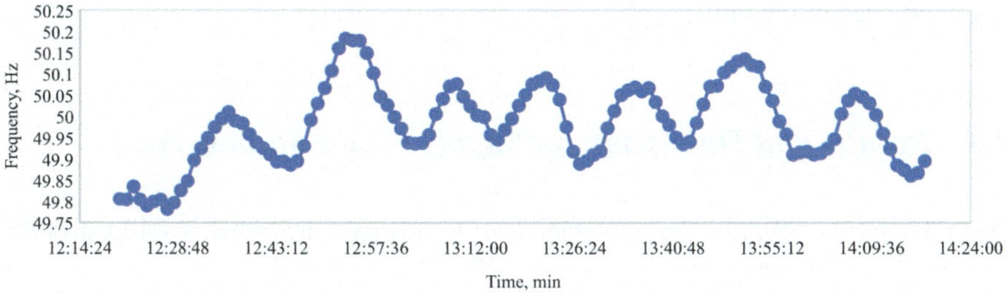

Fig. 14 Time versus frequency

of CEA. However, during the unbalanced (or) base load condition of the power system, the observed I-THDF value of 32–33% is significantly higher than the nominal values. The observed harmonic level is significantly higher than the prescribed limit of 8%, which is the allowable limit of India's Central Electricity Regulatory Commission (CERC).

Figure 14 illustrates the time versus frequency of the coir industry. The system's frequency is a health indicator, which ranges from 49.8 to 50.2 Hz and is maintained under the health condition during full and part load conditions.

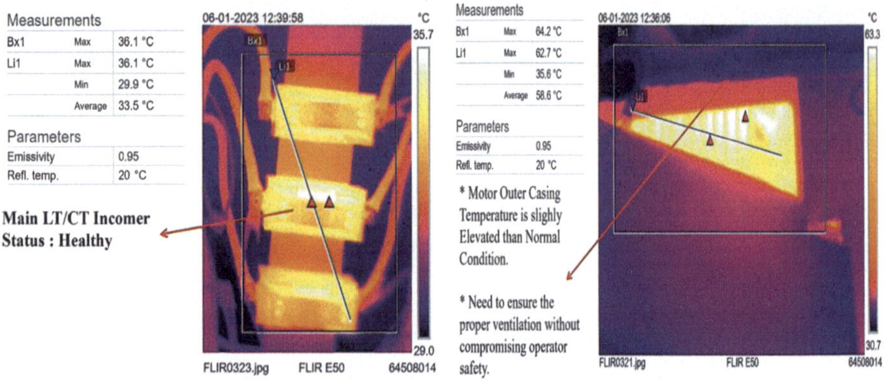

Fig. 15 Infrared thermographic analysis

4.3 Results of Infrared Thermographic Studies

Detailed infrared thermographic studies show that the beater's motor temperature is slightly elevated at 30 hp. Figure 15 (left) illustrates the thermographic analysis of the main incomer of the coir industry, which is maintained under the permissible temperature limit (i.e., 35.7 °C). According to the National Electrical Manufacturers Association (NEMA), the temperature of the motor casing is slightly elevated, and it is recommended that the proper air ventilation be ensured for better motor efficiency. Also, to ensure the health of lubricant, bearing and other fasteners to avoid mechanical anomalies in the unit.

4.4 Results and Discussion on Supply Chain Management

From the view of supply chain management to mitigate the unbalanced load in the coir industry, consider implementing the following solutions;

- *Supplier Diversification:* Work with multiple coir suppliers to reduce reliance on a single source. This ensures a steady supply of raw materials even if one supplier faces issues.
- *Forecasting and Demand Planning:* Utilize historical data and market trends to forecast demand accurately. This allows for better planning of production schedules and raw material procurement, preventing underutilization or overutilization of resources.
- *Inventory Management:* Implement efficient inventory management practices to avoid stockouts or excess inventory. Buffer storage of raw materials could reduce the risk of unbalanced loads.
- *Waste Management:* Implement effective waste management strategies to minimize energy losses from waste disposal and explore opportunities for recycling or repurposing waste materials to generate additional energy or revenue streams.

By implementing these measures, the coir industry can achieve better energy efficiency while ensuring economic viability and environmental sustainability. From the above-mentioned solutions in supply chain management, waste repurposing is one of the key ways to reduce energy revenue and the reliability of the grid power supply by producing electricity using the waste generated in the coir industry, such as baby coir.

- For waste-to-energy conversion, thermochemical techniques such as gasification are most suitable due to low cost and lower time consumption, especially downdraft gasifiers due to tar-free syngas (H_2, CO).
- Due to its tar-free gas generation, purifying the syngas used as fuel for the engine to operate and generating electricity through a generator is easy and cost-effective.
- For electricity generation, it is essential to determine that the waste produced is rich in energy.
- So, several analyses and tests have been done to determine the fuel's energy value, such as proximate analysis, ultimate analysis, and bomb calorimetry.

4.5 Results and Discussion on Proximate Analysis

Based on the proximate analysis results of three samples of coir waste, as illustrated in Fig. 16 and reported in Table 1, several inferences have been made:

Moisture Content: The moisture content across all three samples is relatively low, ranging from 6.66% to 7.44%. This suggests that the coir waste is relatively dry, which could be advantageous for storage and transportation purposes.

Fig. 16 Samples for proximate analysis

Table 1 Proximate parameters of samples under investigation

	Coir waste		
Biomass (%)	Reagent sample 1	Reagent sample 2	Reagent sample 3
Moisture content	6.66	6.83	7.44
Volatile matter	71.63	65.97	72.19
Fixed carbon	19.56	23.74	14.92
Ash content	2.15	3.46	5.45

Volatile Matter: Samples have the highest volatile matter content, up to 72.19%. This indicates that coir waste is rich in volatile compounds, likely contributing to its combustibility and potential use as a fuel source.

Fixed Carbon: Samples have the highest fixed carbon content up to 23.74%. Fixed carbon represents the nonvolatile carbonaceous residue left after volatile matter is expelled during combustion. Higher fixed carbon content suggests better fuel potential and higher energy yield upon combustion.

Ash Content: Samples have the highest ash content up to 5.45%. Ash content reflects the inorganic mineral content present in the biomass. Ash content of coir waste is much lesser than the other biomass. Lower ash content indicates lower impurities or mineral residue levels, which could improve combustion efficiency and reduce ash disposal requirements.

Based on these results, the variation in the results is mainly due to the heterogeneity of the samples in the industry. To address this complex heterogeneity, the stratified sampling technique is suggested. Since the size, source location, and age of the coir, the processing method could be stratified into individual strata for better accuracy of the varied samples in a large energy production process. In addition, systematic integrated grid sampling strategies are proposed for better sampling from the stockpile [30]. These results suggest that coir waste has potential as a fuel source due to its low moisture content, significant volatile matter, fixed carbon content, and low ash content. Additionally, variations in composition between the samples indicate the need for consistent quality control measures in processing coir waste for various large-scale applications.

4.6 Results and Discussion on Ultimate Analysis

Based on the elemental composition analysis results of three samples of coir waste, the following inferences were reported in Table 2,

Carbon Content: Samples have the highest carbon content, as high as 69.5315%. Carbon content is a crucial parameter in determining the energy content of biomass. Higher carbon content indicates higher potential energy yield upon combustion.

Hydrogen Content: Samples have the highest hydrogen content, up to 14.0268%. Hydrogen contributes to the combustibility and energy content of biomass. Higher hydrogen content generally results in higher energy output upon combustion.

Table 2 Ultimate parameters of sample under investigation

	Coir waste		
Biomass (%)	Sample 1	Sample 2	Sample 3
Carbon	69.4032	69.5315	65.6953
Hydrogen	13.0226	14.0268	11.3301
Nitrogen	0.6674	0.7806	0.6562
Oxygen	14.7593	12.2011	16.8684

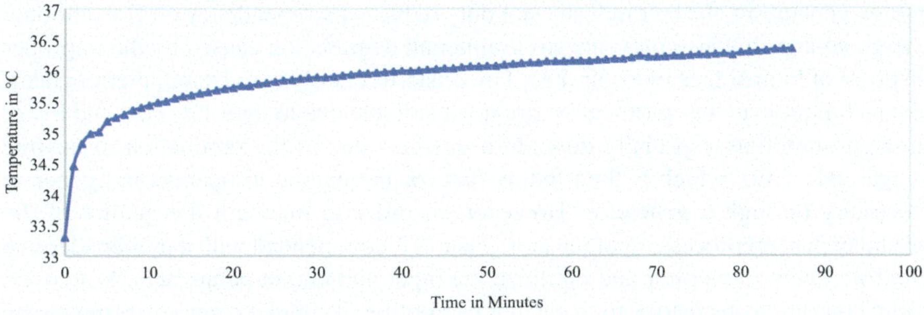

Fig. 17 Time versus temperature of bomb calorimeter analysis of GCV

Nitrogen Content: Samples have the lowest nitrogen content at 0.7806%, followed by Sample 1 at 0.6674% and Sample 3 at 0.6562%. Nitrogen content is essential for assessing the potential emissions of nitrogen oxides during combustion.

Oxygen Content: Samples have the highest oxygen content at 16.8684%. Oxygen content is inversely related to the energy content of biomass. Here, the oxygen content is comparatively lower than other biomass, especially rice husk, indicating higher energy content and a lower likelihood of incomplete combustion.

Overall, these results indicate that coir waste exhibits the highest carbon and hydrogen content, which suggests it may have the highest energy content. The nitrogen content across all samples is relatively low, which is favourable for combustion as it reduces the emission of nitrogen oxides.

Figure 17 shows that the combustion 1 gram sample of baby coir increased the water temperature from 33.30 °C to 36.35 °C. The calorific value of the 1-gram sample of baby coir is 7065.5 calories/gram, which is relatively higher than other biomass. This higher energy content of coir waste indicates its energy richness and is a suitable source of energy production through gasification.

5 Conclusions

Based on the systematic analysis, the findings are summarized and reported to ensure sustainability in the Indian rural industries. The manuscript focused on supply chain optimization and energy management and aligns with the circular economy principles by emphasizing the efficient utilization of resources throughout the production process. By integrating these strategies, the coir industry aims to maintain overall energy efficiency while reducing waste and promoting sustainable practices. Therefore, the manuscript shifts the industry focus from 3 M (Human, Financial, Inventory) to 4 M, including energy management. The paper highlights the significant potential of unhandled coir waste, such as short fibres, peat, and dust, to contribute to both economic sustainability and environmental preservation through resource recycling and a circular economy approach. By effectively utilizing coir waste for energy generation, the coir industry not only reduces its dependency on conventional energy sources but also prevents environmental degradation caused by the improper disposal of bioresidues like coir dust. For waste-to-energy conversion, thermochemical techniques such as gasification are most suitable due to their low cost and lower time consumption, especially downdraft gasifiers due to the production of tar-free syngas (H_2, CO), which is then fed as fuel for the engine to operate and generate electricity through a generator. However, in order to maintain the quality of the producer gas, preprocessing of the coir waste is recommended with a proper adaptive control system to monitor and optimize the input and output parameters. Moreover, using coir dust as feedstock for a gasifier exemplifies a circular economy principle by repurposing waste into a valuable resource. Instead of allowing coir dust to settle over the land, potentially causing harm to water bodies and the environment, the industry harnesses it as a renewable energy source, thereby closing the loop in the production cycle. Techno-economic feasibility study underscores the economic viability of integrating resource recycling measures, emphasizing the circular economy's role in driving sustainable solutions. This study concludes that analysis of the coir dust sample indicates it is rich in energy content; therefore, coir dust of 75–100 kg has the potential to generate 15 kW_e and 45 kW_{th} energy. According to the Ministry of Power (MoP) and Central Electricity Authority (CEA), India's specific power CO_2 emissions are 1.28 kg/kWh. By resource recycling, the overall annual carbon emission reduction of 7008 kg of CO_2 from the industrial operation. From the power quality analysis, poor power quality and power factor could be enhanced by installing an APFC panel of 73–75 kVAr capacity and optimal loading of machines. Also, APFCs need to be purchased with the provision for future capacity addition (i.e., Number of stages) for better future power optimization.

References

1. Stelte W, Reddy N, Barsberg S, Sanadi AR (2023) Coir from coconut processing waste as a raw material for applications beyond traditional uses Wolfgang. Bioresources 18:2187–2212. https://doi.org/10.15376/biores.18.1.Stelte
2. AlNouss A, Parthasarathy P, Shahbaz M, Al-Ansari T, Mackey H, McKay G (2020) Techno-economic and sensitivity analysis of coconut coir pith-biomass gasification using ASPEN PLUS. Appl. Energy 261:114350. https://doi.org/10.1016/j.apenergy.2019.114350
3. Choudhury ND, Bhuyan N, Bordoloi N, Saikia N, Kataki R (2021) Production of bio-oil from coir pith via pyrolysis: kinetics, thermodynamics, and optimization using response surface methodology. Biomass Convers. Biorefinery 11:2881–2898. https://doi.org/10.1007/s13399-020-00630-3
4. Choudhury ND, Saha N, Phukan BR, Kataki R (2021) Characterization and evaluation of energy properties of pellets produced from Coir pith, Saw dust and Ipomoea carnea and their blends. Energy Sources, Part A: Recovery, Utilization Environ Effect:1–18. https://doi.org/10.1080/15567036.2020.1871446
5. Kukharets S, Jasinskas A, Golub G, Sukmaniuk O, Hutsol T, Mudryk K, Čėsna J, Glowacki S, Horetska I (2023) The experimental study of the efficiency of the gasification process of the fast-growing willow biomass in a downdraft gasifier. https://doi.org/10.3390/en16020578

6. Kirubakaran V, Sivaramakrishnan V, Nalini R, Sekar T, Premalatha M, Subramanian P (2009) A review on gasification of biomass. Renew. Sust. Energ. Rev. 13:179–186. https://doi.org/10.1016/j.rser.2007.07.001
7. Nadimuthu LPR, Victor K (2021) Environmental friendly micro cold storage for last-mile Covid-19 vaccine logistics. Environ. Sci. Pollut. Res. 29:23767. https://doi.org/10.1007/s11356-021-17584-2
8. Kopnina H (2019) Green-washing or best case practices? Using circular economy and cradle to cradle case studies in business education. J. Clean. Prod. 219:613–621. https://doi.org/10.1016/j.jclepro.2019.02.005
9. Menon SD, Sampath K, Kaarthik SS (2021) Feasibility studies of coconut shells biomass for downdraft gasification. Mater Today: Proc 44:3133–3137. https://doi.org/10.1016/j.matpr.2021.02.813
10. Thanakam T, Kumsuwan Y (2023) A novel phase-lock-loop algorithm for three-phase grid voltage distortion-connected electric vehicle chargers. AEU-Int. J. Electron. C. 158:154468. https://doi.org/10.1016/j.aeue.2022.154468
11. Lalith Pankaj Raj N, Kirubakaran V (2021) Energy efficiency enhancement and climate change mitigations of SMEs through grid interactive solar photovoltaic system. Int J Photoenergy 2021: 1. https://doi.org/10.1155/2021/6651717
12. Sheeba KN, Babu JSC, Jaisankar S (2009) Air gasification characteristics of coir pith in a circulating fluidized bed gasifier. Energy Sustain. Dev. 13:166–173. https://doi.org/10.1016/j.esd.2009.06.002
13. Rangel N, Li H, Aristidou P (2023) An optimisation tool for minimising fuel consumption, costs and emissions from diesel-PV-battery hybrid microgrids. Appl. Energy 335:120748. https://doi.org/10.1016/j.apenergy.2023.120748
14. Ramadhas AS, Jayaraj S, Muraleedharan C (2006) Power generation using coir-pith and wood derived producer gas in diesel engines. Fuel Process. Technol. 87:849–853. https://doi.org/10.1016/j.fuproc.2005.06.003
15. Ramjani S, Gunasekar JJ, Vijayakumary P, Ramachandran J (2020) Characterization of coir pith for fluidized bed gasification. Int J Chem Stud 8:1120–1124. https://doi.org/10.22271/chemi.2020.v8.i6p.10912
16. Priya AB, Shri GKR, Dineshkumar M, Romi JN, Nepolean AV, Kirubakaran V (2020) Bio char and syngas production from coconut shell by pyrolysis: an experimental study. AIP Conference Proceedings. 2225. https://doi.org/10.1063/5.0005571
17. Yahaya AZ, Somalu MR, Muchtar A, Sulaiman SA, Wan Daud WR (2019) Effect of particle size and temperature on gasification performance of coconut and palm kernel shells in downdraft fixed-bed reactor. Energy 175:931–940. https://doi.org/10.1016/j.energy.2019.03.138
18. Lalith Pankaj Raj Nadimuthu GN, Karthik VM, Mohanraj M, Kirubakaran V (2019) Fast thermal degradation of biomass using scrapped solar cell with special focus on photovoltaic (PV) waste disposal. In: Waste valorisation and recycling. Springer, Singapore, pp 349–361. https://doi.org/10.1007/978-981-13-2784-1_33
19. Wiyono A, Pratiwi TPSI, Priadi CR, Surjosatyo A, Dafiqurrohman H (2017) Investigation of co-gasification characteristics of wood-coconut fibers pellet and rice husk mixtures in a downdraft fixed bed gasifier. Int J Technol 8:1207–1216. https://doi.org/10.14716/ijtech.v8i7.774
20. Martínez JD, Mahkamov K, Andrade RV, Silva Lora EE (2012) Syngas production in downdraft biomass gasifiers and its application using internal combustion engines. Renew. Energy 38:1–9. https://doi.org/10.1016/j.renene.2011.07.035
21. Nadimuthu LPR, Victor K, Basha CH, Mariprasath T, Dhanamjayulu C, Sanjeevikumar P, Khan B (2021) Energy conservation approach for continuous power quality improvement: a case study. IEEE Access. 1:146959. https://doi.org/10.1109/ACCESS.2021.3123153
22. Sreenath C, Siva Sakthi Velan S, Shankar C, Rajakumaran V, Kirubakaran V (2015) Experimental study on optimum area utilisation for maximum power output in solar photovoltaic system. Int J Energy Technol Policy 11:209–219. https://doi.org/10.1504/IJETP.2015.073792

23. Divya AL, Vigneshwari CA, Damini M, Sabeena MA, Kolambikai V, Kirubakaran V (2015) Energy auditing for a farm house. In: 2015 International Conference on Circuits, Power and Computing Technologies [ICCPCT-2015]. pp. 1–6. https://doi.org/10.1109/ICCPCT.2015.7159446
24. Nadimuthu LPR, Victor K (2021) Optimization of energy-intensive process in Ayurvedic medicine manufacturing unit—a case study. Process Integr Optim Sustain 5:975. https://doi.org/10.1007/s41660-021-00194-3
25. Kamaleswaran K, Venkateshwaran M, Harinath P, Mydeen MA, Kirubakaran V (2015) Energy conservation potential in rural industry: a case study on coir industry. IEEE International Conference on Circuit, Power and Computing Technologies, ICCPCT 2015. https://doi.org/10.1109/ICCPCT.2015.7159444
26. Alam MK, Thakur OA, Islam FT (2024) Inventory management systems of small and medium enterprises in Bangladesh. Rajagiri Manage J 18:8–19. https://doi.org/10.1108/RAMJ-09-2022-0145
27. Katuu S (2020) Enterprise resource planning: past, present, and future. New Rev. Inf. Netw. 25: 37–46. https://doi.org/10.1080/13614576.2020.1742770
28. Zhao Z-Y, Hao Y-X, Chang R-D, Wang Q-C (2023) Assessing the vulnerability of energy supply chains: influencing factors and countermeasures. Sustain Energy Technol Assess 56: 103018. https://doi.org/10.1016/j.seta.2023.103018
29. Nunes LJR, Matias JCO, Loureiro LMEF, Sá LCR, Silva HFC, Rodrigues AM, Causer TP, DeVallance DB, Ciolkosz DE (2021) Evaluation of the potential of agricultural waste recovery: energy densification as a factor for residual biomass logistics. Optimization 11. https://doi.org/10.3390/app11010020
30. Stelte W, Barsberg ST, Clemons C, Morais JPS, de Freitas Rosa M, Sanadi AR (2019) Coir fibers as valuable raw material for biofuel pellet production. Waste Biomass Valori 10:3535–3543. https://doi.org/10.1007/s12649-018-0362-2

Microcrystalline Cellulose: A Sustainable Alternative to Combat Heavy Metal Pollution in Soil

Lazar Arun and Evangelin Ramani Sujatha

Abstract Microcrystalline cellulose (MCC) is a natural biopolymer derived from partially depolymerized and purified cellulose, produced by treating alpha cellulose with a large quantity of mineral acids. MCC has a large surface area, is biodegradable, renewable, nontoxic, biocompatible, and has appreciable mechanical properties and economic value. The heavy metal intrusion from various industrial activities like dyeing industries, textiles, chemical industries, manufacturing industries, atomic power plants, etc., into the soil not only contaminates the soil but also pollutes the groundwater. The properties of the soil are adversely affected by heavy metal contamination. In this study, MCC is used to modify soil properties such that it attenuates heavy metal incursion into soil. Lead iron-oxide (Pb_2O_3) is selected as synthetic leachate for the study. The heavy metal removal was observed by adding 2% MCC by weight of dry soil at pH of 4, 6, 8, and 10. The dosage of MCC was selected based on its capacity to enhance the geotechnical properties of the soil. The contact time varied between 0 and 8 h. The investigation showed that the maximum adsorption capacity of 2% MCC at a pH of 10 is 48.46 mg/L while that of soil was 24.13 mg/L. The adsorption increased by nearly two times by treating the soil with 2% MCC. The results of the study show that MCC can be employed to control soil pollution.

Keywords Kaolin · Microcrystalline Cellulose (MCC) · Heavy metal Adsorption

1 Introduction

The performance and integrity of substructures are affected by pollutants that contaminate the soil. Sources of soil pollutants include both domestic and industrial processes, such as waste disposal from industries like ore and mineral processing, waste incinerators, lead-acid battery manufacturers, leather tanneries, power plant

---

L. Arun · E. R. Sujatha (✉)
School of Civil Engineering, SASTRA Deemed to be University, Thanjavur, Tamilnadu, India
e-mail: sujatha@civil.sastra.edu

P. Pathak et al. (eds.), *Advances in Environmental Sustainability, Energy and Earth Science*, https://doi.org/10.1007/978-3-031-73820-3_4

waste disposal, pesticide, sewage, smelting and mining, etc. The treatment and reclamation techniques adopted for various heavy metals depend on their effect on the human population. The toxicity levels for select heavy metals on human health is of the order: Cobalt < Aluminum < Chromium < Lead < Nickel < Zinc < Copper < Cadmium < Mercury [1]. The effect of heavy metal contamination is unique to the heavy metal under consideration; for example, the trivalent form of Cr bound with clay mineral structure as a residual portion can cause chemical reactions depending on the changes in the subsurface conditions. Similarly, aquatic ecosystems can also be severely affected by pollution caused by heavy metal contamination in the soil, which will also affect human health [2]. The percolation of leachate into the soil can alter the physical and chemical characteristics of pore fluids, modify soil structure, and affect the volume change behavior of the soil and its water-holding capacity [3]. Cemented soils are also affected by soil pollution, as the presence of heavy metals can affect their crystalline matrix. The effects of heavy metal soil pollution underline the need for various interventions to attenuate or prevent the migration of heavy metals into the soil.

Traditional techniques for heavy metal removal include physical adsorption, electrokinetic remediation, leaching, electrolysis, flocculation precipitation, phytoremediation, biofuel cell, and biomineralization [4]. Biogenic processes and additives offer an environment-friendly and sustainable alternative over traditional techniques that are used for heavy metal remediation in combating soil pollution caused by heavy metal contaminations. Biomineralization, a biological method that uses microorganisms to induce mineralization, is not only highly efficient and environmentally friendly but also offers long-term stability. It can help in achieving heavy metal remediation through the formation of carbonate invertebrates, silicates in diatoms and algae, carbonate and calcium phosphate in vertebrates, and phosphates through microbial metabolic activities. Heavy metal immobilization using microbial carbonate precipitation (MICP) by urea hydrolysis is now one of the most researched techniques and shows promising results. Similarly, biopolymers can be used as an additive to soil, and they help in modifying the porosity, permeability, and compressibility of the soil, which aids in attenuating the heavy metal migration into the soil. Hydrogels that are formed on the addition of biopolymers to soil help in heavy metal adsorption. The performance of the treated soil is assessed through adsorption studies by varying pH range, concentration of the heavy metal, dosage of additive, and the contact period [5] and used xanthan gum to contain heavy metal contamination in soil for heavy metals Cu, Cd, Pb, and Zn. The authors [5] used a mixing ratio of biopolymer to heavy metal as 1:100 and observed the removal efficiency was maximum for Pb at 80.9% while the least removal efficiency was observed for Zn at 61.7%. Cellulose being abundantly available in nature can be a viable biopolymer to remove heavy metal lead from contaminated soil. Lead contamination in the soil can result from various anthropogenic activities like mining, peeling of lead-coated paints, the battery industry, agriculture, etc. In this study, 2% of MCC is used to attenuate the ingress of synthetic lead leachate into kaolin, a highly compressible clay soil. Adsorption studies by varying contact time and pH are carried out to understand the effect of MCC on lead-contaminated kaolin, and the removal efficiency is estimated for the selected pH range and contact time.

2 Methodology

2.1 Materials

The kaolin clay was procured from Astraa Chemicals, Chennai, Tamil Nadu. Kaolin is white in color with a flaky texture. The molecular formula of kaolinite clay is $Al_2Si_2O_5(OH)_4$, and its molecular weight is 258 g/mol. Microcrystalline cellulose (MCC) powder is a highly refined variant of cellulose powder produced by spray drying to attain exceptional purity, devoid of organic and inorganic impurities. It boasts superior water absorption, swelling, and dispersion characteristics, setting it apart from conventional alpha cellulose. This fine, white powder flows freely and exhibits insolubility in water, weak acids, common organic solvents, and oils. It demonstrates partial solubility and swelling in dilute alkali solutions, remains metabolically inert, offers significant internal surface area and absorbent qualities. MCC was purchased from Amster Microcell in Gujarat, India. It is insoluble in water and does not possess gel-forming capabilities.

The lead stock solution was prepared based on IUPAC standards. Initially, 0.16 g of lead nitrate was dissolved in 100 ml of distilled water, followed by the addition of lead-free nitric acid to the mixture. Upon thorough mixing, distilled water was incrementally added to achieve a final volume of 1000 ml. Lead contamination in soils leads to severe problems for humans, such as chronic and acute toxicity. The extent of toxicity depends on both duration and level of exposure [6]. However, acute lead toxicity is rare.

2.2 Adsorption Isotherms

Adsorption energy reflects the binding between the adsorbate and substrate. It is a vital parameter that determines the interaction between the material and the adsorbate, making this method a popularly adopted method of various treatments to remove or attenuate the movement of pollutants. Here, adsorbate refers to the material that is adsorbed, and adsorbent is the medium (solid/liquid) where adsorbate accumulates. Examples of adsorbents are natural clay, algae, biopolymers, microorganisms, etc., while examples of adsorbate are radionuclides, industrial effluents, and heavy metals [7]. The sorption kinetics of Pb were evaluated at various contact times: 10, 30, 60, 90, 120, 180, 240, 300, 360, 420, and 480 min. The biopolymer was mixed with soil and the heavy metal solution at a ratio of 1:20, with the initial concentration adjusted to 100 mg/l. The slurry containing kaolin and MCC was stirred at 150 rpm and 25°C throughout the experiment. Samples were filtered by allowing them to pass through 180 μm thick filter paper at each specified contact interval. Following Pb concentration analysis, Langmuir and Freundlich sorption isotherms were derived to ascertain the sorption capacity and characteristics of the biopolymer.

Pb uptake per gram was determined using the equation [8].

$$q_e = \frac{(C_0 - C_e)}{m} \times V \quad (1)$$

The % removal of Pb after sorption is

$$\text{Removal\%} = \frac{(C_0 - C_e)}{C_0} \times 100 \quad (2)$$

where q_e is the Pb quantity adsorbed at equilibrium time (mg/L), C_0 is the initial Pb concentration (mg/L), C_e is the equilibrium Pb concentration (mg/L), V is the volume of Pb solution (L), and m is the mass of adsorbents. In addition, K is the distribution coefficient, and n is a correction factor.

3 Results and Discussion

3.1 Sorption Characteristics of Pb onto MCC

The average removal percentages of Pb by the biopolymer at various pH ranges as well as contact periods are 68.8%, 84%, 80.9%, and 96.9% corresponding to the pH ranges of 4, 6, 8, and 10, respectively, as observed from Fig. 1. The biopolymer attained the maximum removal capacity in a short reaction time, signifying that sorption equilibrium occurred swiftly, with an exception at pH 6. The data on sorption kinetics were obtained from experiments with various contact times between MCC and Pb. The Pb sorption at different pH ranges followed a positive correlation coefficient with pseudo-second order kinetics (Fig. 1). In the case of Langmuir isotherm, the correlation coefficients of 2% MCC were about 0.951, 0.936, 0.746, and 0.777 for the pH range of 4, 6, 8, and 10, respectively, whereas the correlation coefficients were 0.988, 0.984, 0.926, and 0.933 in Freundlich isotherm, for the above said pH range.

The surface coverage (θ) of Pb on the monolayer was estimated from the Eq. (3), using the sorption constant (K_L) [5].

$$\theta = \frac{K_L \times C_0}{1 + K_L C_0} \quad (3)$$

where θ is the surface coverage of the biopolymer which increased gradually with increasing pH range. The maximum surface coverage was nearer to 1 at all

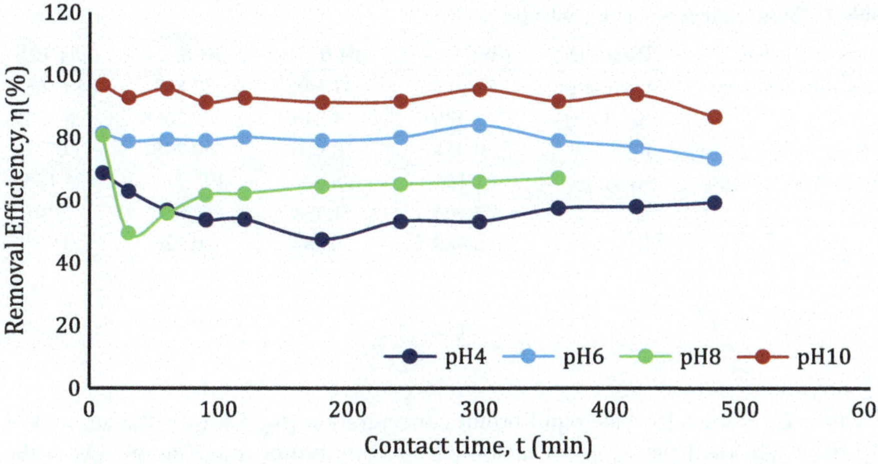

Fig. 1 Removal efficiency of 2% MCC at various pH ranges and contact time

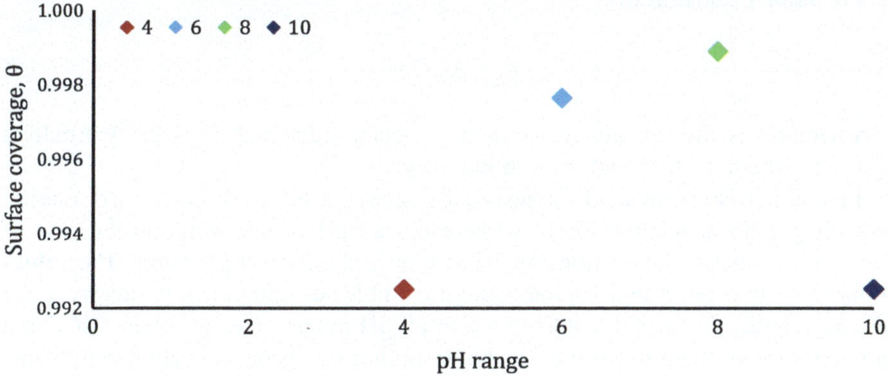

Fig. 2 Effect of pH on surface coverage

investigated pH ranges, indicating that the MCC surface was covered almost completely with a Pb monolayer. Figure 2 shows the surface coverage versus pH relationship.

The Langmuir isotherm models are founded on the assumption of monolayer adsorption occurring on uniform, specific homogeneous surface of adsorbent with a finite number of adsorption sites [8], while Freundlich isotherm is based on the heterogeneous surface area of adsorption, suggesting multilayer adsorption. The isotherm model parameters for lead adsorption are shown in Table 1.

Langmuir isotherm [8].

Table 1 Pb adsorption isotherm model parameters

Isotherm models	Parameters	pH 4	pH 6	pH 8	pH 10
Langmuir isotherm	Q_0 (mg/g)	38.911	31.348	22.883	43.668
	K_L (L/mg)	1.3268	4.2445	9.1648	0.3537
	R^2	0.951	0.936	0.746	0.777
Freundlich isotherm	K_F (L/g)	61.154	385.68	175.2	53.133
	n_F	7.463	1.438	2.080	1.401
	R^2	0.988	0.984	0.926	0.933

$$q_e = \frac{Q_0 K_L C_e}{1 + K_L C_e} \tag{4}$$

where C_e is the adsorbate equilibrium concentration (mg/L), q_e is the amount of adsorbate adsorbed per gram of adsorbent at equilibrium time (mg/g), Q_0 is the amount of Pb adsorbed monolayer capacity (mg/g), and K_L is the Langmuir isotherm constant (L/mg).

Freundlich isotherm [8].

$$q_e = K_F C_e^{1/nF} \tag{5}$$

where K_F is the Freundlich isotherm constant (L/g) and n_F is the Freundlich constant, which relates to adsorption intensity.

Freundlich isotherm model consistently shows a $R^2 > 0.9$ at all pH ranges, indicating a linear relationship at all investigated pH ranges, while in the case of Langmuir isotherm, the relationship is linear only at the lower pH range. The values of n_F range between 1 and 10, indicating favorable adsorption [8]. In this work the nF range indicates favorable adsorption in all pH ranges. The adsorbent MCC not only removes or immobilizes the Pb, but also improves geotechnical properties like shear strength and controls the permeability of the soil with aging. Changes in pH affect mobility and leaching effect. When pH is low, mobility is high [9] due to higher dispersion of the adsorbent [9].

4 Conclusions

The study points out that MCC, a green material derived from refined wood, can be used as a sorption material for the heavy metal Pb. The sorption equilibrium time was determined by the reaction between the heavy metal solution and the mixture of MCC with kaolin. Sorption occurred at 10 min in the pH range of 4, 8, and 10 whereas in pH 6, occurred in 300 min. The sorption capacity of kaolin and MCC-treated kaolin, both contaminated with lead, was studied. Freundlich isotherm model showed a better linear fit at all investigated pH ranges, and the sorption kinetics followed pseudo-second order. The maximum removal efficiencies observed at investigated pH ranges of 4, 6, 8, and 10 were 68.78%, 84.01%, 80.86%, and 96.91%, respectively. The removal efficiency increased with an increase in the pH of synthetic leachate. The sorption experiment suggested that the MCC can be used as a postulant material for the remediation of heavy metals contaminated soil.

References

1. Mansourri G, Madani M (2016) Examination of the level of heavy metals in wastewater of Bandar Abbas wastewater treatment plant. Open J Ecol 6(2):55–61. https://doi.org/10.4236/oje.2016.62006
2. Prasad S et al (2021) Chromium contamination and effect on environmental health and its remediation: a sustainable approaches. J Environ Manag 285:112174. https://doi.org/10.1016/j.jenvman.2021.112174
3. Negahdar A, Shabanian MR, NikGhalbPour M (2018) The effect of heavy metal contaminants on the strength parameters of sandy clay. Amirkabir J Civil Engg 50(5):835–844. https://doi.org/10.22060/ceej.2017.12869.5287
4. Wang S et al (2023) Bio-remediation of heavy metal-contaminated soil by microbial-induced carbonate precipitation (MICP)—a critical review. Sustainability 15(9):7622. https://doi.org/10.3390/su15097622
5. Ko M-S, Jeon Y-J, Kim K-W (2022) Novel application of xanthan gum-based biopolymer for heavy metal immobilization in soil. J Environ Chem Eng 10(5):108240. https://doi.org/10.1016/j.jece.2022.108240
6. Johnson FM (1998) The genetic effects of environmental lead. Mutat Res 410(2):123–140. https://doi.org/10.1016/S1383-5742(97)00032-X
7. Dawoud MMA et al (2021) Overview of environmental pollution and clean management of heavy metals and radionuclides by using microcrystalline cellulose. J Nucl Energy Sci Power Gener Technol 3:2
8. Selvakumar A, Rangabhashiyam S (2019) Biosorption of rhodamine B onto novel biosorbents from Kappaphycus alvarezii, Gracilaria salicornia and Gracilaria edulis. Environ Pollut 255:113291. https://doi.org/10.1016/j.envpol.2019.113291
9. Pikuła D, Stępień W (2021) Effect of the degree of soil contamination with heavy metals on their mobility in the soil profile in a microplot experiment. Agronomy 11(5):878. https://doi.org/10.3390/agronomy11050878

Sustainable Recycling of PPE Biomedical Waste for Eco-Friendly Brick Production

Balendra Mouli Marrapu and Vikram Sai Pavan

Abstract Personal protective equipment's (PPEs) waste is one of the hazardous wastes, which includes surgical gowns, head covers, surgical masks, shoe covers, etc. During COVID-19, a large amount of PPEs waste was generated and was disposed into oceans and surrounding environments (landfill or incineration). PPEs are generally made up of nonwoven fabrics which have been used widely in numerous applications like geotextiles, medical, etc., and they consist of many polymers like polypropylene, polyacrylonitrile, polyurethane, etc. This kind of waste leads to different kinds of environmental pollution like water pollution, air pollution, etc. This paper presents a study on generating construction material (brick) composites from PPEs waste and fly ash. The work (manufacturing of brick) has been carried out with a proportion of PPEs waste (40%) and fly ash (60%). Different properties such as compressive strength, water absorption, efflorescence, and soundness were investigated through standard tests suggested by IS: 3495 (1992). A comparison between the properties of the proposed brick with a standard clay brick reveals that the brick prepared from the PPEs waste may be used in the construction of temporary structures, non-load-bearing walls, and partition walls with satisfactory performance.

Keywords Bricks · PPE (Personal Protective Equipment) · Sustainability · Non-Load Bearing Walls & Partition Walls

---

B. M. Marrapu (✉)
Department of Civil Engineering, Anil Neerukonda Institute of Technology and Sciences, Visakhapatnam, India

V. S. Pavan
School of Construction, NICMAR, Pune, India

P. Pathak et al. (eds.), *Advances in Environmental Sustainability, Energy and Earth Science*, https://doi.org/10.1007/978-3-031-73820-3_5

1 Introduction

The global outbreak of the COVID-19 pandemic has triggered an unprecedented surge in the utilization of personal protective equipment (PPE) kits worldwide [1]. Central Pollution Control Board (CPCB) has been monitoring biomedical waste, including PPE waste. During the peak of the COVID-19 pandemic, India was generating around 101 metric tonnes of COVID-19 related biomedical waste per day. CPCB reported that India generated over 18,000 tonnes of COVID-19 related biomedical waste from June 2020 to March 2021.These kits, encompassing masks, gloves, gowns, and face shields, play a crucial role in protecting frontline workers and the public from the transmission of the virus [2]. However, the extensive usage of PPE kits has led to a substantial increase in their disposal rates, posing significant environmental and public health challenges [3]. The resultant accumulation of PPE waste, predominantly comprising nonbiodegradable materials, not only strains landfill capacities but also presents hazardous risks to ecosystems and human health [4].

On the other hand, fly ash, a by-product of coal combustion in power plants, also poses formidable environmental challenges [5]. Often disposed of in ash ponds or landfills, fly ash poses risks such as groundwater contamination, air pollution, and disruption of habitats [6]. Poor management of ash ponds can lead to the leaching of heavy metals and other harmful substances into soil and water, jeopardizing both ecosystems and human well-being [7]. Additionally, wind dispersion of fly ash particles exacerbates air pollution, adversely affecting respiratory health and environmental integrity [8]. Moreover, the extensive land use for ash pond storage contributes to habitat loss and biodiversity decline, further exacerbating environmental degradation [9].

The traditional practice of clay extraction for brick manufacturing also presents environmental challenges, including soil degradation and habitat destruction [10]. The firing process releases pollutants such as dust and gases, contributing to air pollution and climate change [11]. However, amidst these challenges lies a promising solution—the sustainable recycling of PPE biomedical waste enhanced with effective usage of fly ash for eco-friendly brick production [12]. This innovative approach offers a dual benefit, addressing multiple environmental concerns simultaneously [13]. By incorporating PPE waste into brick production, the demand for clay diminishes, reducing the environmental impact of clay extraction on soil and habitats [14]. Additionally, this reduction in clay brick production results in decreased emissions of greenhouse gases and other pollutants associated with traditional brick firing processes [15].

1.1 Personal Protective Equipment

Personal protective equipment (PPE) is specialized gear worn to establish a barrier between individuals and harmful germs. This barrier effectively diminishes the risk of contact, exposure, and transmission of germs. The utilization of PPE has become paramount in combating the transmission of COVID-19, serving as a prominent and dependable preventive measure. Amidst the COVID-19 pandemic, there has been a significant surge in its usage due to its single-use nature, necessitated by fears of COVID-19 transmission. However, this increased usage has led to significant disposal issues. PPE encompasses a wide range of items, including helmets, gloves, eye protection, gowns, and more. This study specifically focuses on the utilization of nonwoven polypropylene disposable gowns.

Here are some problems associated with the disposal of nonwoven polypropylene. Nonwoven polypropylene protective gowns often end up in landfills, where they accumulate over time. Since polypropylene is not biodegradable, these items can persist in the environment for hundreds of years, taking up valuable space in landfills. Over time, nonwoven polypropylene materials can break down into smaller plastic particles known as microplastics. These microplastics can leach harmful chemicals and pollutants into the soil and water, posing risks to wildlife and ecosystems. Additionally, microplastics can be ingested by marine organisms and enter the food chain, potentially causing harm to human health. Improper disposal of nonwoven polypropylene products can lead to environmental contamination. When these items are littered or improperly disposed of, they can end up in rivers, oceans, and other natural habitats, where they can harm wildlife and disrupt ecosystems. In the present study, PPE waste is collected from the ANIL NEERUKONDA HOSPITAL (ANH), Tagarapuvalasa, Visakhapatnam.

1.2 Fly Ash

Class F fly ash is a type of coal combustion residue that is produced during the combustion of pulverized coal in thermal power plants. Fly ash is composed of fine, glassy particles that are carried away from the burning coal in the flue gas. It has high levels of silica, alumina, and iron. When improperly disposed of, it poses environmental hazards such as soil and water contamination due to its heavy metal content, including arsenic, mercury, and lead. Additionally, the airborne dispersion of fly ash particles during handling and storage can contribute to air pollution and respiratory health issues in nearby communities. In the present study, fly ash is collected from the thermal power plants from Visakhapatnam NTPC.

2 Methodology

This methodology presents a novel approach to address the growing environmental concern of PPE biomedical waste by integrating it into eco-friendly brick production. Through systematic collection, processing, and incorporation of PPE waste alongside sustainable materials, this methodology aims to contribute to both waste management and sustainable construction practices.

2.1 Collection of PPE Gowns

To collect this disposable gown, people must follow proper procedures to ensure safety and hygiene. First, there is a need to collect the discarded PPE kit. That was obtained from a specific bin, separating it from the other waste. Second, keep the bin isolated for the required amount of time, up to 72 h. For this study, PPE gowns were collected from ANH, Tagarapuvalsa, Visakhapatnam. Figure 1 shows the disposed PPE waste in ANH.

2.2 Disinfectant

After that, they can be washed thoroughly using disinfectant. For this process, PPE gowns are soaked in disinfectant for a day, i.e., 24 h. For this process, Dettol disinfectant liquid of 50 ml mixed with 20 liters water is used as a disinfectant solution and can be observed in Fig. 2.

Fig. 1 Disposal gowns collected from ANH, Tagarapuvalasa

Fig. 2 Typical image of soaked PPE waste in 20 liters of disinfectant solution

2.3 Shredding

After the decontamination process, the waste should be exposed to atmospheric conditions, and then the waste should be allowed to densify into smaller pieces of different dimensions 2 cm × 2 cm, 3 cm × 3 cm, 4 cm × 4 cm, 5 cm × 5 cm, 2 cm × 5 cm, 2 cm × 4 cm, and 2 cm × 5 cm by using scissors/shredding machine. A sample of studded pieces is observed in Fig. 3.

2.4 Mixing Fly Ash and PPE Waste

Fly ash is mixed with different sizes of pieces of PPE, like 50% by weight fly ash is mixed with 50% by weight of PPE gown of 2 cm × 2 cm size, next fly ash is increased by 10% and PPE 2 cm × 2 cm reduced to 40% and prepared three bricks each. Similarly, fly ash of different compositions is mixed with different sizes of PPE gown pieces and prepared three bricks of each composition. Bricks size is standard size as per IS: 3495 (1992) [16]. Figure 4a–c shows the mixing of PPE and fly ash, molding of bricks, and prepared brick.

Fig. 3 Smaller pieces of 3 cm × 3 cm PPE gown

(a) (b) (c)

Fig. 4 (**a**) Mixing of PPE gown pieces with fly ash; (**b**) Preparing PPE bricks with the molds; (**c**) Finally prepared PPE brick for testing

2.5 Testing on Bricks

After preparing the bricks, they are left under sunlight for 3 days, as shown in Fig. 4c. Then, the testing of these bricks must be performed as per the code. The following tests are required to identify the class of bricks. Compressive strength test, water absorption, effloresce test, and weight of the brick should be identified. Before performing these tests based on compressive strength, the composition of materials should be identified by performing a compressive strength test on all compositions of bricks using a compressive testing machine (CTM). It was identified that the strength of fly ash 60% + 40% PPE of size 2 cm × 5 cm is getting the optimum value of around 40 Kgf/cm^2. Other composition shear strength results are shown in Fig. 5 below.

3 Results and Discussions

After identifying the composition of brick as 60% fly ash +40% PPE of size 2 cm × 5 cm strip from Fig. 5. Compressive strength test, water absorption, efflouresce test, and weight were performed as per IS: 3495 (1992), and the obtained results were presented in Table 1. From this table, it was identified as a Class-3 Brick as per IS 1077 (1992) [17].

Overall, the sustainable recycling of PPE biomedical waste for eco-friendly brick production presents a holistic solution that not only mitigates the adverse effects of waste accumulation but also contributes to the conservation of natural resources and the reduction of air pollution. In this paper, we delve into the technical processes,

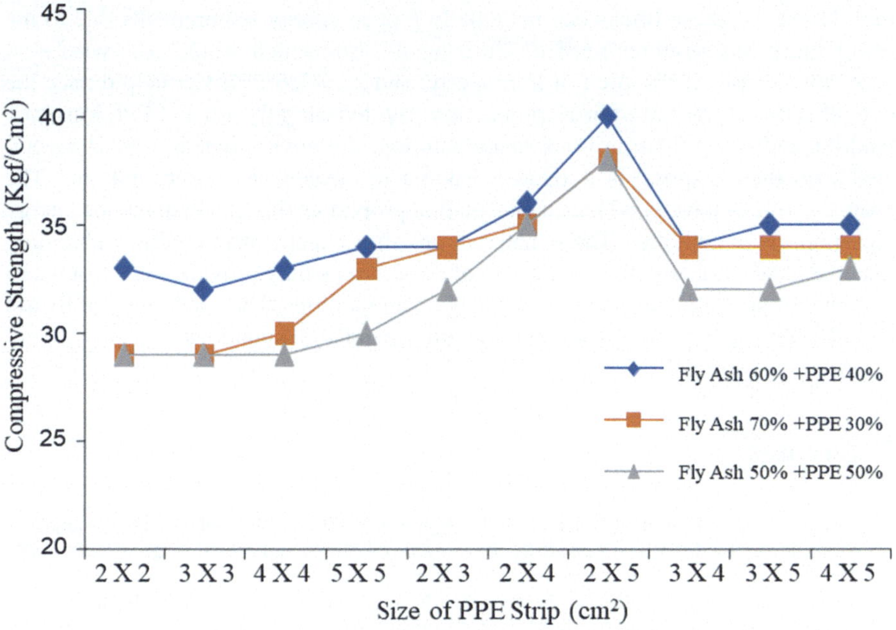

Fig. 5 Compressive strength of different compositions of bricks

Table 1 Physical and mechanical properties of PPE brick

S. N.	Type of test	Result from PPE brick	Class-3 brick as per IS 1077 (1992)
1	Brick size	24 cm × 11.5 cm × 8.5 cm	24 cm × 11.5 cm × 8.5 cm
2	Compressive strength	40 kgf/cm^2	35 kgf/cm^2
3	Water absorption	16%	<20%
4	Efflouresce	Nil	Nil
5	Weight	2.7–2.9 kg	–
6	Brick color	Gray	More variations in color

environmental benefits, and socioeconomic implications of this innovative approach, emphasizing its potential to foster a more sustainable future for both waste management and construction industries.

4 Conclusions

From the results obtained, it is observed that the following advantages can be achieved through PPE bricks over clay bricks. By using fly ash and discarded PPE kits as raw materials, these sustainable bricks reduce the demand for virgin materials such as clay, thereby conserving natural resources. Traditional brick manufacturing processes often require high temperatures, leading to significant energy consumption. However, these bricks can potentially reduce energy requirements due to the lower firing temperatures needed. Utilizing fly ash, which would otherwise emit greenhouse gases if disposed of improperly, and recycled PPE kits help reduce the overall carbon footprint of brick production. By diverting fly ash and PPE kits from landfills and using them in brick manufacturing, the environmental impact associated with their disposal is mitigated, leading to cleaner air, water, and soil. The proposed bricks have good thermal insulation properties that can help reduce energy consumption in buildings. These materials are often cheaper than traditional building materials. Additionally, the production of these bricks may provide a new source of revenue for power plants and waste management facilities. The availability of fly ash and PPE kit waste is comparatively high after pandemic situations.

References

1. Daigle P, Leung V, Yin V, Kalin-Hajdu E, Nijhawan N (2021) Personal protective equipment (PPE) during the COVID-19 pandemic for oculofacial plastic and orbital surgery. Orbit 40(4): 281–286
2. Gondi S, Beckman AL, Deveau N, Raja AS, Ranney ML, Popkin R, He S (2020) Personal protective equipment needs in the USA during the COVID-19 pandemic. Lancet 395(10237): e90–e91
3. Yang S, Cheng Y, Liu T, Huang S, Yin L, Yuepu P, Liang G (2022) Impact of waste of COVID-19 protective equipment on the environment, animals and human health: a review. Environ Chem Lett 20(5):2951–2970
4. Wang Z, Guy C, Ng KTW, An C (2021) A new challenge for the management and disposal of personal protective equipment waste during the COVID-19 pandemic. Sustainability 13(13): 7034
5. Huang TY, Chiueh PT, Lo SL (2017) Life-cycle environmental and cost impacts of reusing fly ash. Resour Conserv Recycl 123:255–260
6. Yao ZT, Ji XS, Sarker PK, Tang JH, Ge LQ, Xia MS, Xi YQ (2015) A comprehensive review on the applications of coal fly ash. Earth Sci Rev 141:105–121
7. Prasad B, Maiti D, Singh KKK (2019) Impact of fly ash placement in an abandoned opencast mine on surface and ground water quality: a case study. Mine Water Environ. 38(1):72–80

8. Hassen W, Hnaien N, Said LB, Albati FM, Ayadi B, Rajhi W, Kolsi L (2023) Air pollution dispersion in hail city: climate and urban topography impact. Heliyon 9(10)
9. Campus, SETU Carlow (2023) Ecological impact assessment
10. Wang S, Gainey L, Mackinnon IDR, Allen C, Yuantong G, Xi Y (2023) Thermal behaviors of clay minerals as key components and additives for fired brick properties: a review. J Build Engg 66:105802
11. Nasr MS, Salman AJ, Ghayyib RJ, Shubbar A, Al-Mamoori S, Al-khafaji Z, Hashim TM, Hasan ZA, Sadique M (2023) Effect of clay brick waste powder on the fresh and hardened properties of self-compacting concrete: state-of-the-art and life cycle assessment. Energies 16(12):4587
12. Thoudam K, Hossiney N, Kumar SL, Alex J, Prakasan S, Chandra S, Urs Y, Arunkumar AS (2023) Recycled surgical mask waste as a resource material in sustainable Geopolymer bricks. Recycling 8(6):93
13. Abbasi S, Sıcakyüz Ç (2024) A review of the COVID-19 pandemic's effects and challenges on worldwide waste management for sustainable development. Int. J. Environ. Sci. Technol.:1–30
14. Akponovo O, Onyebuchukwu LI (2023) A review on recycled materials used in construction. Int J Struct Constr Engg 17(10):408–414
15. Raza IA, Maaze R, Attri GK, Shrivastava S (2023) Comparative life cycle assessment of recycled soil-stabilized bricks and traditional bricks. Mater Today: Proc 80:532–537
16. Standard I (2019) IS: 3495: 1992, methods of Tests of burnt clay building bricks—specification. Bureau of Indian Standards, New Delhi
17. Standard I (1997) IS 1077: 1992, common burnt clay building bricks–specifications. Bureau of Indian Standards, New Delhi

Assessment of Financial Risk Associated with Integrated Project Delivery Using BIM

Aditya Wagh and Gayatri S. Vyas

Abstract In modern trends in construction industries, the paradigm shift towards Integrated Project Delivery (IPD) methodologies and the increasing reliance on Building Information Modeling (BIM) for project execution are transforming the industry. Studies have found the wide use of BIM technology for IPD risk management. In India, a rapidly growing economy with a burgeoning construction sector, the adoption of IPD and BIM presents both opportunities and challenges, influenced by unique socioeconomic dynamics and regulatory landscapes. Despite a growing body of literature emphasizing the positive impact of IPD and BIM on project outcomes, there remains a distinct gap in understanding the specific financial risks associated with IPD projects executed through BIM. This research seeks to address this gap by identifying and assessing the financial risks inherent in the execution of IPD projects using BIM. The objectives include conducting a comprehensive study to identify key risk parameters in IPD, developing a BIM model with various layers for clash detection, validating simulation results, and comparing financial outcomes with conventional building projects. The research aims to contribute valuable insights into the financial risk landscape of IPD projects, aiding stakeholders in making informed decisions for successful project delivery in the dynamic construction industry. To achieve this objective, case studies of residential projects from Pune city, India, are taken to assess the financial risk using BIM with reference to the framework mentioned in previous studies.

Keywords Building Information Modeling (BIM) · Circular Economy · Construction Industry · Financial Risk Assessment · Integrated Project Delivery (IPD) · Project Management

---

A. Wagh (✉) · G. S. Vyas
Department of Civil Engineering, COEP Technological University, Pune, India

P. Pathak et al. (eds.), *Advances in Environmental Sustainability, Energy and Earth Science*, https://doi.org/10.1007/978-3-031-73820-3_6

1 Introduction

The construction industry is undergoing a significant transformation, with a growing trend towards more collaborative procurement systems and the adoption of advanced information and communication technologies [1]. Integrated Project Delivery (IPD) and Building Information Modeling (BIM) [2] have emerged as two of the most widely adopted systems driving this shift. These concepts are expected to become the new industry standard in the near future due to their collaborative nature, ability to implement sustainable procurement strategies, risk and reward sharing basis, and high efficiency of construction by promoting dry construction. One of the key drivers for the adoption of IPD [3] and BIM in the construction industry is the need for improved collaboration among project stakeholders. IPD is an innovative procurement approach that brings together the client, designer, and contractor under a single contract, fostering a collaborative environment. BIM [1], on the other hand, serves as a digital representation of the physical and functional characteristics of a facility, providing a shared knowledge resource for information about a facility and forming a reliable basis for decisions during its life cycle. The integration of IPD and BIM has been shown to improve project outcomes, such as reduced costs, shorter schedules, and higher quality, by supporting the key principles of IPD, such as early involvement of stakeholders, collaborative decision-making, and a focus on project goals rather than individual interests.

Another significant driver for the adoption of IPD and BIM is the increasing demand for sustainable development in the built environment. The construction industry is resource-intensive, and a circular economy (CE) approach is crucial to minimize global impacts and conserve natural resources. BIM has been recognized as a valuable tool for promoting sustainability in construction projects by facilitating the integration of CE principles, such as design for disassembly, use of renewable and recycled materials, and optimization of resource use. The adoption of IPD and BIM can lead to cost savings, innovation in building design and construction techniques [4], and the creation of new job opportunities in the areas of waste management [5], material recovery, and recycling.

The construction industry's shift towards IPD and BIM adoption is also driven by the need for improved project performance and efficiency [6]. BIM has been at the forefront of this technological awakening, promising to improve project quality and efficiency in numerous ways [1]. The integration of BIM faced challenges in the early stages, struggling to gain acceptance due to largely ad-hoc integration strategies within organizations. As technologies continue to develop beyond and in parallel with BIM, there is a lack of tools available to help evaluate, plan, and integrate such advancements into the workflows of the Architecture, Engineering, and Construction (AEC) industry. This highlights the need for a more structured approach to technology integration in the construction sector. Several studies have identified key factors influencing the adoption of IPD and BIM in the construction industry. A case study [2] of BIM adoption in Singapore revealed 26 critical success factors, including strategic and adequate IT infrastructure, availability of standards and guidelines, and government policy. Similarly, a study on IPD adoption in the Middle East found that the main barriers to implementation stem from cultural resistance to the new system and lack of knowledge about it, emphasizing the importance of strategies such as a tendency towards change and value-driven selection criteria [7].

The adoption of IPD and BIM in the construction industry is driven by the need for improved collaboration, sustainability, and project performance. While challenges exist, such as cultural resistance and lack of knowledge, the growing recognition of the benefits associated with these innovative approaches has led to an increasing trend towards their adoption worldwide. As the construction industry continues to evolve, the successful integration of IPD and BIM will play a crucial role in achieving a more sustainable, efficient, and resilient built environment. Based on the provided abstract, guidelines, and search results, here is a detailed section on how the principles of circular economy and resource recycling can be applied to the construction industry and the potential benefits.

1.1 Integrated Project Delivery

The topic of assessing the financial risk associated with IPD using BIM is a critical area of study within the realm of construction project management. IPD is a collaborative approach that involves various stakeholders, including the owner, architect, contractor, and other key participants working together from the project's inception to completion. This method emphasizes teamwork, transparency, and shared risk and reward among the project team. IPD facilitates data sharing, visualization, and analysis, allowing for better decision-making and coordination throughout the project lifecycle. The IPD framework requires a comprehensive understanding of how project dynamics and contractual arrangements impact financial outcomes [8]. IPD redistributes risks and rewards among project participants differently from traditional delivery methods, often requiring a shift in the way financial risks are identified, allocated, and managed.

Key parameters to consider in this assessment include:

1. Collaborative Risk Management: IPD promotes a shared risk environment where all parties have a vested interest in project success. Understanding how risk is distributed and managed collectively becomes crucial.
2. Contractual Framework: IPD contracts typically differ from traditional contracts, emphasizing shared responsibilities and outcomes. Analyzing the contractual framework is essential for identifying financial liabilities and incentives.
3. BIM-Enabled Cost Analysis: BIM offers capabilities for detailed cost estimation, scenario analysis, and forecasting. Leveraging BIM data can enhance financial risk assessment by providing more accurate and real-time insights into project costs and potential risks.

4. Performance-based Compensation: IPD often involves performance-based compensation structures where rewards are tied to project outcomes. Evaluating these compensation models and their impact on financial risk is essential.
5. Technology Adoption Challenges: Implementing BIM within an IPD framework introduces technological complexities and adoption challenges that can influence financial risk. Understanding these challenges is vital for effective risk management.

The integration of IPD with BIM presents unique opportunities and challenges for assessing financial risk in construction projects. This topic underscores the need for a holistic approach that considers collaborative practices, contractual arrangements, technological advancements, and their collective impact on financial risk management.

1.2 Building Information Modeling

BIM plays a crucial role in the assessment of financial risk associated with IPD. This advanced technology offers significant advantages in project planning, execution, and risk management, particularly in the context of IPD [9]. Firstly, BIM facilitates enhanced collaboration and coordination among project stakeholders. In an IPD framework where multiple parties are involved, including architects, engineers, contractors, and owners, BIM serves as a centralized platform for sharing and accessing project information. This integrated approach reduces the likelihood of miscommunication or errors during project phases, ultimately mitigating financial risks arising from rework or delays. Secondly, BIM enables better cost estimation and forecasting. By creating detailed digital models that encompass various project aspects such as design, construction, and scheduling, BIM enhances the accuracy of cost assessments [7]. This capability is crucial in IPD where financial transparency and predictability are essential for effective risk management. Thirdly, BIM supports improved risk identification and mitigation strategies. Through BIM, project teams can simulate and analyze different scenarios, identifying potential risks early in the project lifecycle. This proactive approach allows for the implementation of risk mitigation measures, thereby reducing the likelihood of cost overruns or unforeseen financial challenges. Furthermore, BIM contributes to enhanced project efficiency and productivity. By streamlining workflows and automating repetitive tasks, BIM optimizes resource utilization and project timelines. This efficiency directly impacts financial risk by minimizing operational inefficiencies and associated costs. In summary, the significance of BIM in assessing financial risk within an IPD framework is profound. It empowers project teams with comprehensive tools for collaboration, cost management, risk analysis, and overall project efficiency. Leveraging BIM technologies in IPD not only improves project outcomes but also minimizes financial uncertainties, fostering successful project delivery within budgetary constraints.

1.3 Financial Risks in Construction Projects

In the realm of construction projects, the adoption of BIM within IPD introduces unique financial risks that stakeholders must navigate. Understanding and assessing these risks is critical for project success and financial stability.

Cost Overruns and Budget Management: One significant financial risk associated with IPD using BIM is the potential for cost overruns. While BIM can enhance project visualization and coordination, unforeseen complexities or changes during the construction phase may lead to increased costs. The interconnected nature of IPD requires close collaboration among various parties, and any disruptions or miscommunication can impact cost estimations and budget management. BIM tools can assist in identifying potential clashes or discrepancies early on, but effective risk management strategies [10] are needed to address evolving cost projections and maintain financial discipline.

Data Integration and Compatibility: Another critical financial risk arises from challenges in data integration and compatibility across different BIM platforms and project participants. IPD involves multiple stakeholders with varying levels of BIM proficiency and software preferences. Misaligned data integration can lead to errors, delays, and additional costs in rectifying discrepancies. Investing in standardized protocols and interoperable BIM solutions can mitigate these risks, ensuring seamless data exchange and reducing financial implications related to rework or data translation issues.

Contractual Ambiguities and Liability: IPD contracts often involve complex arrangements that distribute risks and responsibilities among parties differently than traditional project delivery methods. Financial risks can emerge from contractual ambiguities or inadequate risk allocation strategies [11]. With BIM-enabled IPD, defining clear contractual obligations related to data ownership, intellectual property rights, and liability for design errors or omissions becomes crucial. Lack of clarity in contracts can result in disputes, legal costs, and financial liabilities impacting project outcomes.

Technology Adoption and Training Costs: Implementing BIM within an IPD framework requires substantial investments in technology infrastructure and workforce training. Financial risks arise from upfront capital expenditures for BIM, hardware, and IT support, as well as ongoing costs associated with software updates and maintenance. Moreover, the learning curve for stakeholders transitioning to BIM-based IPD workflows can impact productivity and project timelines. Effective risk management entails budgeting for technology adoption and skill development to optimize financial outcomes and project performance. The integration of BIM within IPD introduces both opportunities and financial risks for construction projects. Addressing these risks requires proactive risk management strategies, collaborative contractual frameworks, and investments in technology and skills development. By understanding and mitigating these financial risks, project stakeholders can capitalize on the benefits of BIM-enabled IPD, fostering greater efficiency, transparency, and financial sustainability in construction endeavors.

1.4 Linking IPD and BIM for Risk Assessment

Linking IPD with BIM offers a transformative approach to assess and mitigate financial risks associated with construction projects. IPD fosters collaboration among key stakeholders, including owners, architects, engineers, and contractors, from project inception to completion. BIM, on the other hand, enables the creation and management of digital representations of the physical and functional characteristics of a project. Combining these methodologies enhances risk assessment in the following ways:

Enhanced Collaboration and Communication: IPD promotes early involvement of all parties, fostering a culture of open communication and collaboration [3]. BIM serves as a common platform for sharing project information, allowing stakeholders to access real-time data, identify potential risks, and develop mitigation strategies collectively. This integrated approach minimizes misunderstandings and conflicts, reducing financial risks stemming from miscommunication or delayed information.

Real-time Simulation and Visualization: BIM facilitates the creation of detailed 3D models that can be used for simulation and visualization. By integrating cost and schedule data into these models, stakeholders can simulate various scenarios to anticipate risks related to construction sequencing, resource allocation, or design changes. For instance, clashes or inconsistencies detected through BIM can inform proactive risk management decisions, averting costly rework during construction.

Early Risk Identification and Mitigation: The collaborative nature of IPD coupled with BIM's analytical capabilities enables early identification of potential risks. By conducting virtual walkthroughs of the project using BIM [1], stakeholders can identify safety hazards, logistical challenges, or constructability issues well before physical construction begins. Addressing these risks upfront can prevent costly disruptions and change orders during later stages.

Data-Driven Decision Making: BIM generates a wealth of data that can be leveraged for informed decision-making. By integrating BIM with financial modeling tools, stakeholders can assess the financial implications of design alternatives or project changes in real time. This data-driven approach helps in evaluating trade-offs between project scope, cost, and schedule, thereby optimizing resource allocation and minimizing financial risks [7].

Continuous Monitoring and Control: Throughout the project lifecycle, BIM serves as a centralized repository of project information. This allows stakeholders to continuously monitor project performance against predefined metrics and benchmarks. By tracking key performance indicators such as cost variance or schedule deviations, stakeholders can implement timely corrective actions to keep the project on track financially. Integrating IPD with BIM provides a holistic framework for assessing and managing financial risks associated with construction projects. By fostering collaboration, enabling real-time simulation, facilitating early risk identification, supporting data-driven decision-making, and ensuring continuous monitoring, this approach enhances project predictability, efficiency, and, ultimately, financial performance.

1.5 Developing a BIM Model for Clash Detection and Simulation in IPD Projects

Developing a BIM model for clash detection and simulation within IPD projects is crucial for effectively assessing and mitigating financial risks associated with such collaborative endeavors. In an IPD framework, where multiple stakeholders work closely from project inception to completion, BIM serves as a foundational tool for streamlining communication and coordination [12]. Clash detection, a key feature of BIM, plays a pivotal role in risk management by identifying conflicts or interferences between different building system elements early in the design and construction phases. The development of a BIM model tailored for clash detection involves several essential steps. Firstly, the BIM model integrates 3D models contributed by various project participants, including architects, engineers, and contractors. Each discipline's models are layered and coordinated within the BIM environment, facilitating a comprehensive view of the project's virtual representation. Secondly, clash detection algorithms are applied to the aggregated BIM model to automatically identify clashes or spatial inconsistencies between different building components [13]. These clashes could involve structural elements conflicting with mechanical or electrical systems, for instance. By pinpointing these clashes virtually, project teams can proactively resolve issues before they manifest on-site, thus averting costly rework or delays.

Simulation functionalities within BIM further enhance risk assessment in IPD projects. Through BIM-enabled simulations, project teams can visualize and analyze construction sequences, material flows, and logistical operations. This simulation-driven approach allows stakeholders to identify potential bottlenecks, optimize workflows, and anticipate project risks related to schedule and budget. By developing a BIM model with robust clash detection and simulation capabilities, project teams in IPD settings can achieve significant risk mitigation benefits. Early clash detection minimizes design coordination errors, reducing the likelihood of costly change orders during construction. Simulation-based analysis empowers teams to make informed decisions, enhancing project predictability and overall financial performance. In conclusion, leveraging BIM technology for clash detection and simulation in IPD projects is instrumental in assessing and managing financial risks. The proactive identification and resolution of clashes and the informed decision-making enabled by BIM simulations contribute to improved project outcomes, aligning with the collaborative spirit and risk-sharing ethos of IPD.

2 Literature Review

Elghaish et al. [14] aims to enhance AEC project outcomes by fostering collaboration among stakeholders, involving them early on, and sharing risks and rewards. This paper proposes a framework using BIM tools to develop a cash flow approach without a traditional tender stage. It employs Activity-Based Costing (ABC) to precisely allocate costs to construction processes. The framework provides estimated cost data as cash inflow during the buyout stage, facilitating informed decision-making and improving cost management aligned with BIM and IPD goals. The study fills a gap in utilizing BIM for cash flow analysis within the IPD context, demonstrates its effectiveness through a real-life case study [2], and advocates integrating IPD with BIM to enhance project delivery. However, IPD faces challenges in accurate cost estimation at project initiation due to limited information. This research proposes integrating Target Value Design (TVD), ABC, and Monte Carlo simulation into the IPD cost structure using BIM. A framework is developed to address this, validated with a case study showing significant profit-at-risk variations compared to traditional methods. This approach improves IPD's cost estimation accuracy and practicality.

IPD relies on robust cost management practices that prioritize accurate, timely, and transparent sharing of cost data among project stakeholders [1]. A survey of 50 IPD experts in the USA, UK, and Australia identified key practices for effective cost management. ABC helps identify overhead costs and enhances transparency. Earned value management (EVM), when combined with ABC, facilitates equitable risk-reward distribution through mathematical modeling. Web-based management systems, supported by BIM, foster collaboration that is crucial for IPD success. This research underscores the need to elevate traditional cost management practices to ensure successful IPD delivery [3]. This study compared Design-Build (DB) and IPD methods in constructing two transmission control building facilities. Data on total cost growth, cost per square foot, schedule growth, construction intensity, and change-order factors were analyzed. Results favored IPD due to improved collaboration and BIM usage, contrasting with DB's performance. Future studies should employ more robust statistical analyses for validation, aiding owners in method selection for similar projects [15]. Project management aims to eliminate or minimize risks associated with a project, whether they stem from human activities or natural factors. While some risks are anticipated, others are unforeseen by project managers. Collaborative efforts within project teams, often using tools like, play a crucial role in risk mitigation. BIM facilitates project monitoring, ensuring effective execution and milestone achievement from planning through construction. This research examines BIM's effectiveness in mitigating project risks, particularly within the construction industry [16]. This paper identifies key challenges faced by stakeholders in the AEC industry regarding the adoption of IPD for successful construction project delivery. Challenges were identified through a literature review and a focus group discussion with seven industry professionals. Fourteen critical challenges were categorized as cultural, technological, financial, legal, and others. The study highlights barriers specific to IPD adoption without providing solutions, offering valuable insights for industry stakeholders to inform policy development.

Quebec is moving towards adopting BIM for major infrastructure projects but faces legal and contractual challenges like tendering processes, intellectual property rights, and risk-reward sharing [17]. The study examines how norms impact BIM implementation in Quebec's public infrastructure, proposing IPD to address legal barriers and foster collaboration. It suggests modifying norms to standardize collaborative mechanisms, incorporate two-stage negotiated processes, and implement risk-reward sharing through target costing, encouraging wider stakeholder participation [18]. This paper quantifies barriers to IPD based on input from 115 construction professionals in Malaysia. A conceptual model was developed from global research on key barriers and validated using partial least squares structural equation modeling. The findings reveal crucial insights into the nature and relative importance of these barriers. However, caution is advised when applying these results outside the Malaysian context. Practical recommendations include using change management tools like the European Corporate Sustainability Framework to promote IPD adoption and enhance project performance. This study addresses a research gap in Malaysia and offers guidance to stakeholders in developing countries facing similar challenges [19]. IPD is a collaborative system that promotes efficiency through accurate information and new technologies. A study reviewed IPD principles, including contract, process, information & modeling (I&M), team, and communication, using a qualitative analysis of articles from 2001 to 2018 across scientific databases. Results show a strong focus on collaborative contracts and frameworks for efficiency, increasing use of BIM in I&M, and the importance of well-structured teams. Process and communication areas present growth opportunities for research in IPD. This study provides a foundation for future IPD research.

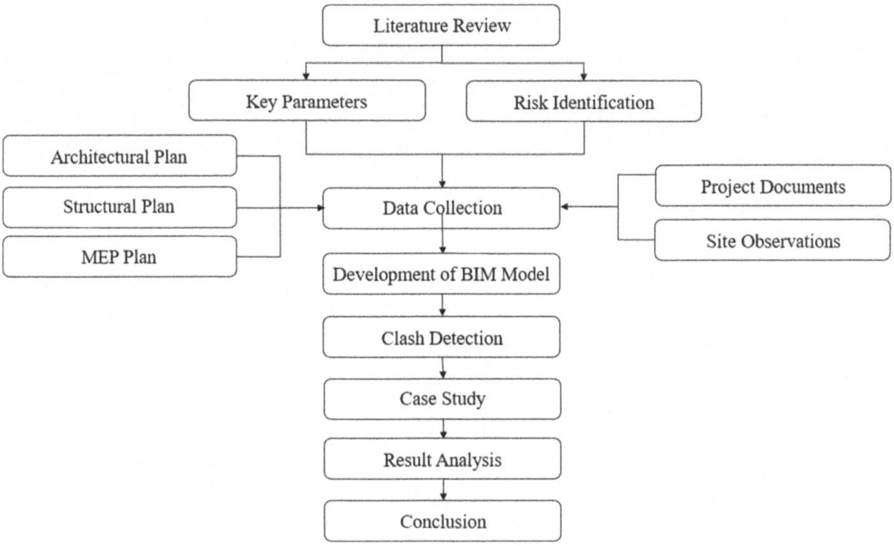

Fig. 1 Methodology flowchart

3 Methodology (Fig. 1)

3.1 Case Study Approach

Research conducted in real time to evaluate how BIM with IPD affects a project that is still in progress. In order to determine how the integration of BIM and IPD affects these crucial variables, the study will examine a number of different topics, including project durations, prices, collaboration, and quality. Understanding the efficacy of contemporary building techniques and their potential to raise industry standards and project outcomes is made possible by this research.

3.1.1 Data Collection Process

To comprehensively assess the financial risk associated with IPD utilizing BIM, a structured data collection process will be implemented. This process aims to gather insights from multiple sources to provide a holistic understanding of project dynamics, challenges, and financial implications. The following methodologies will be employed:

- *Project Documents Review*

  A comprehensive examination of project documents includes contracts, BIM models, financial reports, and related documentation. This review will facilitate the understanding of the project scope, delivery approach, technological integration, and financial performance indicators.

- *Interviews with Project Stakeholders*

  Conducting semi-structured interviews with key stakeholders involved in IPD-BIM projects, including project owners, designers, and contractors. These interviews will capture firsthand experiences, challenges encountered, lessons learned, and perceptions regarding financial risks associated with this project delivery method.

- *Site Visits and Observations*

  On-site visits will be conducted to observe the construction process, assess BIM utilization levels, and gather data on resource management practices and waste reduction strategies. These visits will provide valuable empirical insights into the operational aspects of IPD-BIM projects.

Each of these data collection methods is designed to capture different dimensions of the IPD-BIM approach and its financial implications comprehensively. By triangulating data from multiple sources—project documentation, stakeholder perspectives, direct observations, and industry reports—the study aims to provide a robust foundation for assessing the financial risks inherent in integrating IPD with BIM technologies. This systematic data collection process will ensure the reliability, validity, and richness of the findings, enabling a nuanced evaluation of the financial risk landscape associated with IPD-BIM projects. The integrated analysis of these diverse data sets will contribute towards developing informed strategies for mitigating financial risks and enhancing the overall success of IPD-BIM initiatives in construction projects.

3.2 BIM Model Development

Working with industry specialists, a full BIM model is constructed to produce an advanced conflict detection model. Through the early identification and resolution of potential conflicts or clashes during the design and construction phases, this model will be essential in evaluating changes in the project's cost. The project team may reduce risks and save money on rework by proactively resolving these conflicts, which will result in a more effective and economical project delivery.

3.2.1 Software and Tools Used

The assessment of financial risk associated with integrated project delivery using BIM relies heavily on sophisticated software and tools tailored for collaborative construction environments. Autodesk Revit 2023, a prominent BIM software, serves as the foundational tool for developing detailed 3D models enriched with project data. Its intelligent capabilities enable the visualization and coordination of various project components. To manage project complexities effectively, supplementary tools like Navisworks 2023 are deployed for clash detection and coordination reviews. Navisworks 2023 facilitates the identification and resolution of clashes or conflicts within the BIM models, thus reducing the likelihood of costly rework during construction. Additionally, BIM 360 enhances project collaboration by providing a cloud-based platform for real-time document management, communication, and information sharing among project participants. The integration of these software solutions not only supports the seamless implementation of integrated project delivery but also contributes to mitigating financial risks associated with construction projects.

3.2.2 Modeling Process and Clash Detection

In the assessment of financial risk associated with IPD using BIM, the modeling process and clash detection play crucial roles in managing project efficiency and minimizing financial exposure due to design coordination challenges. The BIM model development process for IPD begins with the creation of the architectural model as the foundational element. This initial step establishes a comprehensive framework upon which subsequent disciplines—such as structural, MEP (mechanical, electrical, and plumbing), and others are progressively integrated. This staged approach ensures a synchronized and cohesive design across all project dimensions. Clash detection within the BIM environment is a pivotal aspect of risk management. It involves continuous evaluations at each modeling phase to identify potential conflicts between disparate building elements. These clashes, if left unaddressed, can lead to costly rework and delays during the construction phase.

The following key points are fundamental in the clash detection process:

- *Clash Categories:* Throughout the project lifecycle, various clash categories are identified, such as structural clashes, MEP clashes, architectural clashes, and so forth. Each category represents a unique set of coordination challenges.
- *Plan Value:* This denotes the projected cost associated with resolving clashes within each category. It is an estimation of potential expenditures based on the severity and complexity of identified clashes.
- *Actual Cost:* The actual cost provides insights into the financial impact of coordination issues on project budgets, reflecting the real expenses incurred in addressing clashes.
- *Total Plan Value and Total Actual Cost:* These aggregate values sum up the estimated and actual costs across all clash categories, respectively, providing an overview of the financial implications of clash resolution efforts throughout the project duration.

By leveraging sophisticated BIM tools like Autodesk Revit 2023, Navisworks 2023, and BIM 360, project teams can establish a collaborative environment conducive to proactive conflict resolution. These tools enable real-time clash detection, facilitating streamlined coordination and reducing the likelihood of costly disruptions during construction. The systematic approach to clash detection within an IPD-BIM framework aims to optimize project efficiency by minimizing rework and addressing coordination challenges early in the design phase. This proactive stance not only enhances project quality but also contributes significantly to cost-effectiveness, mitigating financial risks associated with design coordination issues. The integration of BIM with IPD offers a comprehensive solution to assess and manage financial risks through efficient modeling processes and proactive clash detection strategies. This approach fosters collaboration and enhances project outcomes by reducing uncertainties and potential financial liabilities stemming from design coordination conflicts.

4 Results and Discussion

Clash resolution costs analysis reveals variances between estimated and actual expenses, indicating the effectiveness of clash detection and management processes. Clash severity, impact on schedule, and estimated rework costs are analyzed, guiding project prioritization. Comparing planned costs with actual costs due to clashes unveils discrepancies, highlighting financial impacts and management

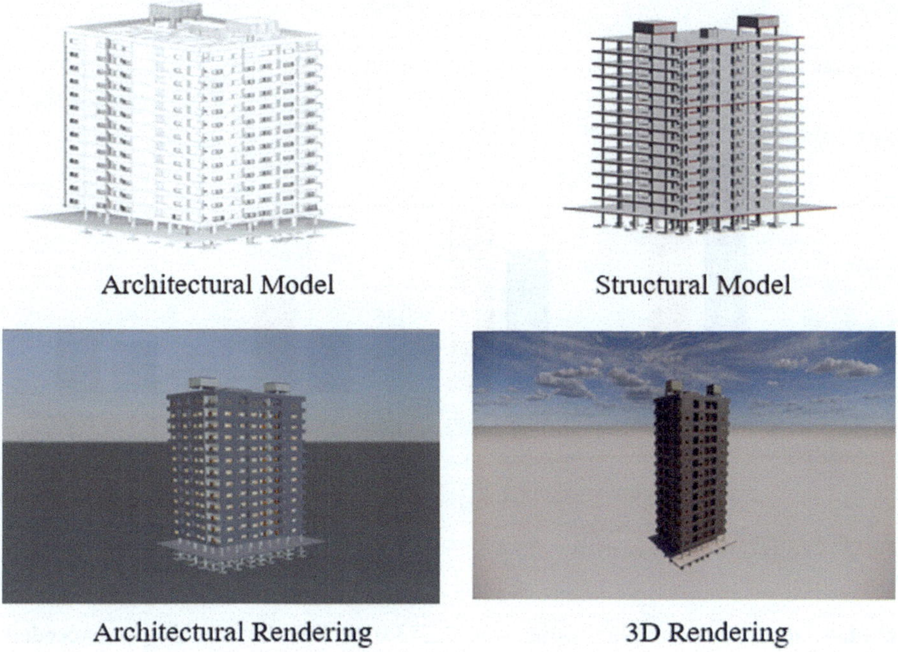

Fig. 2 BIM model

Table 1 Clash activity costs analysis

Clash category	Plan value (estimated cost)	Actual cost
Structural clash	3,671,891	4,904,701
MEP clash	2,253,918	2,779,310
Architectural clash	1,003,134	1,503,448
Interdisciplinary clash	835,945	1,009,540
Total	7,764,888	10,196,999

efficiencies. Understanding these disparities aids in mitigating risks and improving future project planning and execution (Fig. 2).

4.1 Clash Activity Costs Analysis (Table 1)

Figure 3 shows a comparison between estimated plan values and actual costs incurred for resolving various types of clashes in a project. For structural clashes, the estimated plan value was 3,671,891, but the actual cost was higher at 4,904,701. This indicates that the project was less efficient than anticipated in addressing structural issues, resulting in a cost expense of 1,232,810. In the case of MEP

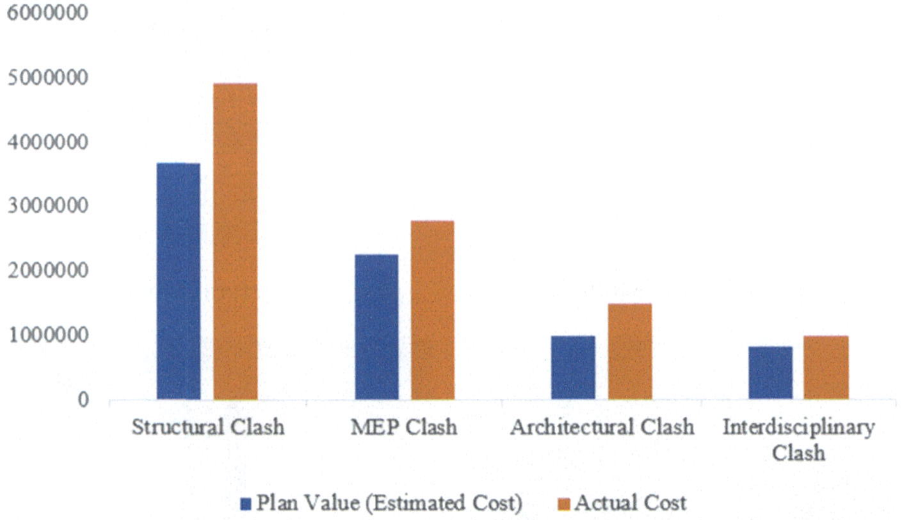

Fig. 3 Clash activity costs analysis

clashes, the estimated plan value was 2,253,918, but the actual cost exceeded expectations at 2779310. This suggests that the complexity or extent of MEP clashes may have been underestimated, leading to higher expenses of 525,392. Similarly, for architectural clashes, the estimated plan value was 1,003,134, and the actual cost incurred was 1,503,448, indicating a variance of 500,314. The interdisciplinary clashes had an estimated plan value of 835,945 and an actual cost of 1,009,540, with a variance of 173,595. Overall, while the total estimated plan value for all clash categories was 7,764,888, the actual expenses totaled 10,196,999, resulting in a variance of 2,432,111. This comparison highlights the effectiveness of clash detection and management processes. Areas, where costs were lower than estimated, indicate successful planning and execution, whereas higher costs reveal potential areas for improvement in estimating and managing clash resolution.

4.2 Clash Detection Activity

Interpreting the clash data provided involves analyzing the severity, impact on schedule, and estimated rework cost associated with each clash between different elements in a construction or engineering project. Clash severity is categorized as either "Hard" or "Soft." "Hard" clashes (Clash1, Clash4, and Clash5) typically indicate critical issues that directly impact the structural integrity or functionality of the elements involved. In contrast, "Soft" clashes (Clash2, Clash3, and Clash6) are less severe, often representing coordination or minor conflicts that can be resolved with less effort, as shown in Table 2 and its detection is shown in Fig. 4.

Table 2 Clash detection activity

Clash ID	Element 1	Element 2	Clash type	Severity	Impact on schedule (days)	Estimated rework cost (Rs.)
Clash1	Beam A	Column H	Hard	High	4	150,000
Clash2	Column B	Beam I	Soft	Medium	2	60,000
Clash3	HVAC duct C	Wall J	Soft	Low	1	55,000
Clash4	Pipe E	HVAC duct K	Hard	High	5	185,000
Clash5	Beam F	Pipe L	Hard	Medium	3	155,000
Clash6	Wall G	Floor M	Soft	Low	1	50,000
Total: 655,000						

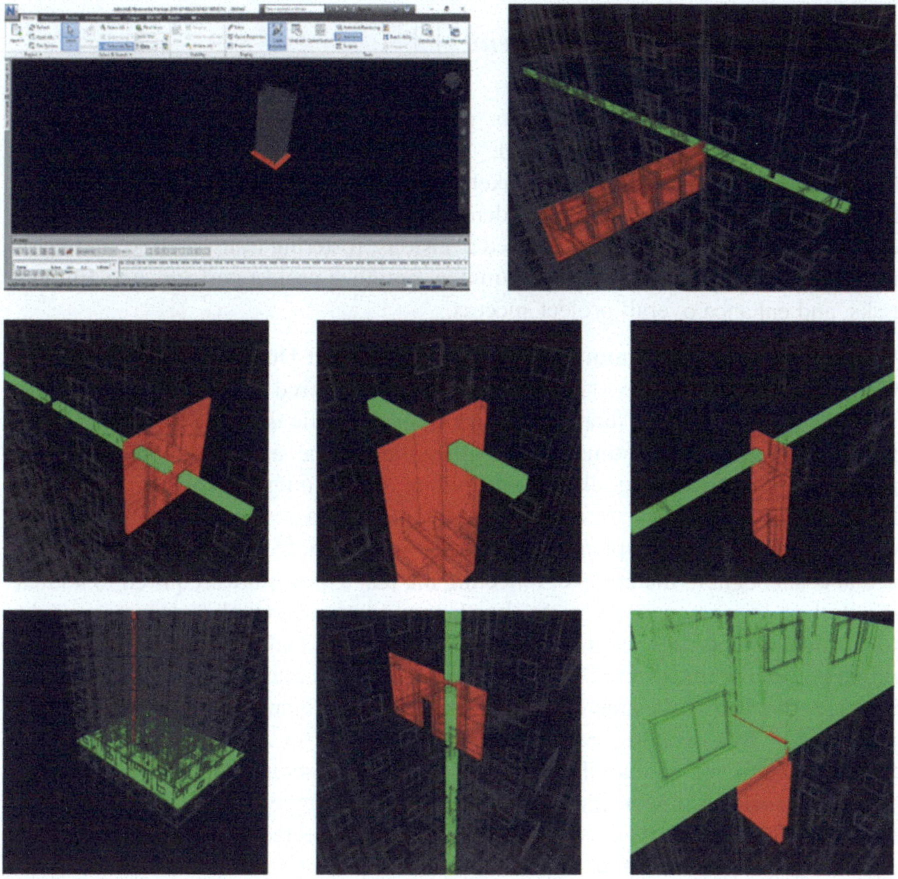

Fig. 4 Clash detection

Impact on Schedule: The clashes have varying impacts on project timelines, measured in days. Hard clashes (Clash1, Clash4, and Clash5) tend to cause more significant delays (4–6 days), affecting project schedules significantly. Soft clashes (Clash2, Clash3, and Clash6) have milder impacts, with delays ranging from 2 to 3 days. Estimated Rework Cost: Each clash also incurs a cost for rework, reflecting the effort and resources required to resolve the issue. Hard clashes (Clash1, Clash4, and Clash5) have higher rework costs (Rs. 1,50,000 to Rs. 2,00,000), indicating more complex and resource-intensive solutions. Soft clashes (Clash2, Clash3, and Clash6) have lower rework costs (Rs. 50,000 to Rs. 80,000), reflecting simpler and less costly corrective actions. Overall, these clash interpretations highlight critical areas of concern within the project, allowing project managers to prioritize resources and interventions based on severity, schedule impact, and cost implications. Addressing these clashes promptly can mitigate risks and ensure the project progresses efficiently and within budget.

4.3 Financial Risk Assessment

Comparing financial risks is vital for gaining deeper insights into the project. It enables proactive risk mitigation, efficient resource allocation, informed decision-making, and transparent stakeholder communication and fosters a culture of continuous improvement. By identifying potential threats early, allocating resources effectively, making informed decisions, fostering transparency, and learning from past experiences, project teams can optimize project performance, mitigate risks, and enhance overall project success.

Comparison between Planned Cost and Actual Cost Due to Clashes
Comparing the planned costs to the actual costs incurred due to clashes provides valuable insights into the financial impact of these conflicts on the project. Initially, the planned costs are estimates or projections of the expenses expected to be associated with resolving clashes, including redesigning, rework, or additional labor. These planned costs are based on thorough assessments and predictions made during the project planning phase. However, as the project progresses, the actual costs begin to materialize, reflecting the real expenditures required to address the clashes. Discrepancies between the planned costs and actual costs may arise due to various factors, such as unforeseen complexities in clash resolution, changes in project scope, or unexpected delays. If the actual costs exceed the planned costs, it indicates that the clashes have had a more significant financial impact than initially anticipated. This scenario can lead to budget overruns, affecting the project's financial health and potentially delaying the project schedule. On the other hand, if the actual costs are lower than the planned costs, it suggests that the clashes were resolved more efficiently or at a lower cost than expected. This outcome reflects effective management of resources and successful mitigation of financial risks

Table 3 Comparison between actual cost including clashes and actual cost after removing clashes

Category	Actual cost including clashes (Rs.)	Actual cost after removing clashes (Rs.)	Variance (Rs.)
Total estimated rework cost (Rs.)	6,55,000	39,000	−6,16,000

associated with clashes. However, it is essential to investigate the reasons behind the cost variances to identify areas for improvement in future projects (Table 3).

BIM models are essential for managing financial risks in IPD projects, providing a detailed, accurate, and collaborative platform for visualization, coordination, and decision-making. A clash detection activity costs analysis reveals the effectiveness of these processes, with lower costs indicating successful planning and execution. Clash interpretations help prioritize resources and interventions based on severity, schedule impact, and cost implications. Proactive clash detection and resolution are crucial in construction projects to minimize potential capital loss and ensure efficient and cost-effective project delivery.

5 Conclusions

The financial risk assessment of IPD utilizing BIM incorporates various methodologies to evaluate the project's financial viability and potential risks. Among these methods, the clash detection approach stands out for its consideration of the time value of money. In assessing the financial risks associated with clashes, the BIM model reveals crucial insights. A comparison between the actual cost, including clashes, and the cost after resolving clashes illustrates significant variances. Notably, the total estimated rework cost amounted to Rs. 6,16,000, but after effective clash resolution, this cost was mitigated entirely. This drastic reduction underscores the financial benefits of efficient clash detection and resolution processes within integrated project delivery using BIM. Analyzing clash resolution costs further highlights the financial implications of various clash categories. While some clashes incurred costs lower than estimated, others exceeded initial projections, indicating areas for improvement in estimating and managing clash resolution. Clash detection activities provide additional insights into financial risk assessment by examining severity, schedule impact, and estimated rework costs associated with each clash. Variations in severity, impact on schedule, and rework costs between hard and soft clashes underscore the diverse financial risks inherent in integrated project delivery using BIM. Comparing planned costs to actual costs due to clashes reveals critical financial impacts on the project. Discrepancies between these costs highlight potential budget overruns or cost savings, indicating areas for improvement in clash resolution management and financial risk mitigation strategies. The BIM model, combined with a thorough analysis of clash resolution costs and detection activities, facilitates a comprehensive assessment of financial risks associated with IPD using BIM. This analysis enables project stakeholders to make informed decisions to optimize financial outcomes and mitigate risks effectively.

Despite the insightful information gained from employing clash detection and BIM to analyze financial risks in IPD, there are restrictions. These include potential inaccuracies in cost data, a focus on clash-related costs over broader financial risks, limited generalizability, challenges in capturing long-term trends, reliance on assumptions about mitigated costs, complex financial modeling, and external factors affecting project finances. Improving the validity and applicability of financial risk assessments in IPD requires a comprehensive approach addressing data accuracy, considering a wider range of risks, validating assumptions, and acknowledging external influences.

To enhance financial risk assessments in IPD utilizing clash detection and BIM, it is recommended to prioritize data accuracy by improving cost tracking for clash resolution and broader financial risks. Additionally, conducting comprehensive risk assessments beyond clash-related costs is essential to capture long-term trends and external influences accurately. Validating assumptions about cost mitigation and enhancing financial modeling are crucial steps for ensuring more reliable and comprehensive IPD risk assessments.

References

1. Khanna M, Elghaish F, McIlwaine S et al (2021) Feasibility of implementing IPD approach for infrastructure projects in developing countries. J Inf Technol Constr 26:902–921. https://doi.org/10.36680/J.ITCON.2021.048
2. Elghaish F, Abrishami S, Hosseini MR et al (2019) Integrated project delivery with BIM: an automated EVM-based approach. Autom Constr 106:102907. https://doi.org/10.1016/j.autcon.2019.102907
3. Adamtey SA (2021) A case study performance analysis of design-build and integrated project delivery methods. Int J Constr Educ Res 17:68–84. https://doi.org/10.1080/15578771.2019.1696903
4. Heigermoser D, García de Soto B, Abbott ELS et al (2018) BIM-based last planner system tool for improving construction project management. Autom Constr 104:246–254. https://doi.org/10.1016/j.autcon.2019.03.019
5. Sherif M, Abotaleb I (2023) Application of integrated project delivery (IPD) in Egypt: implementation and challenges. Lect Notes Civ Eng 239:73–84. https://doi.org/10.1007/978-981-19-0503-2_7
6. Bapat H, Sarkar D, Gujar R (2021) Evaluation of key performance indicators of integrated project delivery and BIM model for infrastructure transportation project in Ahmedabad, India through decision-making approach. J Inst Eng Ser A 102:995–1011. https://doi.org/10.1007/s40030-021-00568-3
7. Ma Q, Li S, Cheung SO (2022) Unveiling embedded risks in integrated project delivery. J Constr Eng Manag 148:1–13. https://doi.org/10.1061/(asce)co.1943-7862.0002212
8. Manata B, Garcia AJ, Mollaoglu S et al (2021) The effect of commitment differentiation on integrated project delivery team dynamics: the critical roles of goal alignment, communication behaviors, and decision quality. Int J Proj Manag 39:259–269. https://doi.org/10.1016/j.ijproman.2020.12.003

9. Turk Ž, Sonkor MS, Klinc R (2022) Cybersecurity assessment of bim/cde design environment using cyber assessment framework. J Civ Eng Manag 28:349–364. https://doi.org/10.3846/jcem.2022.16682
10. Su G, Hastak M, Deng X, Khallaf R (2021) Risk sharing strategies for IPD projects: interactional analysis of participants' decision-making. J Manag Eng 37:1–12. https://doi.org/10.1061/(asce)me.1943-5479.0000853
11. Mohammadi S, Aibinu AA, Oraee M (2024) Legal and contractual risks and challenges for BIM. J Leg Aff Disput Resolut Eng Constr 16:1–13. https://doi.org/10.1061/jladah.ladr-1040
12. Buk'hail RS, Al-Sabah RS (2022) Exploring the barriers to implementing the integrated project delivery method. J King Saud Univ - Eng Sci. https://doi.org/10.1016/j.jksues.2022.04.003
13. Ibrahim MW, Hanna A, Kievet D (2020) Quantitative comparison of project performance between project delivery systems. J Manag Eng 36:1–13. https://doi.org/10.1061/(asce)me.1943-5479.0000837
14. Elghaish F, Abrishami S, Hosseini MR (2020) Integrated project delivery with blockchain: an automated financial system. Autom Constr 114:103182. https://doi.org/10.1016/j.autcon.2020.103182
15. El Asmar M, Hanna AS, Loh W-Y (2013) Quantifying performance for the integrated project delivery system as compared to established delivery systems. J Constr Eng Manag 139:1–14. https://doi.org/10.1061/(asce)co.1943-7862.0000744
16. Roy D, Malsane S, Samanta PK (2018) Identification of critical challenges for adoption of integrated project delivery. Lean Constr J:1–15
17. Jobidon G, Lemieux P, Beauregard R (2021) Building information modeling in Quebec's procurement for public infrastructure: a case for integrated project delivery. Laws 10:2. https://doi.org/10.3390/laws10020043
18. Durdyev S, Hosseini MR, Martek I et al (2020) Barriers to the use of integrated project delivery (IPD): a quantified model for Malaysia. Eng Constr Archit Manag 27:186–204. https://doi.org/10.1108/ECAM-12-2018-0535
19. Viana ML, Hadikusumo BHW, Mohammad MZ et al (2020) Integrated project delivery (IPD): an updated review and analysis case study. J Eng Proj Prod Manag 10:147–161. https://doi.org/10.2478/jeppm-2020-0017

Part II
Green Technologies for Sustainable Solutions

Feasibility Study on Utilization of Incinerated Solid Waste Ash for Mitigating Swell–Shrink and Consolidation Characteristics of Expansive Soil

Ramakrishna Bollam, Sangeetha Sundar, Shaik Abdul Rahman, and Hari Krishna Padavala

Abstract The disposal of municipal solid waste is a critical issue in urban environments. In recent times, waste-to-energy plants have been installed and operated across the globe to reduce the volume of trash. In Hyderabad, India, there are currently five such waste-to-energy plants operating in Dundigal, Jawaharnagar, Bibinagar, Pyaranagar, and Yacharam. These energy production units produce incinerated ash as a by-product. In the current study, an attempt was made to determine the feasibility of using incinerated ash from domestic waste to mitigate the swell-shrink behavior of locally available expansive soil. Domestic waste was collected from a municipal solid waste dump yard, and ash was made by burning it. Tests such as shrinkage limit test, free swell index test, and oedometer tests were performed as per IS 2720 to study the swell–shrink and consolidation characteristics of locally available soil and the soil treated with incinerated ash. The shrinkage limit of virgin soil, which was initially 15%, increased to 38%, 52%, and 55% by the inclusion of 20%, 30%, and 50% of incinerated ash by weight of soil, respectively. By incorporating 25% of burned ash into the soil mass, the free swell index of virgin soil was dropped by 23%, which was 30% initially. Consolidation characteristics were found for virgin soil, incinerated ash, and soil mixed with the optimum quantity of ash. An increase of 75% in the rate of consolidation was observed for the soil mixed with the optimum quantity of ash. Based on the test results, the ideal amount of incinerated ash required to mitigate swell-shrink and settlement characteristics of expansive soil can be determined.

R. Bollam (✉) · S. Sundar · S. A. Rahman
Department of Civil Engineering VNR Vignana Jyothi Institute of Engineering & Technology, Hyderabad, India

H. K. Padavala
Department of Civil Engineering, NIT, Warangal, India

P. Pathak et al. (eds.), *Advances in Environmental Sustainability, Energy and Earth Science*, https://doi.org/10.1007/978-3-031-73820-3_7

Keywords Waste-to-energy plants · Incinerated ash · Expansive soil · Swell–shrink characteristics · Shrinkage limit

1 Introduction

Municipal solid waste (MSW) poses significant challenges in waste management due to its large-scale generation and limited disposal options. Daily, approximately 143,449 metric tons (MT) of MSW is generated across the country, with only 111,000 MT collected by various municipalities. However, merely 35,602 MT of MSW undergoes treatment [1]. Focusing on Telangana, the daily generation of MSW is reported to be 7371 MT, which accounts for approximately 5% of the total daily waste generation in the country. Within Telangana, the city of Hyderabad contributes significantly to this figure, generating 2187 MT of MSW daily [2]. These statistics underscore the potential for utilizing MSW incinerated ash in soil stabilization efforts. Though there are many possible ways of assessing MSW, as reported [3–5], by incineration, the volume is reduced significantly, and the waste is converted into heat energy, resulting in the formation of ash. Incineration is primarily accomplished through two methods: controlled interaction at waste-to-energy plants and mass burn incineration at dump yards. In India, uncontrolled burning is used more than controlled incineration at energy plants [6]. Uncontrolled burning poses serious environmental and health risks due to the release of harmful pollutants into the atmosphere. Therefore, promoting and implementing controlled incineration methods at waste-to-energy plants can mitigate these risks while also producing ash that can be used for soil stabilization. In general, managing MSW can become more environmentally friendly and sustainable by implementing controlled incineration techniques into waste management procedures.

One common issue in the construction of geotechnical infrastructures is the presence of expansive soils, which can cause problems such as shrinkage, swell, and prolonged consolidation. Expansive soils undergo significant volume changes in response to moisture variations, leading to differential settlement. These volume changes can have detrimental effects on the foundations of structures. This nonuniform movement can exert uneven pressure on the building, causing cracks in walls, floors, and foundations, compromising the structural integrity and safety of the entire structure. Shrinking or swelling of the soil causes more than twice as much damage as natural disasters. Shrinking and swelling cost nearly $2.3 billion annually in damages to homes, buildings, roads, and pipelines [7]. It is, therefore, crucial to address these soil-related challenges in construction projects, as they can have far-reaching consequences. By understanding the behavior of expansive soils and implementing appropriate measures, engineers can mitigate the risks associated with these unpredictable soil movements. Sridharan and Rao investigated kaolinite's volume change behavior in solvents and pore fluids. The results show repulsive forces in water and benzene, while electrostatic attraction and Van der Waals forces are not significant. The volume change behavior was primarily controlled by frictional forces and clay fabric.

A comprehensive outline of the global issue of expansive soil and the progress made in its optimization were provided here [8, 9]. In a recent review study [10], it was explained about swelling and shrinkage properties, including mechanics, factors, effects, and methods for assessing them. Their reviews offer thorough mitigation techniques and emphasize the use of different admixtures as soil stabilizers, which have demonstrated encouraging outcomes in reducing the difficulties caused by expansive soil. The admixtures include marble waste powder, stone waste fly ash, lime powder, and eggshell powder. Each of these materials has unique properties that contribute to soil stabilization. For example, fly ash and marble waste powder are industrial by-products that improve the mechanical and physical characteristics of the soil while also helping to manage waste sustainably. Industrial by-products can be used as soil stabilizers, improving soil properties and reducing waste. This process also helps in lowering carbon footprints and promoting circular economy principles. These environmental benefits point out the advantages of soil stabilization techniques.

The stability of structures built on soil relies heavily on their behavior under load. A crucial test to assess this is the consolidation test, which helps us understand how effectively the soil handles water pressure and subsequent settlement. Fiber incorporation into the soil can significantly improve the consolidation parameter [11]. It was discovered that the admixture-reinforced soil had a compression index 1.8 times greater than the nonreinforced soil, but the void ratio had decreased. The consolidation characteristics of clayey soil, when mixed with nano-bentonite in one dimension [12], found that incorporating nano-bentonite into the clayey soil sped up its drainage consolidation process.

It is noted that the addition of Municipal Solid Waste Incineration Ash (MSWIA) reduced the maximum dry density of soil [13]. The addition of MSWIA alone to clayey soil caused a reduction in the soil's plastic and liquid limits. Furthermore, the Optimum Moisture Content (OMC) value also showed a decrease with the addition of MSWIA. Another study states that adding MSWIA content to clayey soil increased permeability and unconfined compressive strength values while soil differential swell decreased [14]. A study on the feasibility of using fly ash and bottom ash as lightweight embankment or backfill materials found that most samples using fly ash and bottom ash are viable options for meeting the Road Development Authority (RDA) standards [15]. Bottom ash from municipal waste incineration produced the highest strength values in compression and bending tests with soil [16]. Another study stated that using fly ash would improve particle size distribution, resulting in better characteristics for pavement design and performance [17]. As the fly ash content increases, a slight reduction in hydraulic conductivity in the mixtures of compacted ash was observed. This results in enhanced shear strength and increased resistance to water flow through voids between particles [18].

Overall, the studies signify that incorporating MSWIA into clayey soil can have both positive and negative effects on soil properties. Utilization of MSWIA can improve permeability and unconfined compressive strength, but it results in the reduction of the density of soil. This study investigates the swelling, shrinking, and consolidation behavior of expansive soil collected from a nearby borrow area, with MSWIA added as a partial soil replacement. These results imply that a potential topic for further investigation is the use of waste incineration ash in soil stabilization.

This could potentially lead to the development of more sustainable solutions for waste volume reduction and subsequently improve the geotechnical characteristics of problematic soil.

2 Materials and Methods

2.1 Raw Materials

2.1.1 Soil

One of the goals of this study was to reduce the swell–shrink and consolidation behaviors of expansive soil by using domestic solid waste incinerated ash. Freshly excavated soil from the Bollaram Industrial Area in Hyderabad, India, was collected for the experiment. Before being tested, the soil was oven-dried at 100°C and sieved through a 425-μm sieve. The sieve size of 425 μm was chosen to account for the plastic nature of the soil, which causes swelling, shrinking, and consolidation. Existing studies show that soil particles smaller than 425 μm exhibit greater cohesion. As per Indian standards, grain size distribution, Atterberg limits, and specific gravity properties were determined. All of the experimental results are the averages of three to four samples. It was found that the specific gravity was 2.4, which was less than that of normal soil. This is due to the presence of carbon in the soil under consideration. Additional preliminary tests, including hydrometer analysis (Fig. 1) and Atterberg limit tests, indicate that the soil is silt, according to the IS

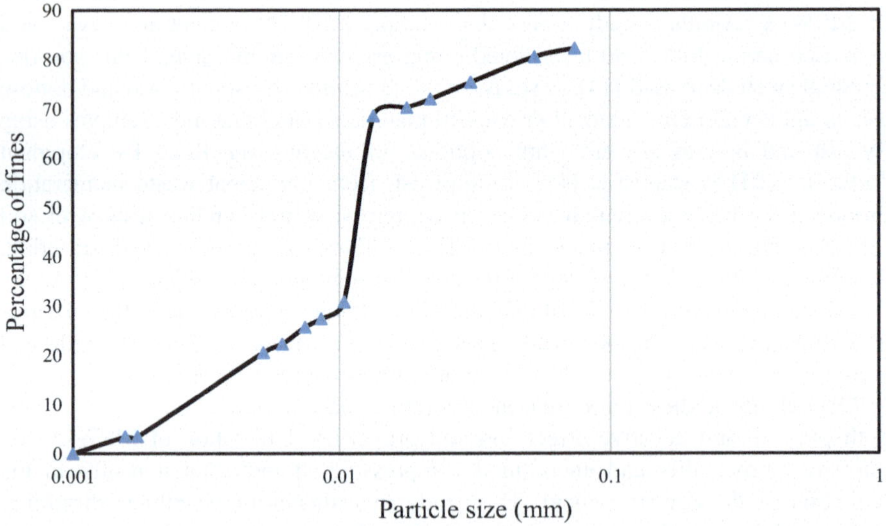

Fig. 1 Particle size distribution based on sedimentation analysis

Table 1 Consistency limits and soil classification as per Indian standards

Test name	Code	Result and inference
Hydrometer analysis	IS 2720-4	%5 of clay and 95% silt
Liquid limit (LL)	IS 2720-5	LL = 46%; intermediate compressibility (I)
Plastic limit (PL)	IS 2720-5	PL = 28%
Plasticity index (PI)	IS 1498	Value lies below A-line; hence the soil is classified as silt of intermediate plasticity

classification, falls into the category of intermediate compressibility based on the liquid limit analysis, and has a plasticity limit of 28%. Table 1 shows the details of the soil properties.

2.1.2 Ash

The problem of ash being generated at waste-to-energy plants due to incineration is also observed at domestic incinerators in residential communities. Such practices of incineration of domestic waste are also in wide practice at independent houses. In a recent attempt, the geotechnical properties of MSWIA and its incorporation with polyfibers were investigated [19]. The waste was collected as per the prescribed composition (2) and oven-dried at 100°C for easy ignition. The dried waste was then burned in the laboratory to generate ash from the domestic waste, as shown in Fig. 2.

After complete incineration of the waste, ash was collected from it in a tray. To ensure uniformity for soil testing, the ash was also sieved through a 425-μm sieve. Further testing of the ash was done on the ash passing through a 425-μm sieve. The specific gravity of the ash material was determined to be 1.3, which is similar to that of fly ash made from burnt coal and used in construction projects [20].

2.2 Test Compositions

2.2.1 Shrinkage Limit

The shrinkage limit is one of the Atterberg limits and a fundamental soil characterization parameter. The soil sample ratios used in the study to measure the shrinkage limit test are listed below. The samples were designated as S1, S2, S3, and S4, and their compositions are mentioned in Table 2. These ratios were chosen to investigate how the addition of ash affects the shrinkage limit of the soil samples.

Fig. 2 Lab-scale incineration of the collected domestic waste in a semi-closed environment

Table 2 Test compositions at various ash replacements of soil, adopted for the shrinkage limit study

Designation	Compositions
S1	Soil (100%)
S2	Soil (80%) + ash (20%)
S3	Soil (70%) + ash (30%)
S4	Soil (50%) + ash (50%)

Table 3 Sample designations considered for the investigation of swelling and compressibility behavior of soil, ash, and soil–ash composition

Designation	Contents
M1	Soil (100%)
M2	Ash (100%)
M3	Soil (75%) + ash (25%)

2.2.2 Swell Index and Consolidation Test

Through the shrinkage limit study, the quantity of ash required to replace soil was optimized. The ideal ratio for improving the shrinkage limit was found to be 25% ash as a soil replacement. Swell index and consolidation studies were performed for the following samples, as mentioned in Table 3.

2.3 *Test Methods*

2.3.1 Carbon Content

Carbon content, which is also called loss on ignition, was done according to the IS 1760-1 guidelines. The average value of the carbon content is reported with three repetitions. This test was conducted in the muffle furnace, as revealed in Fig. 3, at two different test temperatures, 450°C and 900°C, for 4 h to find the organic carbon content and total carbon content.

The organic and inorganic carbon contents of the sample were calculated based on the weight loss observed during the heating process. This technique is frequently employed to assess the organic matter content in soil or sediment samples. The temperature and duration of heating are crucial factors in ensuring accurate results, as they can affect the decomposition of organic material. Figure 4 shows the ash after keeping in the muffle furnace for the stipulated time.

2.3.2 Shrinkage Limit

Shrinkage limit test was carried out for the test combinations mentioned in Table 2, following IS 2720-6. When more moisture is lost, and further volume reduction is no longer possible, the water content at this state is known as the shrinkage limit. The volume of the metal container and dry pat of soil after oven drying was measured using lab-grade mercury, as depicted in Fig. 5.

Fig. 3 Muffle furnace used to find the carbon content in ash and soil samples at lab scale and at various temperatures

Fig. 4 Residual ash sample after burning the lab-scale incinerated domestic waste at 900°C in the muffle furnace

Fig. 5 Container filled with mercury to measure the volume of container and volume of dried soil pat in shrinkage limit testing

2.3.3 Swell Index

The swell index test was conducted as per IS 2720-40 to determine the potential volume increase of soil when exposed to water. As the ash was floating in the water, as observed in Fig. 6, when it was experimented with to find its swell index, the testing was a failed attempt to measure the swell index of only ash (M2).

2.3.4 Consolidation Test

Consolidation testing was carried out in compliance with IS 2720-15. At intervals of 2, 4, 9, 16, 25, 36, and further square minutes up to 1440, dial gauge readings were recorded. This report presents the average values of the consolidation results obtained from three samples each of designation M1, M2, and M3. In Fig. 7, the experimental setup is displayed.

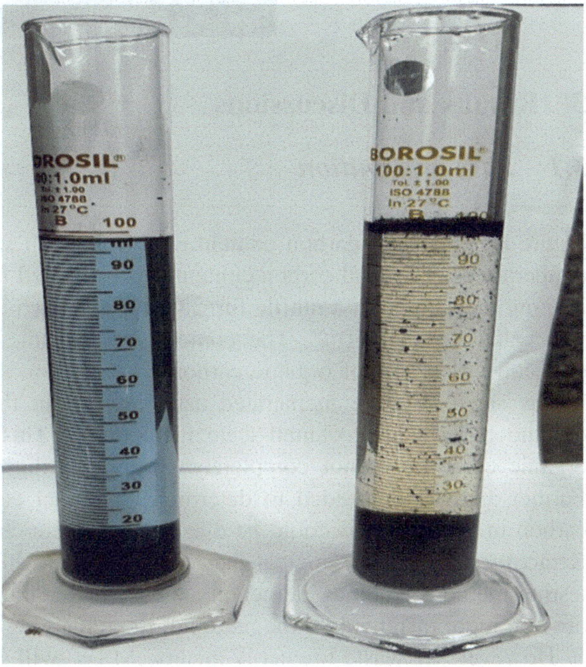

Fig. 6 Soil–ash composition used for the study shows no expansion in kerosene (towards left) and slight expansion in water (towards right), and also some amount of ash floating on water

Fig. 7 Arrangement of soil, ash, and soil–ash compositions in an odometer to investigate their compressibility behaviors

3 Results and Discussions

3.1 Loss on Ignition

Figure 8 depicts the carbon content of soil and ash samples at 450°C and 900°C temperatures. The total carbon content was estimated from the mass loss caused by igniting the sample in a muffle furnace at 900°C, while the organic carbon content was calculated at 450°C. The content of inorganic carbon was determined by deducting the value of organic carbon content from the total carbon content. The results show that the incinerated ash prepared in the laboratory contains more organic carbon due to small-scale incineration. This indicates that some of the organic carbon was not completely burned off during the incineration process. Further analysis is needed to determine the exact composition of the remaining carbon in the ash. This could be due to incomplete combustion or the presence of refractory organic compounds that are resistant to high temperatures. Additional testing methods, such as elemental analysis or scanning electron microscopy, may provide more insights into the nature of the remaining carbon.

However, in waste-to-energy plants, MSW will be incinerated at very high temperatures, leaving less organic carbon in the ash. Therefore, it is important to consider the incineration conditions when interpreting the organic carbon content in ash samples. The presence of carbon determines the quality of the ash produced during the incineration process. Organic carbon content in soils is undesirable for geotechnical applications. Most of the organic carbon in soils is due to organic

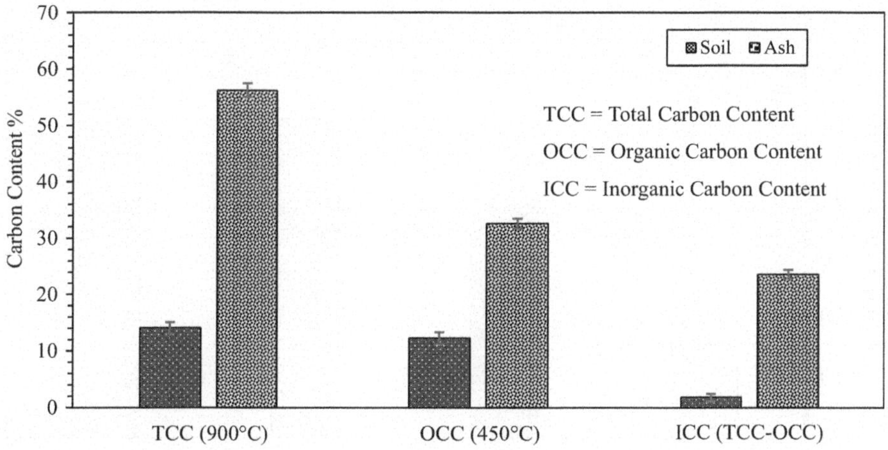

Fig. 8 Total, organic and inorganic carbon contents found by incineration of soil and ash samples in a muffle furnace at 450°C and 900°C

materials existing at shallow depths or on the surfaces of the soil. Hence, soil from deeper depths is preferred for backfilling in construction projects to avoid potential issues with organic carbon content. It is also important to conduct thorough testing and analysis of soil samples for carbon content before using them in backfilling to ensure stability and reduce compressibility during post-construction stages.

3.2 *Rate of Shrinkage*

The presence of ash content in soil influences the shrinkage characteristics of the soil. The data presented in Fig. 9 clearly shows that the shrinkage limit increases as the ash component of the soil–ash composition increases. The positive correlation highlights the possibility of ash as a useful replacement for soil stabilization. The increase in shrinkage limit with higher ash content aligns with the principles of pozzolanic reactions. Ash, being rich in pozzolanic materials, interacts with the calcium hydroxide present in the soil–water system, forming additional cementitious compounds. These compounds contribute to the densification of the soil matrix, thereby reducing its susceptibility to shrinkage. This mechanism effectively addresses the root causes of soil shrinkage, which are mainly related to soil structure and the presence of clay minerals.

MSW incinerated ash is found to be an effective soil stabilizer due to the notable increase in shrinkage limit that occurs when ash content rises from 0% to 50%. The shrinkage limit was 16% for the natural soil, which improved significantly with the addition of ash, reaching 56% at 50% ash as soil replacement. This noteworthy rise emphasizes how ash can change the characteristics of soil and how it can be used to address shrinkage problems in soil. Additionally, a larger rate of rise in the shrinkage

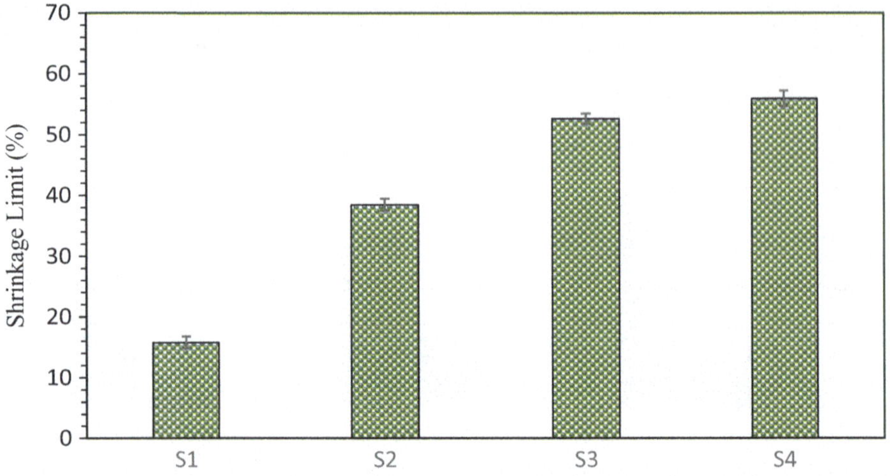

Fig. 9 Shrinkage limit of various test compositions showing insignificant change in shrinkage limit after 30% replacement (S3-S4)

limit was noted between 20% and 30% soil replacement. There was no such significant change in the shrinkage limit by increasing the ash content beyond 30%. Therefore, subsequent research was limited to 25% soil replacement with ash.

3.3 Swelling

Free swell index (FSI) is a crucial metric in construction and engineering, as it quantifies the potential of soil to expand or swell when it comes into contact with water. This expansion can lead to severe structural damage if not properly managed. FSI is determined by comparing the volume of soil in its dry state to its volume when fully saturated with water, refer to Fig. 10. A higher FSI signifies greater soil expansion, necessitating preventative measures. Previous studies have shown that adding fly ash or rice husk ash to expansive clays can significantly reduce their FSI [21].

In the current study, the FSI of natural soil was found to be 30%, indicating that it falls under the category of moderate expansiveness according to IS 1498 classification of expansive soils. However, when MSW incinerated ash was incorporated into the soil, the FSI decreased to 23%. This reduction, which represents a nearly 30% decrease, suggests that the soil–ash compositions could be beneficially used as backfill material. The decrease in FSI can be attributed to the unique properties of the ash, refer to Table 4. The possibility of swelling is limited because the ash particles occupy the spaces that would otherwise allow for the absorption of water by the soil. The pozzolanic nature of the ash might have reduced the availability of

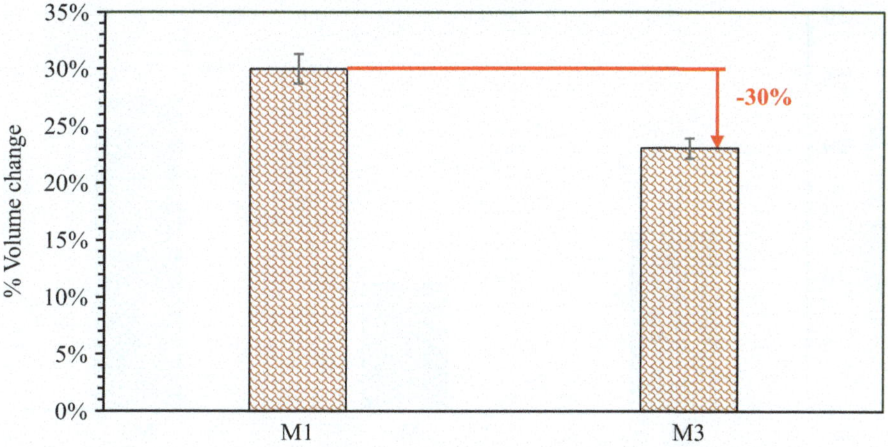

Fig. 10 Reduction in free swelling of the soil–ash composition (M3) compared to the virgin soil (M1)

Table 4 Classification of soil based on swelling behavior according to Indian standards

Properties of soil	Results observed	IS code range for the degree of expansiveness	Interpretation
Liquid limit	46%	40–60%	The soil considered for the study falls under the category of moderate degree of expansiveness
Plastic limit	28%	25–50%	
Shrinkage limit	16%	8–18%	
Free swell index	30%	20–35	

water required to form a double layer around the clay mineral of the natural soil, thus reducing swelling phenomena.

3.4 *Compressibility Behaviour*

The impact of ash inclusion on the soil's consolidation properties was investigated in this study. Important characteristics for soil, ash, and a soil–ash mix were measured under various load scenarios. These included the coefficient of consolidation (C_v), compression index (C_c), coefficient of compressibility (a_v), and coefficient of volume change (m_v). One significant factor that regulates the pace at which pore water pressure in the soil diffuses and causes settlement is the coefficient of consolidation

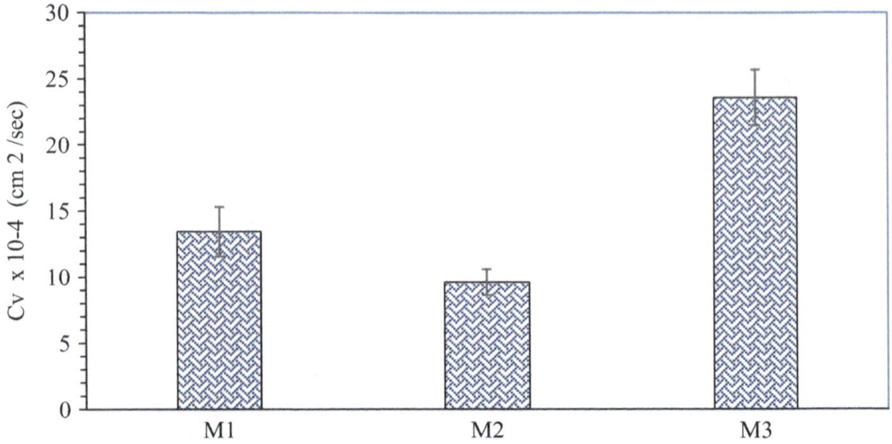

Fig. 11 Coefficient of consolidation of adopted test samples revealing better geotechnical property of soil–ash composition (M3)

(C_v). The square root of time and the logarithm of time are the two techniques used to compute soil C_v. The square root method was used for C_v calculations in this study. The soil sample showed a slower rate of consolidation, with a C_v of 13.46×10^{-4} cm^2/sec. Ash has a lower C_v value of 9.6×10^{-4} cm^2/sec. The soil–ash composition significantly improved the consolidation rate with a C_v of 23.55×10^{-4} cm^2/sec, a 75% increase over the base soil. Ash particles, with their larger grain size and increased permeability, can form interconnected pathways within a composite material, allowing for faster pore water evacuation and consolidation. They also have a lower void ratio, which means less compressible pore space and a stiffer mix that consolidates more quickly. The type of ash can affect the pore water of soil chemistry by acting as a pozzolanic material, combining with water and calcium to form cementitious compounds that bind soil particles together, stiffen the soil matrix, and improve consolidation behavior. The variations in the rate of consolidation of samples are portrayed in Fig. 11.

The study also investigated the compressibility characteristics of soil, ash, and soil–ash compositions by determining their compression index (C_c) using pressure vs. void ratio graphs. This parameter reflects the soil's propensity to compress under increasing load. The base soil had a moderate compressibility (C_c) of 0.048, indicating noticeable deformation under pressure. Ash had a lower C_c value of 0.023, suggesting a more rigid material compared to soil. The soil–ash mix had a similar C_c value as only ash (0.023), indicating that the addition of ash significantly reduced the compressibility of the soil composite. This suggests that the ash particles' size distribution or internal structure may influence the compressibility of the soil. The base soil had a coefficient of compressibility (a_v) of 0.01261 m^2/kN, indicating medium compressibility. Ash had a slightly higher a_v than soil, measuring 0.0131 m^2/kN. This suggests that, when loaded, ash may experience a slightly larger volume change than base soil. The a_v value of the

Table 5 Compressibility parameters of the mixes adopted for investigation

Samples	Coefficient of compressibility a_v [m^2/kN]	Compression index C_c	Coefficient of volumetric compression m_v [m^2/kN]	Coefficient of consolidation $C_v \times 10^{-4}$ [cm^2/sec]	Permeability $k \times 10^{-6}$ [cm/sec]
M1	0.0208	0.048	0.01261	13.43	1.661
M2	0.0213	0.23	0.0131	9.6	1.233
M3	0.1665	0.23	0.0111	23.55	2.564

soil–ash mix was the lowest, measuring 0.0111 m^2/kN. Table 5 shows the consolidation of characteristics of mix compositions considered for the study.

4 Conclusions

The findings suggest that incorporating ash into the soil at an optimum ratio of 25% replacement is beneficial for backfill materials. Further testing will be necessary to explore the long-term effects of using ash as a soil replacement in various applications. The results of this study could potentially lead to more sustainable practices in the geotechnical infrastructure. To sum up, this study offers an important new understanding of how ash content affects soil–ash mixture shrinkage limits. The observed positive correlation between ash content and shrinkage limit highlights the efficacy of ash as a remedial agent for soil stabilization. Future research avenues may involve investigating the long-term performance and environmental implications of using ash in soil engineering applications, thereby fostering sustainable infrastructure development.

- The study suggests that the use of incinerated ash as a partial replacement for backfill improves the overall sustainability of waste management practices.
- The optimal quantity of ash to be added to the soil was found to be 25%. This quantity increased the shrinkage limit (S.L.) by 188%, reduced the FSI by 30%.
- The study proved the sustainable utilization of waste incinerated by-products as a partial filler material in geostructures. This approach can reduce land contamination, providing a pathway for sustainability.
- The use of ash resulted in reduced settlements due to improved swell, shrink, and consolidation characteristics. It also enhanced permeability and drainage by about 54%.
- The rate of consolidation was 75% higher compared to soil, which can moderate the overall construction time.
- The study presents a novel remedial technology that addresses emerging contaminants. The utilization of ash as a soil stabilizer can improve the geotechnical aspects of problematic soil.
- To learn more about the mechanical properties and microlevel physiochemical interactions between natural soil material and burnt ash, more research may be

carried out. This could provide valuable insights for future applications of this technology. Furthermore, research into the mechanical characteristics of soil–ash compositions, such as their compressive and shear strengths, might be beneficial in understanding how well they function as a whole.

References

1. Thomas P, Soren N (2020) An overview of municipal solid waste-to-energy application in Indian scenario. Environ Dev Sustain 22:575–592. https://doi.org/10.1007/s10668-018-0235-7
2. Kumar A, Agrawal A (2020) Recent trends in solid waste management status, challenges, and potential for the future Indian cities—a review. Curr Res Environ Sustain 2:100011. https://doi.org/10.1016/j.crsust.2020.100011
3. Jadhao SB, Shingade SG, Pandit AB, Bakshi BR (2017) Bury, burn, or gasify: assessing municipal solid waste management options in Indian megacities by exergy analysis. Clean Techn Environ Policy 19:1403–1412. https://doi.org/10.1007/s10098-017-1338-9
4. Bansal D, Gupta G, Ramana GV, Datta M (2024) Environmentally responsible disposal and reuse of MSW incineration bottom ash: assessment from two Indian plants. Clean Techn Environ Policy 26:1439–1454. https://doi.org/10.1007/s10098-023-02550-y
5. Joshi R, Ahmed S (2016) Status and challenges of municipal solid waste management in India: a review. Cogent Environ Sci 2:1139434. https://doi.org/10.1080/23311843.2016.1139434
6. Gupta G, Datta M, Ramana GV, Alappat BJ (2021) MSW incineration bottom ash (MIBA) as a substitute to conventional materials in geotechnical applications: a characterization study from India and comparison with literature. Constr Build Mater 308:124925. https://doi.org/10.1016/j.conbuildmat.2021.124925
7. Barman D, Dash SK (2022) Stabilization of expansive soils using chemical additives: a review. J Rock Mech Geotech Eng 14:1319–1342. https://doi.org/10.1016/j.jrmge.2022.02.011
8. Dave TN, Siddiqui AK (2020) A review of expansive soil—effects and mitigation techniques. pp 519–527
9. Zada U, Jamal A, Iqbal M, Eldin SM, Almoshaogeh M, Bekkouche SR, Almuaythir S (2023) Recent advances in expansive soil stabilization using admixtures: current challenges and opportunities. Case Stud Construc Material 18:e01985. https://doi.org/10.1016/j.cscm.2023.e01985
10. Ito M, Azam S (2010) Determination of swelling and shrinkage properties of undisturbed expansive soils. Geotech Geol Eng 28:413–422. https://doi.org/10.1007/s10706-010-9301-0
11. Ali T, Khan MA, Ali K (2021) Consolidation behavior of clayey soil reinforced with geofiber. pp 119–129
12. Cheng G, Zhu H-H, Wen Y-N, Shi B, Gao L (2020) Experimental investigation of consolidation properties of Nano-bentonite mixed clayey soil. Sustainability 12:459. https://doi.org/10.3390/su12020459
13. Singh P, Boora A, Gupta AK (2021) Geotechnical characteristics of clayey soil admixed with municipal solid waste incineration ash, cement and polypropylene fiber. Innov Infrastruct Solut 6:193. https://doi.org/10.1007/s41062-021-00547-4
14. Singh P, Boora A, Gupta AK (2022) Sub-grade characteristics of clayey soil incorporating municipal solid waste incineration ash and marble dust. J Eng, Design Technol 20:712–726. https://doi.org/10.1108/JEDT-08-2020-0347
15. Gimhan PGS, Disanayaka JPB, Nasvi MCM (2018) Geotechnical engineering properties of Fly ash and bottom ash: use as civil engineering construction material. Eng: J Ins Eng, Sri Lanka 51: 49. https://doi.org/10.4038/engineer.v51i1.7287

16. Reddy CS, Mohanty S, Shaik R (2018) Physical, chemical and geotechnical characterization of fly ash, bottom ash and municipal solid waste from Telangana state in India. Int J Geo-Eng 9:23. https://doi.org/10.1186/s40703-018-0093-z
17. Townsend SW, Spreadbury CJ, Laux SJ, Ferraro CC, Kari R, Townsend TG (2020) Blending as a strategy for reusing municipal solid waste incinerator ash in Road-Base construction. J Environ Eng 146. https://doi.org/10.1061/(ASCE)EE.1943-7870.0001788
18. Kim B, Prezzi M, Salgado R (2005) Geotechnical properties of Fly and bottom ash mixtures for use in highway embankments. J Geotech Geoenviron 131:914–924. https://doi.org/10.1061/(ASCE)1090-0241(2005)131:7(914)
19. Zekkos D, Kavazanjian E, Bray JD, Matasovic N, Riemer MF (2010) Physical characterization of municipal solid waste for geotechnical purposes. J Geotech Geoenviron 136:1231–1241. https://doi.org/10.1061/(ASCE)GT.1943-5606.0000326
20. Pandian NS Fly ash characterization with reference to geotechnical applications
21. Phanikumar BR, Sharma RS (2007) Volume change behavior of Fly ash-stabilized clays. J Mater Civ Eng 19:67–74. https://doi.org/10.1061/(ASCE)0899-1561(2007)19:1(67)

Evaluating Bitumen Mix with a Blend of RHA and Synthetic Material: A Comparative Perspective

Malathi Narra, Dharmika Chowdary Karusala, Sai Satya Sreya Godi, Raja Sankhar, and Galla Yaswanth

Abstract Bituminous surface roads are rapidly degrading due to growing traffic loads and climate change. Therefore, the management of fracture and permanent deformations in flexible pavements has been the subject of research in recent years. Various polymer modifiers are used to alleviate fatigue, rutting, and low-temperature cracking concerns. One kind of substance that has a lot of fine glass fibres in it is called glass powder. A key component in enhancing the performance of flexible pavements is additive. Glass powder has been used to improve asphalt mixture performance by preventing fatigue cracking and permanent deformation, as well as improving automatic characteristics. Compared to previous methods, it was discovered that adding glass powder to the bituminous pavement mix as a low-cost modifier improved performance. To assess the performance characteristics of a bitumen mixture, a combination of glass powder and rice husk ash (RHA) is introduced in this study.

Keywords Bitumen · Rice Husk Ash · Glass powder · Roads · Concrete mix

1 Introduction

The properties of the components (binder, aggregate, and additive) and the ratio of these components in the overall mix constitute a few of the elements that affect how well flexible courses work. Bitumen can also be altered by including various kinds of additives. As a result, the quality of illuminance might not be uniform, and some parts may not receive proper lighting. In general, the inclusion of polymers makes bitumen more rigid and enhances its temperature sensitivity. Stiffness increases allow the use of somewhat softer base bitumen, which improves low-temperature performance and increases the mixture's resilience to rutting in hot temperatures.

M. Narra (✉) · D. C. Karusala · S. S. S. Godi · R. Sankhar · G. Yaswanth
Department of Civil Engineering, Velagapudi Ramakrishna Siddhartha Engineering College, Vijayawada, India
e-mail: malathi@vrsiddhartha.ac.in

P. Pathak et al. (eds.), *Advances in Environmental Sustainability, Energy and Earth Science*, https://doi.org/10.1007/978-3-031-73820-3_8

Several studies have demonstrated that because of their high strength, ductility, and durability qualities, synthetic materials are useful for enhancing the mechanical performance of asphalt mixtures.

In this research, we're looking at how much glass powder to add to the bituminous mix to find the ideal number of additives and, ultimately, forecast an improvement in the bituminous concrete mix's performance characteristics. In our experiment, bituminous concrete mix is combined with rice husk ash (RHA) as a moderator. The aim is to provide an overview of the impact of glass powder reinforcement on the asphalt mixture. The purpose of this research is to investigate the properties of bitumen mix additive mixtures with different amounts of glass powder and RHA. Aggregates are graded following MORTH requirements to make mixes. An ideal bitumen percentage is 6%; 3% of RHA and 0.1%, 0.15%, 0.2%, 0.25%, and 0.3% of glass powder are added to the mixture overall. The ubiquitous network of asphalt roads that crisscross the globe stands as a testament to their reliability and durability. However, traditional asphalt mixtures face increasing scrutiny due to their dependence on nonrenewable resources and limited lifespan. As environmental concerns gain prominence, researchers are embarking on a new era of asphalt innovation, seeking sustainable solutions that enhance performance while minimising environmental impact. This pursuit has unlocked the exciting potential of waste and synthetic materials as viable alternatives, paving the way for a future where roads are not just arteries of transportation but also champions of sustainability.

One such avenue lies in harnessing the power of waste high-density polyethene (HDPE). Studies conducted by Hinislioglu and Agar in 2004 demonstrated that incorporating shredded HDPE as a modifier significantly increases the Marshall stability and quotient of asphalt concrete, crucial measures of resistance to deformation [1]. This translates to pavements that can withstand heavier traffic loads and endure longer, reducing the need for frequent repairs and replacements. By diverting waste materials like HDPE from landfills and incorporating them into asphalt mixtures, we not only contribute to a circular economy but also extend the lifespan of these valuable resources.

Another sustainable solution emerges from the transformation of an unlikely hero: fly ash, a by-product of coal combustion. Research by Mirkovic and Mladenovic revealed that incorporating fly ash as a filler in asphalt mixtures can replace a portion of the bitumen, the traditional binding agent derived from petroleum [2]. This not only minimises our reliance on nonrenewable resources but also improves the shear resistance of the asphalt, making it less susceptible to cracking and deformation. By finding a valuable use for fly ash, a previously problematic waste product, we not only lessen the environmental burden but also tap into a readily available resource for asphalt improvement.

The world of synthetic fibres offers another exciting chapter in the story of asphalt innovation. Jia and shen work highlights the potential of polyester fibres in reinforcing asphalt mixtures [3]. Their research showcases that adding polyester fibres significantly improves drainage, fatigue resistance, and rutting resistance—key factors in ensuring smooth, durable roads that can handle heavy traffic without succumbing to wear and tear. By acting as a microscopic reinforcement network, these fibres enhance the overall performance and longevity of asphalt pavements, leading to roads that stand the test of time and traffic.

The concept of fibre-reinforced asphalt mixtures extends beyond simply adding these materials. Chen conducted research that delved into the design and evaluation aspects of such mixtures [4]. Their work paves the way for the wider implementation of fibre-reinforced asphalt by establishing methodologies for their creation and performance assessment. This allows for the optimization of fibre type, quantity, and overall mixture composition to achieve the desired properties.

Understanding how fibre-reinforced asphalt mixtures behave under stress and strain is crucial for their successful implementation. Wu and Ye focused on the rheological and fatigue properties of these mixtures containing polyester fibres [5]. Their research provides valuable insights into the way these mixtures respond to various loading conditions. This knowledge is essential for predicting the long-term performance of pavements constructed with fibre-reinforced asphalt.

The properties of the components (binder, aggregate, and additive) and the ratio of these components in the overall mix constitute a few of the elements that affect how well flexible courses work. Bitumen can also be altered by including various kinds of additives. As a result, the quality of illuminance might not be uniform, and some parts may not receive proper lighting. In general, the inclusion of polymers makes bitumen more rigid and enhances its temperature sensitivity. This can lead to increased brittleness, especially at lower temperatures. Synthetic fibres, known for their high strength, ductility, and durability qualities, have emerged as promising options to address these challenges and further enhance the mechanical performance of asphalt mixtures [6]. This study investigates the combined effects of RHA and synthetic fibres, like polyester, to further improve the performance and sustainability of asphalt pavements.

Even familiar materials like glass fibre are finding their niche in the realm of asphalt innovation. The effectiveness of glass fibre as a stabiliser in Stone Matrix Asphalt, a type of mixture known for its exceptional strength and resistance to deformation [7, 8]. Adding glass fibre to this mix minimises the "drain down" effect, where the bitumen seeps out of the mixture, and improves its creep characteristics, making it less prone to permanent deformation under heavy loads. This not only enhances the performance of Stone Matrix Asphalt but also broadens the range of applications where it can be effectively utilised, paving the way for more versatile and resilient roads. These advancements in asphalt technology offer a glimpse into a future where sustainability and performance go hand in hand. By embracing waste and synthetic materials, we can create asphalt mixtures that are not only robust and long-lasting but also environmentally responsible. With continued research and innovation, the road ahead for asphalt pavements is paved with exciting possibilities, promising a future where every lane leads towards a more sustainable and resilient world.

2 Methodology

2.1 Components

This section covers the composition, what components were used, their materialistic specifications, and their properties.

1. Aggregates: A coarse aggregate of nominal size 10 mm and a fine aggregate of nominal size 8 mm are used in the composite mixture, and the characteristics are shown in Table 1.

2. Bitumen: The bitumen used in the mixture needs to have an appropriate viscosity and properties, as shown in Table 2.

3. Glass Powder: In this study, powdered glass fibre (Fig. 1) was utilized as an ingredient in the asphalt mixture to increase the asphalt's toughness and fatigue characteristics, as shown in Table 3. The varying percentages of glass fibres were occupied by the weight of bitumen.

4. RHA: One of the many benefits of using RHA, a waste byproduct of milling rice, in bituminous highways, is that it reduces environmental deterioration and saves natural resources and building expenses. There is not much research on the use of this substance in asphalt mixture, though. examining the effects of RHA as an asphalt modifier on hot mix asphalt was the aim of this investigation. Testing for penetration grade, ductility, and softening point were among the characteristics of asphalt binders that were assessed. Additionally, an evaluation was conducted on the mechanical characteristics of asphalt mixes, encompassing Marshall stability, stiffness modulus, resistance to rutting, and fatigue behaviour. The findings demonstrated that the use of RHA improved bitumen's rheological characteristics. Furthermore, RHA modification had positive impacts on the Marshall stability, stiffness modulus, rutting strength, and

Table 1 Physical characteristics of aggregates

Index	Aggregate
Specific gravity(gm/cm^3)	2.64
Aggregate crushing value (%)	23.43
Aggregate impact value (%)	7.95
Flakiness and elongation (%)	14.64, 8.64

Table 2 Properties of bitumen

Index	Aggregate
Specific gravity(gm/cm^3)	0.99
Penetration of bitumen	97.66
Flashpoint	280

Fig. 1 Glass powder used in mixture

Table 3 Properties of the glass powder

Test properties	Glass powder
Colour	White
Tensile strength	4700–4800 Mpa
Density	2.46 gm/cc
Base	S- base

fatigue performance of asphalt mixtures. Bitumen mix blends with 2%, 4%, 6%, and 8% RHA modifiers.

2.2 *Preparation of Marshall Mix Design*

In this research, the preparation of the asphalt mix adhered to stringent ASTM-D standards, ensuring accuracy and consistency. The process involved several key steps:

2.2.1 Aggregate Preparation

Crushed aggregate was carefully separated into two size fractions: coarse (diameter exceeding 2.36 mm) and fine (passing through a 2.36-mm sieve), refer to Fig. 2. Further, to remove any moisture content that could impact the mix, both fractions were completely dried in an oven for 24 h at a high temperature between 170°C and

Fig. 2 Segregation of materials according to required weights

190°C. Precise amounts of each fraction were then measured for each specimen: approximately 108 g of coarse aggregate and 617 g of fine aggregate.

2.2.2 Asphalt Heating and Mixing

The asphalt binder, the crucial glue holding the aggregate together, was heated to a temperature between 130°C and 140°C to ensure proper fluidity for mixing. Further, the preheated aggregates were combined with the appropriate amount of hot asphalt at a temperature of 200°C. This high temperature ensures thorough blending and optimal bonding between the asphalt and the aggregate particles; refer to Fig. 3a, b.

2.2.3 Modification for Control and Treated Mixes

The control and modified mixes incorporating glass powder and polyester additives followed the same binder and rice husk concentration. In the modified mixes, the specified amounts of glass powder or polyester were premixed with the heated aggregate and filler before the hot asphalt was added. This step ensures even distribution of the modifier throughout the entire mix.

2.2.4 Compaction and Sample Formation

The thoroughly mixed asphalt mixture was carefully placed in a preheated mould. To achieve uniform density and optimal performance, the mix was compacted using a standardised protocol of 75 blows on each side of the mould. Finally, the

(a) (b)

Fig. 3 (**a**) Heating the materials to 170 °C, and (**b**) materials after heating

Fig. 4 A moulds prepared in this study

compacted mix was extracted from the mould using a dedicated sample extractor, ready for further testing and analysis, as shown in Fig. 4.

It provides an understanding of the meticulous steps involved in preparing the asphalt mix for this study. By adhering to rigorous standards and employing precise temperature control and mixing techniques, the researchers ensured the consistency and quality of the asphalt mixtures, laying the foundation for reliable and accurate data collection and analysis.

2.3 Observations

The prepared composite mixture was tested, and its properties were analysed, as presented in Tables 4, 5 and 6. The schematic graph presented in Figs. You can find the graphs below:

The following observations were found based on the Figs. 5, 6 and 7 of the Marshall stability and flow value that increased strength, compared to the conventional mix, moulds reinforced with polyester exhibited higher Marshall stability values. This suggests that the addition of polyester enhanced the asphalt's ability to resist deformation under load. Further, it enhanced workability. The polyester-reinforced moulds also displayed improved Marshall flow values, as observed on the flow graph. This indicates the modified asphalt offered better workability and compaction during pavement construction. There are also brittleness concerns; while strength and workability improved, observations after testing revealed that the moulds became more brittle. This suggests a potential trade-off between

Table 4 Marshall stability and flow values of RHA reinforced bituminous concrete moulds

S.N.	Composition of RHA (%)	Marshall stability (KN)	Flow value (mm)
1	2	13.06	2.30
2	3	14.84	1.90
3	4	14.86	2.10
4	6	15.23	2.12
5	8	11.52	2.18
6	10	11.78	2.13
7	12	7.38	3.50

Table 5 Marshall stability and flow values of polyester reinforced bituminous concrete moulds

% of polyester	Marshall stability (KN)	Flow value (mm)
0.1	14.15	0.74
0.15	18.21	0.98
0.2	16.0	0.34
0.25	15.55	0.40

Table 6 Marshall stability and flow values of glass Powder reinforced bituminous concrete moulds

% of glass powder	Marshall Stability (KN)	Flow Value (mm)
0.5	20.2	1.21
1	19.6	1.201
1.5	22.3	1.52
2	19.11	1.26
2.5	18.4	0.4
3	20.18	0.27

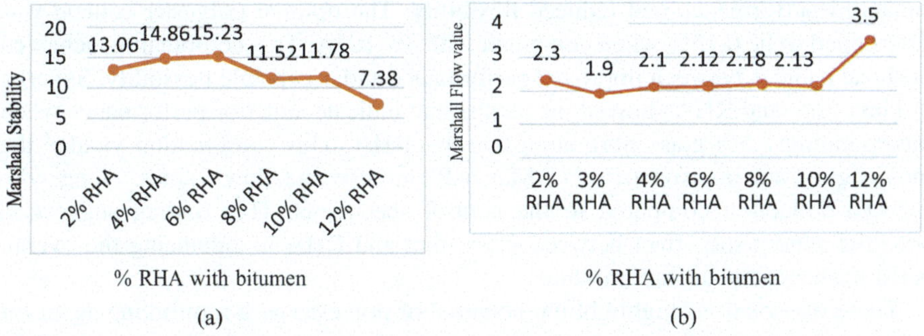

Fig. 5 (**a**) Graph measuring Marshall stability of RHA reinforced bituminous concrete moulds, (**b**) Graph measuring Marshall flow values of RHA reinforced bituminous concrete moulds

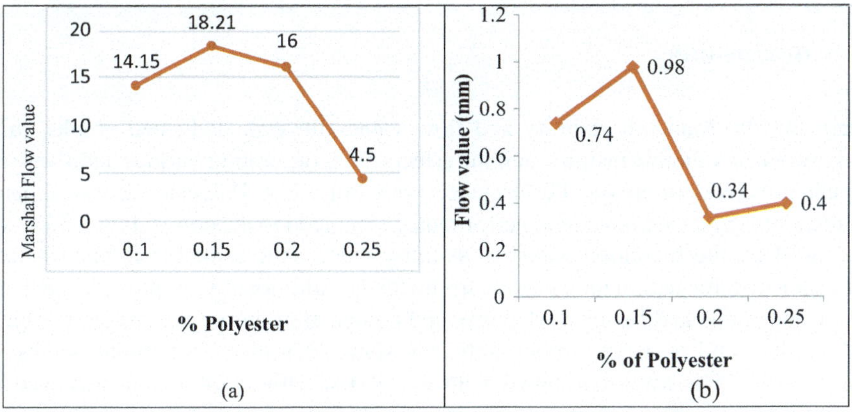

Fig. 6 (**a**) Graph measuring Marshall stability of polyester reinforced bituminous concrete moulds, (**b**) Graph measuring Marshall flow values of polyester reinforced bituminous concrete moulds

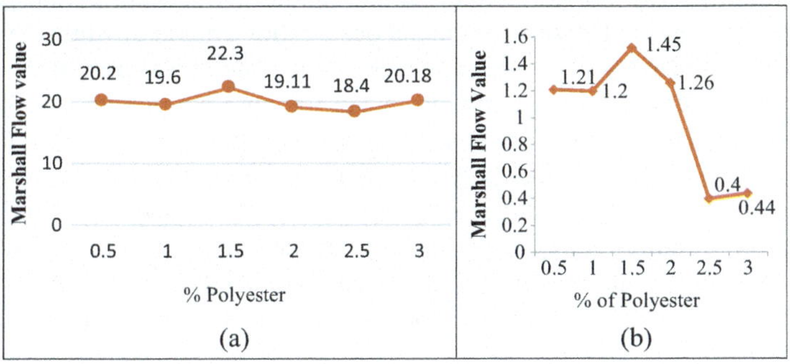

Fig. 7 (**a**) Graph measuring Marshall stability of glass powder reinforced bituminous concrete moulds, (**b**) Graph measuring Marshall flow values of glass powder reinforced bituminous concrete moulds

enhanced performance and reduced flexibility. The optimal polyester content was determined to be 0.15% when combined with 3% RHA. This combination achieved the ideal balance between improved performance and acceptable flexibility. Synergy of glass fibre and RHA showed the results that indicate superior performance when incorporating 2.5% glass fibre alongside 3% RHA. This combination yielded the most significant improvements in Marshall stability and flow values, along with reduced deflection compared to the control specimens. This finding suggests a potential synergistic effect between glass fibre and RHA in enhancing the overall performance of the asphalt mixture.

These observations highlight the potential of polyester as a reinforcing agent for asphalt mixtures but also emphasise the need for further investigation to address the brittleness concern and identify additional modifiers that can maintain workability without compromising strength.

3 Conclusions

Analyzing the Marshall stability and flow values through dedicated graphs, this study revealed valuable insights into the effects of incorporating various additives on asphalt mixtures. Improved Performance with Polyester: Polyester reinforcement significantly enhanced pavement performance compared to the conventional mix. As evidenced by the increased Marshall stability values, the asphalt exhibited better resistance to deformation under heavy traffic loads. Additionally, improved Marshall flow values indicated better workability and compaction during construction. Optimal Content and Synergy: Importantly, the study identified an optimal polyester content of 0.15% when combined with 3% RHA. This combination achieved a desirable balance between improved performance and acceptable flexibility. Furthermore, the results revealed a potentially synergistic effect when incorporating 2.5% glass fibre alongside 3% RHA. This combination yielded the most significant improvements in Marshall stability and flow values, demonstrating enhanced strength and flexibility compared to the control and polyester-reinforced specimens. Addressing Brittleness: However, a crucial observation emerged: while strength and workability increased with some additives, some modifications also exhibited increased brittleness. This highlights the need for further research to address this trade-off, potentially through exploring alternative modifiers or optimizing the content of specific additives like polyester. In conclusion, this research underscores the potential of utilizing sustainable materials like RHA and synthetic fibres to develop more durable and high-performing asphalt mixtures. By optimizing the type, content, and combinations of these additives, future research can aim to achieve the ideal balance: strong, flexible, and workable pavements for the future.

References

1. Hinislioglu S, Agar E (2004) Investigation of the influence of HDPE-modified binder on the Marshall stability, flow, and Marshall quotient. Build Environ 39(10):1200–1207
2. Mrema AH, Noh SH (2020) Performance of glass wool fibres in asphalt mixtures. Constr Build Mater 261:120095
3. Jia H, Sheng Y, Wu S, Liu G (2023) Enhancing mechanical properties of asphalt mixtures with synthetic fibers: a review. Constr Build Mater. 389:129934
4. Chen H, Xu Q, Chen S, Zhang Z (2009) Evaluation and design of fibre-reinforced asphalt mixtures. Mater Des 30(7):2595–2603. https://doi.org/10.1016/j.matdes.2008.10.004
5. Wu S, Ye Q (2007) Investigation of rheological and fatigue properties of asphalt mixtures containing polyester fibres. Constr Build Mater 21(10):1872–1879
6. Alnadish AM, Singh N, Singh S, Alawag AM (2023) Applications of synthetic fibres in asphalt mixtures. Materials 15(4):1004. https://doi.org/10.3390/ma15041004
7. Ayoub EM (2019) Use of glass fibres in bituminous concrete and stone matrix asphalt mixes. Transp Res Rec 2670(1):29–37
8. Mirkovic K, Mladenovic G (2019) Influence of fly ash content on the performance of hot mix asphalt mastic. Constr Build Mater 206:84–93

Demand Forecasting of Highway Construction Materials Using Machine Learning Model

Rahul V. Wasekar and Gayatri S. Vyas

Abstract The global construction sector stands at a crucial juncture, necessitating eco-conscious practices to minimize environmental impact and enhance resource efficiency, with highway construction playing a significant role. In the Indian context, the urgency to adopt sustainable supply chain management practices in highway construction is pronounced, given the country's vast infrastructure projects and environmental challenges. India's ambitious infrastructure goals necessitate innovative approaches to ensure sustainability and resilience in highway construction projects. This study explores integrating sustainable supply chain management into highway construction, focusing on material demand forecasting. Leveraging machine learning techniques, specifically random forest regression, the research aims to transform material demand forecasting in highway construction, addressing issues such as overstocking and stockouts. By providing stakeholders with valuable insights into material demand forecasting, the study seeks to contribute to the timely completion and sustainable development of highway infrastructure projects, fostering environmentally conscious and resilient transportation networks worldwide.

Keywords Demand Forecasting · Machine Learning · Highway Construction · Random Forest Regression · Supply chain management

1 Introduction

The demand for highway construction materials poses significant challenges to sustainable development and environmental conservation [1]. Efficient resource management is essential to mitigate environmental impact and optimize economic outcomes [2]. Highway construction generates substantial material demand, including concrete, asphalt, gravel, and more [3]. However, the dynamic nature of construction projects, coupled with varying environmental factors and economic

R. V. Wasekar (✉) · G. S. Vyas
COEP Technological University, Pune, India

P. Pathak et al. (eds.), *Advances in Environmental Sustainability, Energy and Earth Science*, https://doi.org/10.1007/978-3-031-73820-3_9

conditions, complicates accurate forecasting of material demand [4]. Traditionally, material demand forecasting in highway projects has relied on manual methods and historical data analysis, which often lack the precision and adaptability required to address the dynamic nature of construction projects [5]. These conventional methods can lead to inefficiencies, such as overordering or underordering materials, resulting in financial losses and environmental degradation. With advancements in technology, particularly in machine learning, there is a growing opportunity to revolutionize material demand forecasting in highway construction [6]. Machine learning (ML), a subset of artificial intelligence (AI), presents promising prospects for improving demand forecasting in highway construction materials [7]. By harnessing computational algorithms and data analytics techniques, ML can extract valuable insights from various datasets, encompassing factors such as weather conditions, economic indicators, traffic volume, and material specifications [8]. Additionally, ML algorithms enable real-time monitoring and decision making, facilitating proactive strategies for optimizing material procurement and resource allocation in highway construction projects [9]. The adoption of ML techniques not only enhances accuracy in demand forecasting but also fosters innovation in the construction industry [10]. Integrating these advanced technologies into traditional forecasting models can significantly reduce waste, improve cost-efficiency, and support sustainable construction practices [11]. As the industry moves towards more data-driven decision-making, the importance of sophisticated forecasting tools becomes increasingly evident [12]. This research aims to develop a tailored ML prediction model designed to forecast the demand for highway construction materials. By leveraging the unique characteristics of the dataset, this research seeks to contribute to a more sustainable and environmentally conscious approach within the highway construction industry.

2 Methodology

This section details the processes of data collection, preprocessing, algorithm selection, model development, and model evaluation. Section 2.1 covers the collected data and preprocessing methods, while Sect. 2.2 introduces and applies supervised ML algorithms. In Sect. 2.3, the model development process is explained, followed by the evaluation of ML models in Sect. 2.4. A schematic of prediction is shown in Fig. 1.

2.1 Data Collection and Preprocessing

Data collection, a pivotal aspect of this research, involved gathering data from 30 highway construction projects within the Maharashtra (India) region. The focus was on engaging various stakeholders associated with highway development projects. The datasets encompass a diverse array of variables comprising both

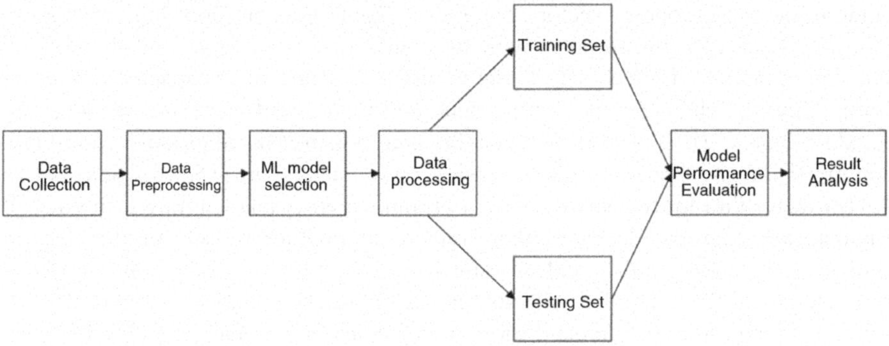

Fig. 1 Research methodology for the development of prediction model

Table 1 Variables with its description considered to develop the RF model

Variables		Code	Parameter
Independent	Continuous	WT	Weather temperature
		WP	Weather precipitaion
		EGDP	Economic GDP %
		TV	Traffic volume
		TD	Transport distance
	Categorical	MT	Material type
		SL	Supplier location
Dependent	Continuous	DE	Demand

categorical and continuous factors, refer to Table 1. Categorical variables encompass supplier locations and material types, while continuous variables include factors such as traffic volume, transport distances, and economic indicators.

Data preprocessing was applied to the collected dataset to improve the performance of the ML model. The preprocessing steps adopted in this study involved label encoding, which converts categorical input variables into numerical ones. Subsequently, feature scaling was conducted using the normalization technique, refer to Eq. 1. It is a crucial preprocessing step in ML, particularly when working with datasets containing features with different scales. It ensures that all features contribute equally to the learning process.

$$Normalized\ value = \frac{Original\ value - Minimum\ value}{Maximum\ value - Minimum\ value} \quad (1)$$

2.2 Machine Learning Algorithm Selection

In this process, selecting the appropriate ML model holds significant importance. Various ML algorithms were assessed, including linear regression, decision trees, random forest, support vector regression, and gradient-boosting regression [13, 14]. Each algorithm comes with its own set of advantages and drawbacks. The determination of which ML model to utilize depends on its capacity to manage complexity, the characteristics of the data, and the desired level of accuracy. The Random Forest (RF) model, a versatile and potent ML algorithm commonly employed for classification and regression tasks, stands out [15, 16]. Selected as the ML algorithm for this study, RF is a prominent ensemble learning technique. It amalgamates multiple decision trees trained on random subsets of the data to enhance prediction accuracy and robustness. Its capability to handle both numerical and categorical data, coupled with its feature importance analysis, renders it well-suited for forecasting the demand for highway construction materials [17, 18].

2.3 Development of Prediction Model

The ML model for predicting the demand for highway construction materials is developed using Python within a Jupyter Notebook from Anaconda Navigator. The Jupyter Notebook, an open-source web application, acts as a platform for Python code execution, facilitating the creation and execution of ML prediction models. To initiate the development process, a range of essential Python libraries are imported and utilized within the Jupyter Notebook environment. Among these, "pandas" is indispensable for data manipulation and analysis, enabling diverse operations on data structures and complex data processing tasks. Furthermore, "scikit-learn" (commonly imported as sklearn) serves multiple purposes: "sklearn.preprocessing" handles categorical variable encoding, "sklearn.modelselection" facilitates dataset splitting, and "sklearn.ensemble" houses the RF regressor, a robust method for regression tasks that constructs multiple decision trees to enhance prediction accuracy and mitigate overfitting. Additionally, "sklearn.metrics" is instrumental in evaluating model accuracy and furnishing metrics such as R-squared and Mean Squared Error. For visualization needs, "matplotlib" and "seaborn" come into play, generating informative graphics like performance plots and feature importance visualizations within the notebook. Each of these libraries assumes a pivotal role in the ML development workflow, spanning from data preprocessing to model evaluation and visualization.

2.4 Model Performance Evaluation

To assess the effectiveness of ML models for predicting material demand, a range of statistical metrics in regression analysis were used. These metrics included Mean Absolute Error (MAE), Mean Squared Error (MSE), Root Mean Squared Error (RMSE), and R-squared (coefficient of determination), which are computed in the following equations:

$$MSE = \sum_{i=1}^{n} \frac{(vi - wi)^2}{n} \tag{2}$$

$$RMSE = \sqrt{\sum_{i=1}^{n} \frac{(vi - wi)^2}{n}} \tag{3}$$

$$R^2 = 1 - \frac{\sum_{i=1}^{n} (vi - wi)^2}{\sum_{i=1}^{n} (vi - \bar{wi})^2} \tag{4}$$

where *wi* represents the observed demand for construction materials and *vi* represents the predicted demand for construction materials, $\bar{wi}$ denotes the mean observed demand, $\bar{vi}$ indicates the average predicted demand, and n denotes the number of samples.

3 Result Analysis and Discussion

3.1 Model Performance Result

The RF model's performance was tested using several statistical metrics like R^2, RMSE, and MSE. R^2 is a measure of how well the independent variables explain the variation in the dependent variable. RMSE tells us the average size of errors between what the model predicts and the actual values. Similarly, MSE quantifies the average squared difference between the actual and predicted values of the target variable. The model performance metrics obtained are as in Table 2.

3.2 Sensitivity Analysis

Figure 2 shows the sensitivity analysis plot. The sensitivity analysis of the demand forecasting model reveals that variations in different input features significantly impact the accuracy of demand predictions. Notably, changes in traffic volume have the most substantial effect, leading to a considerable decrease in mean squared error (MSE). Conversely, alterations in transport distance show a notable impact, resulting in an increase in MSE. Supplier location exhibits sensitivity to geographical distribution, with changes leading to a decrease in MSE. Furthermore, economic GDP growth and weather precipitation also influence demand predictions, with

Table 2 Performance evaluation metrics

Performance metric	RF values
MSE	0.0257
RMSE	0.1604
R^2	0.6355

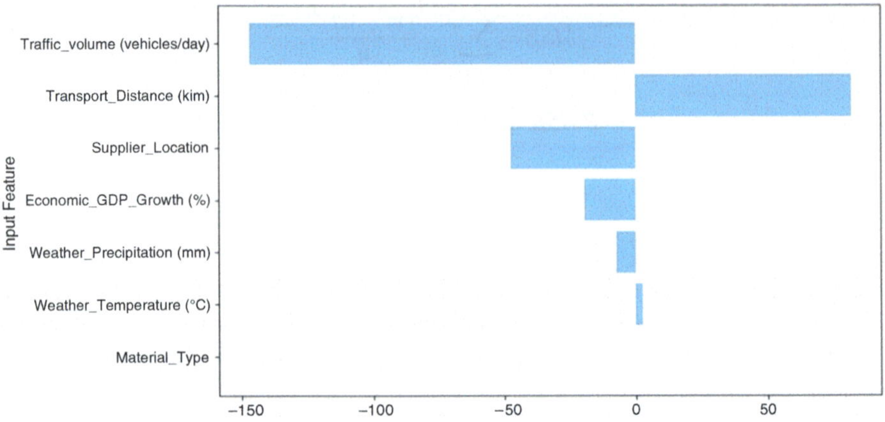

Fig. 2 Sensitivity analysis plot of the input parameters

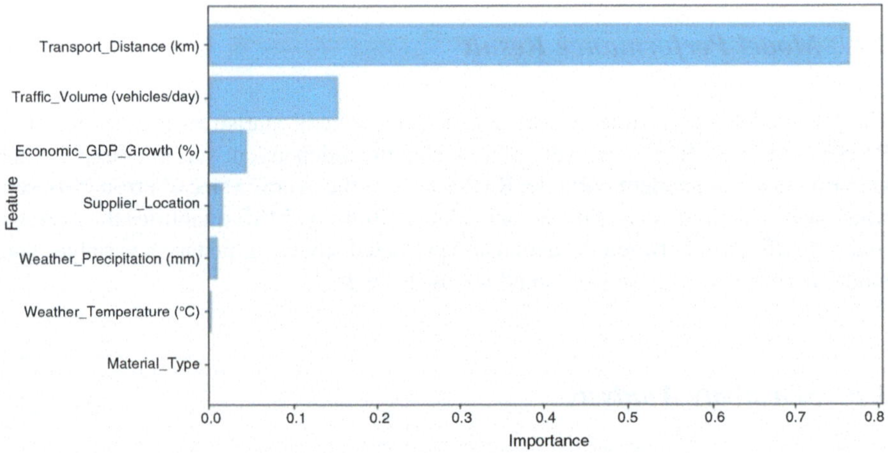

Fig. 3 Relative importance of input parameters in the Random Forest model

variations causing decreases in MSE. Weather temperature displays a slight increase in MSE, indicating relatively lower sensitivity. Material type demonstrates minimal sensitivity.

3.3 *Feature Importance*

Figure 3 shows a feature importance plot, which displays the relative importance of different features in predicting the target variable, which in this case is demand for highway construction materials.

Transport distance emerges as the most influential factor for demand forecasting of highway construction materials, given its highest importance score. The traffic volume is closely followed, indicating its significant role in determining the forecasted demand. Economic GDP growth also exhibits importance, albeit to a lesser degree, implying its modest influence on demand forecasting. In contrast, features such as supplier location, weather precipitation, weather temperature, and material type demonstrate relatively lower importance, suggesting their limited impact on demand prediction in this model. Overall, the feature importance analysis underscores the critical role of transport distance and traffic volume in accurately forecasting demand for highway construction materials, highlighting the importance of logistical factors in SCM.

3.4 Predicted Values vs. Actual Value Scatter Plot

The scatter plot in Figs. 4 and 5 showcases the correlation between the actual demand for construction materials, measured in units, and the corresponding predictions generated by the ML model. On the x-axis, we have the actual demand values, while the y-axis represents the predicted values by the model. The dashed diagonal line represents a perfect agreement, where predicted and actual values are equal. The variance observed in the scatter plots between the training and validation sets presents an opportunity for constructive analysis and improvement in model performance evaluation. While the training set exhibits a close alignment between

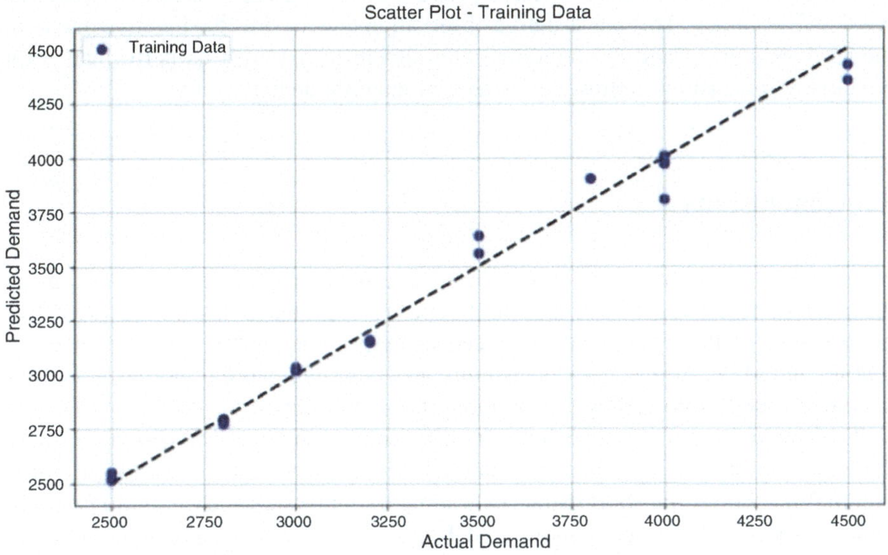

Fig. 4 Scatter-plot diagram for actual vs. predicted demand for training set

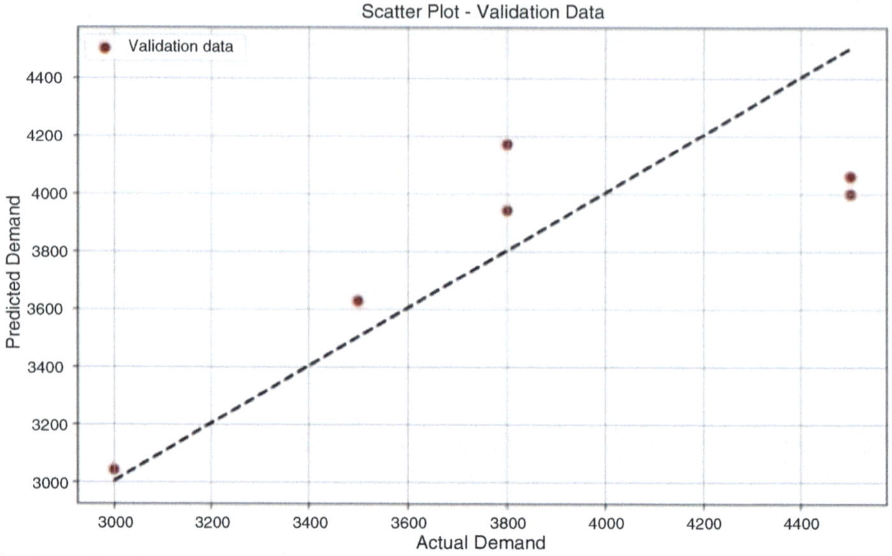

Fig. 5 Scatter-plot diagram for actual vs. predicted demand of validation set

actual and predicted demand values, indicative of the model's foundational accuracy, the validation set displays a slightly broader deviation from the ideal scatter center line. This deviation, although noticeable, signifies the model's capacity to capture essential trends and patterns in demand forecasting. Despite the discernible difference, it highlights areas where the model can be further refined to enhance its predictive accuracy. This constructive observation underscores the iterative nature of model development and the ongoing commitment to optimize its performance for effectively forecasting highway construction material demand.

4 Conclusions

In this study, the highway material demand forecasting model was developed using the random forest regression algorithm due to a limited dataset. The predictive performance of the model was evaluated using several statistical metrics like R^2, RMSE, and MSE. Additionally, sensitivity analysis and feature importance analysis were performed, followed by the creation of a scatter plot.

The sensitivity analysis revealed that traffic volume and transport distance exerted the most significant influence on material demand forecasts, with lower traffic volume and shorter transport distances leading to higher demand predictions. Additionally, supplier location, economic GDP growth, and weather precipitation moderately affected forecasts, while weather temperature and material type exhibited minimal impact. The feature importance plot highlighted that transportation distance and traffic volume are the two most crucial factors influencing material demand forecasting. These findings are consistent with the expectation that logistical considerations and economic trends play significant roles in determining demand levels along highway construction projects.

References

1. Prakash A, Mohanty RP (2015) Understanding construction supply chain management for road projects. Int J Logist Syst Manag 22(4):414–435. https://doi.org/10.1504/IJLSM.2015.072747
2. Han C, Zhang Q (2021) Optimization of supply chain efficiency management based on machine learning and neural network. Neural Comput Applic 33(5):1419–1433. https://doi.org/10.1007/s00521-020-05023-1
3. Ni D, Xiao Z, Lim MK (2020) A systematic review of the research trends of machine learning in supply chain management. Int J Mach Learn Cybern 11(7):1463–1482. https://doi.org/10.1007/s13042-019-01050-0
4. Tirkolaee EB, Sadeghi S, Mooseloo FM, Vandchali HR, Aeini S (2021) Application of machine learning in supply chain management: a comprehensive overview of the main areas. Math Probl Eng 2021:1. https://doi.org/10.1155/2021/1476043
5. Larrea-Gallegos G, Vázquez-Rowe I (2022) Exploring machine learning techniques to predict deforestation to enhance the decision-making of road construction projects. J Ind Ecol 26(1):225–239. https://doi.org/10.1111/jiec.13185
6. Gong H, Sun Y, Shu X, Huang B (2018) Use of random forests regression for predicting IRI of asphalt pavements. Constr Build Mater 189:890–897. https://doi.org/10.1016/j.conbuildmat.2018.09.017
7. Speiser JL, Miller ME, Tooze J, Ip E (2019) A comparison of random forest variable selection methods for classification prediction modeling. Expert Syst Appl 134:93–101. https://doi.org/10.1016/j.eswa.2019.05.028
8. Meharie MG, Shaik N (2020) Predicting highway construction costs: Comparison of the performance of random forest, neural network and support vector machine models. J Soft Comput Civ Eng 4(2):103–112. https://doi.org/10.22115/SCCE.2020.226883.1205
9. Seyedan M, Mafakheri F (2020) Predictive big data analytics for supply chain demand forecasting: methods, applications, and research opportunities. J Big Data 7(1). https://doi.org/10.1186/s40537-020-00329-2
10. Simić N, Petronijević P, Devedžić A, Ivanović M (2022) Preliminary quantity estimation in construction using machine learning methods. Савремена Теорија И Пракса У Градитељству 15(1):83–91. https://doi.org/10.7251/stp2215083s
11. Kovačević M, Antoniou F (2023) Machine-learning-based consumption estimation of prestressed steel for prestressed concrete bridge construction. Buildings 13(5):1187. https://doi.org/10.3390/buildings13051187
12. Silva RV, de Brito J, Dhir RK (2017) Availability and processing of recycled aggregates within the construction and demolition supply chain: a review. J Clean Prod 143:598–614. https://doi.org/10.1016/j.jclepro.2016.12.070
13. Tayefeh Hashemi S, Ebadati OM, Kaur H (2020) Cost estimation and prediction in construction projects: a systematic review on machine learning techniques. SN Appl Sci 2(10):1–27. https://doi.org/10.1007/s42452-020-03497-1
14. Akinosho TD et al (2020) Deep learning in the construction industry: a review of present status and future innovations. J Build Eng 32:101827. https://doi.org/10.1016/j.jobe.2020.101827
15. Suveka V, Shanmuga Priya T (2016) A review on prediction of material prices in construction projects. Int Res J Eng Technol:757–760. [Online]. Available: www.ir-jet.net

16. Liu Y, Wang Y, Lyu P, Hu S, Yang L, Gao G (2021) Rethinking the carbon dioxide emissions of road sector: integrating advanced vehicle technologies and construction supply chains mitigation options under decarbonization plans. J Clean Prod 321(June):128769. https://doi.org/10.1016/j.jclepro.2021.128769
17. Zhang Y, Ling C (2018) A strategy to apply machine learning to small datasets in materials science. npj Comput Mater 4(1):28–33. https://doi.org/10.1038/s41524-018-0081-z
18. Mohammadnazari Z, Ghannadpour SF (2021) Sustainable construction supply chain management with the spotlight of inventory optimization under uncertainty, vol 23. Springer, Netherlands, p 7. https://doi.org/10.1007/s10668-020-01095-0

A Review on Liquefaction Analysis Based on Stress and Energy Based Approaches Using SPT

P. Aditya Sai Teja, B. Vaishnavi, K. Chandra Vishal, and V. Ramya Krishna

Abstract Liquefaction phenomenon occurs in completely saturated, cohesive-less and loose soils that are subjected to periodic loading in undrained conditions. Because of the gradually rising pore pressure during liquefaction, the soil loses its shear strength when the effective stress resulting to zero. Even if a structure is intended to resist earthquakes, liquefaction may lead to foundation failure and subsequent structure collapse. Liquefaction is dependent on the properties of the soil below the ground surface. This paper presents the methods for the analysis of liquefaction potential and aims at the comparison of stress and energy-based methods, which rely mainly on Standard Penetration Test (SPT) values. This paper reviews some of the liquefaction hazards from previous earthquakes as well as case data from field research. This paper also studies remedial strategies employing different soil mitigation techniques that might be applied based on soil conditions at that site to reduce the severe damage caused through liquefaction.

Keywords Earthquake · Energy-based method · Factor of Safety · Liquefaction · Mitigation techniques · Standard Penetration Test · Stress-based method

1 Introduction

Liquefaction is a phenomenon characterized by the diminished strength of saturated and cohesion-less soils [1], resulting from gradually increasing pore pressure and consequent decrease in the effective stress caused by cyclic loading. This occurrence leads to a reduction in the soil's strength and stiffness, commonly induced by seismic activity or other rapid loading events [2]. Saturated soils, where water fills the spaces between the particles, experience liquefaction. Pore pressure, which is applied by this water to the soil particles, becomes disrupted and causes liquefaction [3].

P. A. S. Teja · B. Vaishnavi · K. C. Vishal · V. R. Krishna (✉)
Department of Civil Engineering, VNR Vignana Jyothi Institute of Engineering and Technology, Hyderabad, Telangana, India
e-mail: ramyakrishna_v@vnrvjiet.in

P. Pathak et al. (eds.), *Advances in Environmental Sustainability, Energy and Earth Science*, https://doi.org/10.1007/978-3-031-73820-3_10

1.1 Cause of Liquefaction

Since the occurrence of two significant earthquakes in 1964, researchers globally have dedicated their efforts to studying liquefaction. Subsequently, the modern field of engineering has seen the introduction of numerous terms, conceptual insights, processes, and analytical techniques. Evaluating soil liquefaction and the resulting ground failures associated with earthquakes stands as a vital aspect of geotechnical engineering practice [4]. When an earthquake takes place, the release of stored energy generates waves that traverse the Earth's crust, causing vibrations [5].

India has experienced approximately 74 earthquakes to date. The initial recorded earthquake in India transpired on June 6, 1505, within the Karnali zone near Saldang. Earthquakes possess the potential to inflict severe and catastrophic damage. The environmental repercussions of earthquakes encompass geological faults, tectonic uplift and downlift, tsunamis, soil liquefaction, ground vibrations, landslides, and ground collapse, all of which can be induced either directly or indirectly [6]. Recent seismic activity in various regions has underscored the significance of constructing earthquake-resistant structures. The escalating urban seismic risk is attributed to factors such as high population density, the proliferation of diverse building types (including mud and RCC frames), substandard construction practices, and the prevalence of dense and unpredictable traffic patterns. The term "liquefaction" originated in the aftermath of two seismic events: the 7.5-intensity Nigata earthquake in Japan and the 9.2-magnitude Good Friday earthquake in 1964 in Alaska. These earthquakes led to various consequences such as slope failure, bridge piers sinking, leaning of residences, and the revelation of the structures present in the subsurface. India also experienced notable liquefaction-related damages after the 7.6-magnitude earthquake that struck Bhuj on January 26, 2001.

1.2 Past Records of Liquefaction

Case Study 1
The 45-km San Andreas fault located beneath the Mountains named Santa Cruz in California was struck by the Lorna Prieta earthquake. Significant liquefaction was caused by this seismic event, which had an earthquake moment magnitude 7, in a number of locations close to Monterey Bay [7] and Moss Landing in California on October 17, 1989. The entire 150–300 m broad coastline peninsula known as Moss Landing spit showed signs of liquefaction, which were characterized by deformations.

Case Study 2
On January 26, 2001, Gujarat state in India experienced a powerful 7.6-magnitude earthquake (Mw 7.6), commonly known as the Bhuj Earthquake. The mainshock, located near Bachau. This earthquake, one of the most devastating in Indian history, resulted in the near-complete collapse of Chang Dam primarily due to the

liquefaction of its shallow foundation soils. While Kaswati Dam and Fatehgadh Dam suffered comparatively less severe damage, direct observations of liquefaction were not reported near these dams. However, localized liquefaction of foundation soils contributed to the post-quake damage observed in these areas [8].

Case Study 3

Christchurch, New Zealand, experienced a series of potent local earthquakes from September 2010 to December 2011, leading to considerable destruction and intense ground upheavals. The impact was especially severe during the magnitude 6.2 earthquake on February 22, 2011, causing significant harm to the city and a tragic toll of 185 lives. This particular seismic event is believed to have triggered one of the most extensive and severe instances of liquefaction in natural soils ever recorded, affecting approximately one-third of the city [9].

1.3 Methods for Evaluation of Liquefaction Potential

In recent decades, scientists have put forth various methodologies for identifying soil susceptibility to liquefaction and enhancing our understanding of the liquefaction process. These methods can be broadly categorized into three groups, as outlined by Green in 2001: energy-based, strain-based, and stress-based techniques. Among these, two authors introduced the stress-based method which is widely employed for assessing liquefaction [10]. Typically, this approach relies on empirical data gathered from both field and laboratory test results. An energy-based method (EBM) that operates on the principle that the energy released from the seismic waves in a unit volume of soil is directly correlated with the accumulation of pore pressure was introduced [11].

2 Stress-based Method

In the stress-based method, both the soil layer's seismic demand, marked as CSR, and the soil's ability to withstand liquefaction, denoted as CRR, must be calculated or estimated in order to use the stress-based approach for evaluating the ability of soils against the liquefaction. Various writers have ascribed different symbols to the cyclic stress ratio required to produce liquefaction [12]. In order to reduce ambiguity and separate the mobilized liquefaction resistance from generated cyclic shear stresses, this study uses the term CRR, which is also suggested for engineering practice, to indicate the soil's ability against the liquefaction.

Step 1

The water table's location and SPT N blow counts should be included in the soil data used to evaluate liquefaction susceptibility. By comparing the ability of soil to withstand liquefaction (CRR) to the soil layer's seismic demand (CSR), the

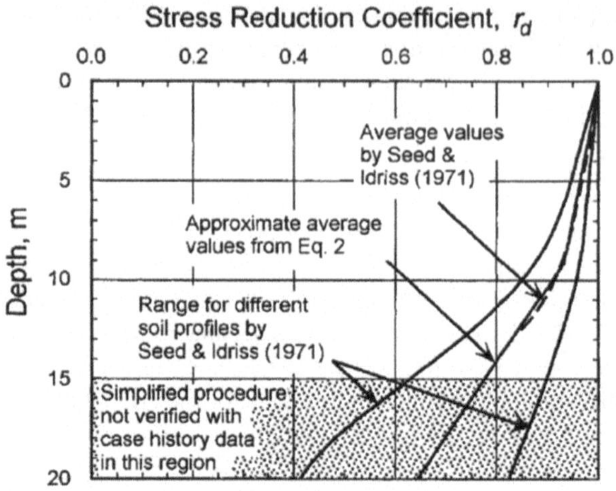

Fig. 1 Stress reduction coefficient versus depth curves [10]

stress-based method determines the liquefaction potential of the given soil material at that site.

Step 2—Evaluation of total vertical overburden stress:

Assess the total vertical overburden pressure (σ_{vo}) and effective vertical overburden pressure (σ_{vo}') at varying heights for all layers within the unit soil volume that have the potential for liquefaction.

Step 3—Evaluation of factors for stress reduction:

$$r_d = e^{[\alpha(z)+(\beta(z)\times M)]} \quad (1)$$

$$\alpha(z) = -1.012 - 1.126\, sin[5.133 + z/11.73] \quad (2)$$

$$\beta(z) = 0.106 + 0.118\, sin[5.142 + z/11.28] \quad (3)$$

where M = earthquake's magnitude.

In this case, z stands for the depth in meters below the surface of the earth. Even though alternative equations for estimating r_d at further depths are presented [13], it's crucial to remember that liquefaction evaluation at these deeper depths occurs outside of the range where the streamlined process has been confirmed. As usual, applications are recommended within this verified depth range. The range and mean of values suggested are shown visually in Fig. 1, along with the r_d mean values that were computed using the equation.

Step 4—Evaluation of cyclic stress ratio CSR generated by the dynamic load using:

Calculation of the cyclic stress ratio has been established through the formulation of the below equation:

Table 1 Seismic zone factor [10]

Zone	II	III	IV	V
Z	0.10	0.16	0.24	36

$$CSR = (\tau_{av}/\sigma_{v0}') = 0.65(a_{max}/g)(\sigma_{v0}/\sigma_{v0}')r_d \quad (4)$$

In this equation, a_{max} represents the peak ground acceleration resulting from seismic waves; g denotes the acceleration due to gravity; σ_{vo} and σ_{v0}' were correspondingly the total and effective vertical overburden pressure; and r_d stands for the coefficient of stress reduction. This coefficient accommodates the soil profile's flexibility.

In case of unavailability of the PGA value, it is acceptable to consider the ratio of a_{max}/g as equivalent to the zone factor due to seismic waves, which is represented as Z. The seismic zone factors are shown in Table 1.

Table 1 gives an idea of the various zones of seismicity with factors. Zone factors are given on the basis of the expected intensity of the earthquake in different zones [14].

Step 5—Evaluation of liquefaction resistance (cyclic resistance ratio).

Various field tests are commonly employed to assess the ability of soil against liquefaction using the various subsurface exploration methods like the standard penetration test, shear-wave velocity measurements, cone penetration test, and Becker penetration test [15]. Various authors diligently worked to establish correlations to analyze the liquefaction criteria derived from field tests, aiming for consistent results regardless of the specific test utilized. Standard penetration and cone penetration tests are typically chosen mostly because of their enormous databases and historical data; however, other tests might be useful in regions with gravelly sediment or restricted access to large machinery.

The criteria used to assess the resistance of soil against liquefaction using the SPT had been proven to be reliable compared with other tests [16]. Most of these criteria may be seen in the cyclic resistance ratio vs. $(N_1)_{60}$ blow count graph shown in Fig. 2. The SPT value from the blow count is corrected after normalizing to an overburden pressure of approximately 100 kPa, and an efficiency of hammer energy 60% is represented by $(N_1)_{60}$ in this instance. The above normalizations were all explained in depth in the paper labeled "Other Corrections." A graph with computed cyclic resistance ratio, which was correlated with $(N_1)_{60}$ data from various sites that either experienced or did not experience liquefaction after being hit by the seismic waves with magnitudes nearly 7.5, is shown in Fig. 1. The graph consisting of cyclic resistance ratio curves are purposefully arranged to separate regions with data suggesting liquefaction from those showing no liquefaction. Specifically, curves were progressed for soils of granular type [17] with the quantity of fines present at ≤5%, 15%, and 35%, as depicted on the graph [18]. The cyclic resistance ratio curve for the number of fines present <5% serves as the basic criteria of penetration for this stress-based approach, denoted as the curve based on clean sand base regarding SPT values. Notably, the cyclic resistance ratio curves in Fig. 1 are applicable specifically

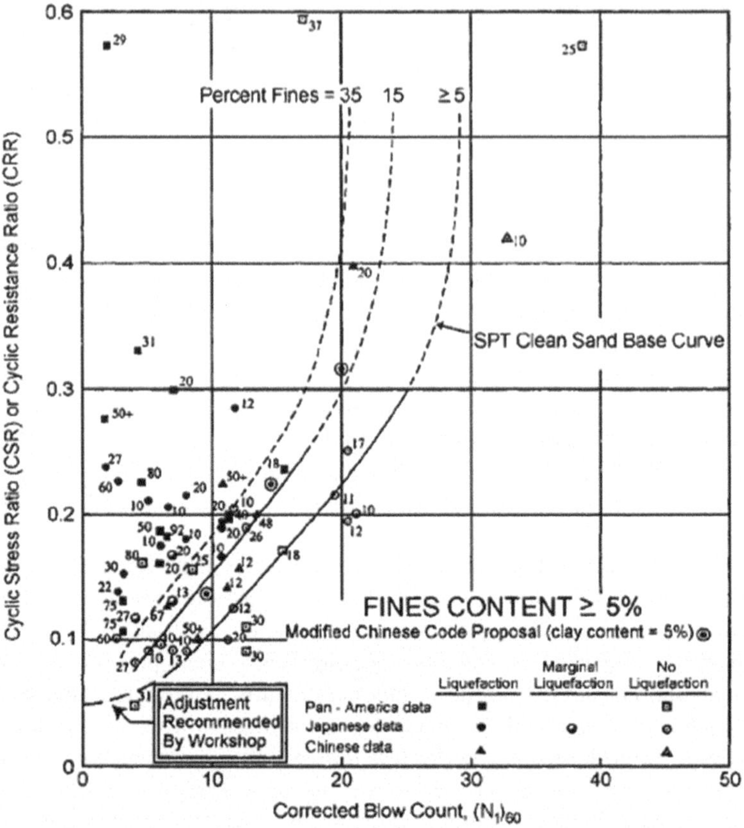

Fig. 2 SPT clean-sand base curve based on 7.5 earthquake magnitude [17]

to 7.5 magnitudes, and correction factors for adjusting these curves to magnitudes other than 7 are given by the empirical formula.

A. F. Rauch, in 1998, provided an approximation for the base curve of clean-sand depicted in Fig. 2 through the below equation:

$$\mathrm{CRR}_{7.5} = \frac{1}{34 - (\mathrm{N1})60} + \frac{(\mathrm{N1})60}{135} + \frac{50}{[10.(\mathrm{N1})60 + 45]^2} - \frac{1}{200} \tag{5}$$

Valid for $(\mathrm{N1})_{60}$ values less than 30, the equation applies to situations where clean granular soils, with $(\mathrm{N1})_{60}$ values equal to or greater than 30, were considered more denser to undergo liquefaction, which were categorized under nonliquefiable soils.

This equation is not only useful for implementation in spreadsheets and various analytical methods, but also serves as a practical tool in evaluating the base curve of clean sand for standard engineering calculations.

Correction for Fine Content

When considering corrections individually on fines content, it is essential to apply engineering judgment and exercise caution. Both Seed and Idriss, with the collaboration, evaluated the below equations to adjust $(N_1)_{60}$ to a corresponding value of clean sand, denoted as $(N1)_{60cs}$ [19].

$$(N_1)_{60cs} = \alpha + \beta (N_1)_{60}$$

The coefficients α and β are established by meeting any of the following conditions:

$$\alpha = 0 \text{ for F.C} \leq 5\%$$

$$\alpha = \exp\left[1.76\text{–}\left(190/\text{F.C}^2\right)\right] \text{for } 5\% < \text{F.C} < 35\%$$

$$\alpha = 5.0 \text{ for F.C} \geq 35\%$$

$$\beta = 1.0 \text{ for F.C} \leq 5\%$$

$$\beta = \left[0.99 + \left(\text{F.C}^{1.5}/1000\right)\right] \text{for } 5\% < \text{F.C} < 35\%$$

$$\beta = 1.2 \text{ for F.C} \geq 35\%$$

All the above equations are suitable for calculating soil's ability against liquefaction. A back-calculated curve for a quantity of fines of 35% closely aligns with the curve corresponding in Fig. 2. Similarly, the curve which is back-calculated for a quantity of fines of 15% is positioned to the right of the 15% original curve.

Overburden Correction

Table 1 outlines various factors, beyond characteristics of grain and quantity of fines, that impact results of SPT. The corrections for the above factors were incorporated [20].

$$(N_1)_{60} = C_N N_{60} \tag{6}$$

where C_N = factor which is a common reference effective over burden pressure to normalize Nm.

As the SPT N-values tend to rise with the augmentation of effective overburden stress, there is a factor applied for the correction [21]. The above factor was typically determined using the below equation, which was formulated by Whitman and Liao:

$$C_N = \left(P_a/\sigma'_{VO}\right)^{0.5} \tag{7}$$

Step 6—Evaluation of Safety Factor

$$\mathrm{FOS} = \frac{\mathrm{CRR}}{\mathrm{CSR}} \tag{8}$$

where FOS = safety factor, CRR = ratio of cyclic resistance, and CSR = ratio of cyclic stress.

Step 7—Analyzing the liquefaction potential of soil [22].

If obtained FOS≤1, then the soil is said to be liquefiable. If not, the soil has no chance of liquefaction.

3 Energy-based Method

Davis and Berrill introduced the energy-based method (EBM) by postulating that the pore pressure accumulation was directly correlated with seismic energy dissipation in soil per its unit volume. This involves estimating seismic energy reaching a site based on the magnitude of the earthquake and the epicenter distance. Another perspective on defining seismic energy was presented by Kayen and Mitchell, who employed Arias Intensity as a metric for assessing the severity of earthquake shaking in order to evaluate liquefaction potential [23].

The study employed an evaluation of liquefaction potential through energy-based method (EBM) on a homogeneous sand model subjected to seismic motions recorded at various locations during earthquakes of different magnitudes [24]. Additionally, the EBM is used as an application to real case studies of liquefaction in the city of Urayasu in the year 2011 Tohoku earthquake with 9 magnitude and in Tanno-cho after 2003 with 8 magnitude Tokachi-oki earthquake. Throughout the above assessments, the obtained results were correlated with those derived from the presently employed stress-based method under geotechnical and seismic conditions.

The findings revealed that the energy-based method produced findings comparable to those of the stress-based method for various recent earthquakes, offering a more straightforward application without the need for additional parameters. In the city of Urayasu, both methods gave mostly regular results by incorporating a suitable coefficient in the stress-based method for an earthquake of 9 magnitude. However, it was noted that both methods tended to overestimate actual liquefaction performance, possibly due to the omission of effects of aging and plasticity on the in-situ strength of liquefaction. In Tanno-cho, the energy-based method successfully assessed the real performance of liquefaction resulting from the motion of short acceleration during a distant high-earthquake magnitude, a task that the stress-based method was unable to accomplish.

Step 1—Evaluation of confining pressure

Confining pressure (σ'_c)

$$\sigma'_c = \frac{\sigma'_V(1 + 2K_o)}{3} \quad (9)$$

where σ'$_v$ = effective stress in the vertical direction (Kpa) and k_0 = earth pressure coefficient.

Normalization of N values using effective overburden pressure at each depth from the below equation.

$$N_1 = \frac{(N^* 1.7)}{\left(\frac{\sigma'_V}{P_o} + 0.7\right)} \quad (10)$$

Here, P_o was specified as 98 kPa, and the parameters C_1 and C_2 vary based on the fines content Fc according to the following relationship:

To evaluate C_1 value

$$\text{If fines content} < 10 \text{ then } C_1 = 1$$

$$\text{If fines content} < 60 \text{ then } C_1 = \frac{FC + 40}{50}$$

$$\text{If fines content} > = 60 \text{ then } C_1 = \frac{FC}{20} - 1$$

To evaluate C_2 value

$$\text{If fines content} < 10 \text{ then } C_2 = 0$$

$$\text{If fines content} > = 10 \text{ then } C_2 = \frac{FC - 10}{18}$$

Step 2—Evaluation of cyclic resistance ratio ($\mathbf{R_{L20}}$)

The R_{L20} value, initially examined in intact soils obtained through an in-situ soil freezing technique, is commonly derived in engineering applications by utilizing the empirical formula based on the SPT N_1-value.

$$\text{If } N_a < = 14 \text{ then } R_{L20} = 0.0882^* \sqrt{\frac{Na}{1.7}}$$

$$\text{If } N_a > 14 \text{ then } R_{L20} = 0.0882^* \left(\sqrt{\frac{Na}{1.7}} + \left(1.6^* 10^{-6}\right) * (N_a - 14)^{4.5}\right)$$

Where N_a is given by the next equation:

$$N_a = C_1 N_1 + C_2 \quad (11)$$

Step 3—Evaluation of normalized dissipated energy

Within the framework of EBM, the initial step involves the identification of the dissipated energy essential for soil liquefaction. Drawing from a collection of undrained triaxial tests [25] subjected to cyclic loading detailed from prior work of Kokusho, $\Delta w/\sigma'_c$, the normalized dissipated energy exhibited a nearly exclusive correlation with R_{L20} across diverse soils featuring distinct quantities of fines and relative densities [26].

$$\Delta w/\sigma'_c = 0.032 - 0.48R_{L20} + 2.40R^2{}_{L20} \quad (12)$$

Step 4—Evaluation of normalized strain energy

The externally applied strain energy, which was normalized, denoted as w/σ'_c, normalized with σ'_c, was distinctly associated with internal dissipation of energy, which got normalized. $\Delta w/\sigma'_c$ correlation remains consistent irrespective of variations in the quantity of fines and relative densities.

$$\Delta w/\sigma'_c = 5.4 \times 10^{1.25 \times \log\left(\Delta w/\sigma'_c\right)} \quad (13)$$

Step 5—Evaluation of energy capacity

$$WH = \left(\frac{w}{\sigma'_V}\right) * \sigma'^{*}_C 1.5 \quad (14)$$

The calculation of upward energy (E_u) involves employing a one-dimensional linear equivalent wave propagation analysis of the soil model. Subsequently, the maximal energy till the end of shaking (E_{uf}) is then calculated for every unit separately.

Rising shear waves at a time interval, $t = 0\sim t_1$, which represents the length of significant shaking as

$$E_u = \rho Vs \int_0^t u^2\, dt \quad (15)$$

Step 6—Evaluation of accumulated energy ratio

Direct comparison between liquefaction energy capacity (WH) and upward rising energy (E_u) is conducted from computing an energy ratio, WH/E_u, for every unit of soil mass. A unit of soil mass exhibiting a lower ratio of energy, WH/E_u, indicates a greater and earlier liquefaction potential compared to other units within a profile of the soil. However, the ultimate liquefaction capacities of each soil mass will ultimately determine the overall assessment.

$$AER = \sum \frac{WH}{E_u} \tag{16}$$

Sequential numbering of energy ratios for individual soil units begins from unit ratio i = 1, progressing towards larger values. These ratios are then summed up as $\Sigma_i(WH/E_u)_i$, represented as the accumulated energy ratio.

It thus transpires in a sequential manner within specific soil units when AER = $\Sigma_i(WH/E_u)_i$ less than or equal to one. This is due to the rising upward energy having the potential to induce liquefaction in every soil mass in the specified sequence till it gets completely exhausted by the energy capacity of the liquefaction.

When one layer undergoes liquefaction in specific soil mass, transmission of energy from that layer to subsequent units experiences a significant reduction. This reduction can impact the precision and predictability of liquefaction for subsequent units, which results in liquefying after the initial one. The above inherent limitation was present in all simplified procedures based on the concepts of stress, including stress-based method. In the above context, the sequence of liquefaction outlined in the energy-based method offers various perceptions for a more realistic interpretation of evaluation results.

4 Techniques for Improvement of Soil Profile against Liquefaction

There are several soil modification approaches that can be used as crucial pre-disaster management tactics to reduce the likelihood of liquefaction. Over time, significant advancements have occurred in this field, leading to the evolution, experimentation, and proven effectiveness of new soil improvement techniques. The choice of a specific technique is predominantly influenced by factors such as the soil type, improvement costs, feasibility of application, and others. These techniques primarily operate through mechanisms such as densification, drainage, reinforcement, and confinement. The following section provides a review of various soil improvement techniques applicable to mitigating liquefaction.

(i) *Vibratory Methods*

Primarily, these techniques operate on the principle of soil mass densification through induced vibrations coupled with varying degrees of drainage, reinforcement, and confinement. The vibratory methods can be broadly categorized into three main approaches. They are

a. Stone columns
b. Vibro-concrete columns
c. Sand compaction piles

(ii) *Compaction Grouting*

Compaction grouting is a technique used to make loose soils more denser involving the injection of slow slump grout under pressure to form a growth bulb, gradually densifying the soil. Effective for depths over 6 m, it addresses structural settlements and improves loose soil. Cost-effective for soils with deep isolated liquefiable layers, it minimizes disturbances during operation and poses minimal risk of catastrophic failure. However, potential damage to underground utilities may occur.

(iii) *Deep Dynamic Compaction*

This method densifies cohesionless soils by repeatedly dropping a huge weight of between 9000 and 27,000 kg from a height between 15 and 40 meters. It's a cost-effective technique for resisting liquefaction in various porous soil types, effective up to around 10 m depth under ideal conditions. However, it's not recommended beyond 10 m for improving soil under existing structures. This method is primarily useful for enhancing better conditions of cohesionless soils prior to construction, reducing void spaces, and dissipating pore pressure to increase effective stress so as to reduce the liquefaction damage.

(iv) *Jet Grouting*

High pressure jets are used in grouting to mix and erode soil, creating a soilcrete column for liquefaction mitigation [27]. It's versatile, forming shear walls or serving as underpinning. Different techniques, like mono, dual, and triple fluid systems, adapt to site conditions.

(v) *Deep Soil Mixing*

It involves the rotation of overlapping augers in a row to the target depth. The binder material is sent through auger tips while they rotate and lift, creating a soilcrete column [28]. The arrangement of soilcrete columns depends on the extent of required reinforcement. This technique is suitable for depths exceeding 30 m and is effective with fine-grained material.

(vi) *Earthquake Drains*

Earthquake drains, made of a pipe which is corrugated with openings and filter fabric, manage excess pore water pressure in liquefied soil layers. Installation methods include conventional drilling and casing. Effective up to approximately 25 m depths, their performance relies on soil characteristics, earthquake shaking attributes, and drain characteristics like spacing, hydraulic resistance, and radius.

(vii) *Dewatering*

The drainage technique aims to prevent liquefaction by controlling pore pressure through continuous dewatering. However, it is costly and challenging to maintain pumps for extended periods. Gillette and Bliss (1997) deemed it unsuitable for a permanent solution due to high operational costs and potential issues with perched water in target drawdown elevation. It also falls short in mitigating liquefaction under continuous recharge conditions.

(viii) *Grand Flex Mole Technology*

This innovative technique involves chemical grouting, developed by [29] in Japan (2012), which offers efficient and straightforward soil reinforcement beneath existing structures. Notably, it eliminates the need for additional vertical shafts, distinguishing it from conventional chemical grouting methods. This advantage makes it particularly suitable for mitigating liquefaction in already developed regions where proactive soil improvement techniques were not applied in advance. Moreover, it has demonstrated cost-effectiveness for depths exceeding 3 m compared to traditional permeation grouting.

The expanding urbanization raises concerns about earthquake-induced liquefaction, emphasizing the importance of effective soil improvement techniques to bolster shearing resistance. Review studies suggest that sand compaction and granular piles are both efficient and not much costlier. While other alternatives may be more effective, their high costs limit practicality. The recently developed Grand Flex Mole Technology shows promise but is not widely adopted. There's a growing need for innovative methods to assess liquefaction potential in disaster response and also accurate modeling using numerical techniques to simulate potentiality under periodic loading [30, 31].

5 Conclusions

The EBM provides valuable insights into assessing liquefaction potential by examining the balance between energy capacity and supply, an aspect not fully addressed by the traditional Seed and Idriss Simplified Method (SBM). The dissipated energy is a critical parameter for understanding pore-pressure buildup and induced strain. A significant challenge in energy-based liquefaction evaluation is accurately determining upward energy for comparison with energy capacity. While this study dealt compared to its complete contribution to density, silt particles in silty sands only partially contribute to liquefaction resistance and cone resistance (qc1N) and cyclic resistance ratio (CRR). This means that silty sands may not have as strong liquefaction resistance and cone resistance as clean sands. The upward seismic energy through dynamic soil response analyses, simpler methods based on the magnitude of the earthquake, energy source distance, and seismic impedance ratios of that particular site are also viable. Further research, particularly incorporating actual case studies, is crucial to improve the applicability and reliability of energy-based approach in engineering design.

References

1. Almani ZA, Ansari KAMRAN, Memon NA (2013) Liquefaction potential of silty sand in simple shear. Mehran Univ Res J Eng Technol 32(1):85–94
2. Kramer SL (2018) Past, present, and future developments in liquefaction hazard analysis. Developments in earthquake geotechnics, 51–60

3. Vaid YP, Thomas J (1995) Liquefaction and post liquefaction behavior of sand. J Geotech Eng 121(2):163–173
4. Figueroa JL, Saada AS, Liang L, Dahisaria NM (1994) Evaluation of soil liquefaction by energy principles. J Geotech Eng 120(9):1554–1569
5. Andrus RD, Stokoe KH (1999) liquefaction resistance based on shear wave velocity
6. Zhang W, Goh AT, Zhang Y, Chen Y, Xiao Y (2015) Assessment of soil liquefaction based on capacity energy concept and multivariate adaptive regression splines. Eng Geol 188:29–37
7. Montalva G, Ruz F, Escribano D, Bastías N, Espinoza D, Paredes F (2022) Chilean liquefaction case history database. Earthquake Spectra 38(3):2260–2280
8. Jayakrishnan V, Beena KS (2021). Liquefaction studies and its mitigation in india–focusing Kerala coast—a review. Advances in sustainable construction materials: select proceedings of ASCM 2020, 321–329
9. Ansary MA, Ansary M, Ahamed S (2024) Liquefaction potential assessment of the DMDP area of Bangladesh using CPT-based methods. Nat Hazards 120:1–32
10. Seed HB, Idriss IM (1971) Simplified procedure for evaluating soil liquefaction potential. J Soil Mech Found Div 97(9):1249–1273
11. Berrill JB, Davis RO (1985) Energy dissipation and seismic liquefaction of sands: revised model. Soils Found 25(2):106–118
12. Bolton Seed H, Tokimatsu K, Harder LF, Chung RM (1985) Influence of SPT procedures in soil liquefaction resistance evaluations. J Geotech Eng 111(12):1425–1445
13. Youd TL, Idriss IM (2001) Liquefaction resistance of soils: summary report from the 1996 NCEER and 1998 NCEER/NSF workshops on evaluation of liquefaction resistance of soils. J Geotech Geoenviron 127(4):297–313
14. Whitman RV, Lambe PC (1985) Liquefaction of Soils during Earthquakes. Committee on earthquake engineering. National Research Council, National Academy Press, Washington, DC
15. Tatsuoka F, Iwasaki T, Tokida KI, Yasuda S, Hirose M, Imai T, Kon-No M (1980) Standard penetration tests and soil liquefaction potential evaluation. Soils Found. 20(4):95–111
16. Franke KW, Wright AD (2013) An alternative performance-based liquefaction initiation procedure for the standard penetration test. In Geo-congress 2013: stability and performance of slopes and embankments III (pp. 846–849)
17. Sawicki A, Mierczyński J (2006) developments in modeling liquefaction of granular soils, caused by cyclic loads, vol 59, p 91
18. Jas K, Dodagoudar GR (2023) Liquefaction potential assessment of soils using machine learning techniques: a state-of-the-art review from 1994–2021. Int J Geomech 23(7):03123002
19. Idriss IM (1999) An update to the Seed-Idriss simplified procedure for evaluating liquefaction potential. Proc., TRB Worshop on New Approaches to Liquefaction, Pubbl. n. FHWA-RD-99-165, Federal Highway Administation
20. Liao SS, Whitman RV (1986) Overburden correction factors for SPT in sand. J Geotech Eng 112(3):373–377
21. Seed HB, Idriss IM, Arango I (1983) Evaluation of liquefaction potential using field performance data. J Geotech Eng 109(3):458–482
22. Monkul MM, Gültekin C, Gülver M, Akın Ö, Eseller-Bayat E (2015) Estimation of liquefaction potential from dry and saturated sandy soils under drained constant volume cyclic simple shear loading. Soil Dyn Earthq Eng 75:27–36
23. Simcock KJ, Davis RO, Berrill JB, Mullenger G (1983) Cyclic triaxial tests with continuous measurement of dissipated energy. Geotech Test J 6(1):35–39
24. Sahua MD, Tiwari MK Methods of liquefaction potential evaluation–a review
25. Seed HB, Lee KL (1967) Undrained strength characteristics of cohesionless soils. J Soil Mech Found Div 93(6):333–360
26. Polito C, Green RA, Dillon E, Sohn C (2013) Effect of load shape on relationship between dissipated energy and residual excess pore pressure generation in cyclic triaxial tests. Can Geotech J 50(11):1118–1128

27. Alqawasmeh H, Alzubi Y, Mahamied A (2023) State-of-the-art review: fiber-reinforced soil as a proactive approach for liquefaction mitigation and risk management. J Eng 2023(1):8737304
28. Johari A, Pourbeirak S, Amjadi AH (2024) Comparison of fine-grained soil liquefaction using numerical modeling and susceptibility criteria: a case study. Geotech Geol Eng:1–15
29. Alavi AH, Gandomi AH (2012) Energy-based numerical models for assessment of soil liquefaction. Geosci Front 3(4):541–555
30. Sharma M, Satyam N, Reddy KR (2021) State of the art review of emerging and biogeotechnical methods for liquefaction mitigation in sands. J Hazard Toxic Radioact Waste 25(1):03120002
31. Baziar MH, Jafarian Y, Shahnazari H, Movahed V, Tutunchian MA (2011) Prediction of strain energy-based liquefaction resistance of sand–silt mixtures: an evolutionary approach. Comput Geosci 37(11):1883–1893

Value Addition of Medicinal Herbs Using Sustainable Solar Dryer: A Case Study of Sirumalai Hills, Tamil Nadu

Tharani James Jayabal, Solai Dharani Murugaiah, Sutharsanan Viknesh Machappuli, Lalith Pankaj Raj Nadimuthu, and Kirubakaran Victor

Abstract A paradigm shift towards Siddha medicines was rapidly increasing due to several health benefits. The utilization and demand for natural medicines to cure diseases has been increased rapidly. Natural medicines require conditioning and preprocessing (i.e., removal of moisture content) to maintain the quality and consistency of the medicine. A sustainable drying process is essential to maintain an effective supply chain. The sustainable solar drying process is improving the shelf life and quality of medicinal plants. Also, it promotes density reduction of the medicinal plants and supports competitive positions through value addition. Sirumalai is a mountain range situated in the Dindigul district in Tamil Nadu. It is known for its rich biodiversity and unique medicinal herbs. Many of these herbs have been traditionally used in Ayurvedic and Siddha medicine for their therapeutic properties. This paper attempts to analyze a passive solar dryer integrated with a solar water heater based on 24×7 drying of herbs to improve the quality of herbs. Solar PV-based force circulation fan with air heat exchanger has been employed for backup heating during night time. The performance analysis has been carried out and reported. Also, the socioeconomic analysis has been carried out and reported in this paper.

Keywords Medicinal herbs · Passive solar system · Combined system efficiency · Cost-benefit analysis

1 Introduction

The study area is located at the downhills of the Sirumalai Hills. An average summer temperature of 28°C and a winter temperature of 10°C encompass a 52 km^2 hill range with a diverse spectrum of flora and animals. A survey of the literature showed

T. J. Jayabal · S. D. Murugaiah · S. V. Machappuli · L. P. R. Nadimuthu · K. Victor (✉)
Centre for Rural Energy, The Gandhigram Rural Institute-Deemed to be University, Gandhigram, Tamil Nadu, India

P. Pathak et al. (eds.), *Advances in Environmental Sustainability, Energy and Earth Science*, https://doi.org/10.1007/978-3-031-73820-3_11

that little ethnobotanical research had been done in the Sirumalai highlands of southern India, particularly in relation to the Paliyan tribes' traditional knowledge, which hinders this region as a whole [1]. Preprocessing and preconditioning are essential stages in enhancing the shelf life of medicinal plants when they become Siddha medicine [2, 3]. The preconditioning and preprocessing stages involve drying under the solar dryer to eliminate the moisture content from crops, vegetables and fruits. Solar dryers are composed of two types: direct and indirect types [4]. The selective black coating is carefully chosen for effective heat absorption through solar radiation in the direct-type solar dryers [5]. So, the area inside the dryer gets heated up. The effective moisture removal from the product is carried out through the convective air circulation-based heat transfer. The greenhouse-type dryer efficiently works during the active sunshine hours. To retain its same efficiency at night, the solar dryer can integrate with an evacuated type water heater, which can collect and store heat in hot water during the daytime and convert it into hot air through a heat exchanger during the night time. Warm water is collected and stored in an evacuated tube collector water heater during the day, and heated air is produced by a heat exchanger at night. Several parallel, clear glass tubes joined to the header pipe make up the evacuated tube collector. Strong, temperature-resistant, and highly transmitting of sun radiation, borosilicate or soda lime glass makes up each tube. Reducing the heat loss is the vacuum's insulating effect. The metal heat pipe within each tube is connected to a copper fin, which heats the water during the day by connecting it to a common point and then to a storage tank. Nighttime usage of the hot water is possible, or the next day, due to the insulating properties of the tank. The heat exchanger is the system that exchanges hot water into hot air. The medicinal herbs used in this paper are *Solanum trilobatum* [purple fruited plant (PFP)], used for the treatment of respiratory problems, liver health, proper blood circulation, sinus, lung disease, and treatment of tuberculosis. *Azardirachta indica* (neem leaves) is an antimicrobial agent that cures leprosy, optic nerve disorders, epistaxis, intestinal worms, anorexia, skin ulcers, and heart and circulatory disease. *Vintex negundo* (Chaste tree) helps in the treatment of sinusitis, headache, and muscle aches and is effective against eczema, skin infections, liver disorder, spleen enlargement, rheumatic pain, and backache. *Origanum vulgare* (Oregano), an antioxidant, has a cholesterol-lowering effect, highly reduces inflammation, and is effective against stomach parasites. *Psidium guajava* (Guava leaves) are highly beneficial in lowering blood sugar levels, boosting heart health, acting as an anticancer effect, and relieving painful mensuration symptoms. Decentralized renewable energy utilization promotes rapid climate change mitigation [6–8] in rural areas by ensuring sustainable solutions. The proposed system encapsulates solar photovoltaics; a greenhouse-type solar dryer and evacuated tube collector ensure the combined system efficiency [9]. The photovoltaics is a solid-state design, which can ensure the electrical demand of the proposed system in the stand-alone mode of operation. The last mile of energy connectivity is ensured through solar photovoltaics and thermal applications for the sustainable value addition and profit-making capacity of the rural community [10].

2 Literature Review

Direct or indirect drying of both organic and nonorganic products is possible with solar energy. External disasters like weather, insects, and animals might impact the items. Several greenhouse solar (GHS) dryers were presented as a solution to the drawbacks of open solar drying. The newly improved greenhouse dryer's data is presented, and the combined thermal system performance is proven [11, 12]. Also, several simulation and numerical studies ensure the significance of the combined system efficiency [13].

An indirect-type solar drying system was created to increase the value of bananas while also improving their sustainability and nutritional worth. The solar dryer comprises a solar smooth air surface heating with three layers of insulating material, a chamber for drying, and an air circulating system with a regulation. The dryer has two alternative airflow arrangements (at the top flow plus bottom moving), generated pressures at 0–3 meters per second, and two mounting skewers [14].

The hybrid solar–gas dryer arrangement has been developed for efficiency testing. An indirect air heater has been proposed instead of straight heating up, with exhaust gases kept outside the chamber to prevent diffusion. The mean temperature of drying and humidity level ranged from 25 to 80°C and 31.3–6.2%. The most incredible dryer efficiency was found to be around 42% for solar-powered drying process, 37% for gas mode drying and 40% for hybrid systems [15]. The Sabah snake grass and Java tea applications are intended for solar greenhouse driers. The solar greenhouse drier operates more efficiently during the day than at night, when the ambient air's relative humidity becomes anomalous. Heat pumps may be used with solar greenhouse dryers to lower relative humidity by 10–15%. The heat pump system's dry air supply, which has a magnitude of 0.25 to 0.50 m/s, contributes to both improving the pace at which the herbs dry and reducing the fluctuations (temperature and relative humidity) inside the greenhouse-type dryer [16].

A detailed experimental analysis was carried out to compare the performance of conventional ambient drying with that of greenhouse-type innovative solar dryers. An evacuated tube integrated solar dryer is utilized for the thermal performance analysis using drying characteristics of *Phyllanthus emblica* (Anvla), *Aloe Vera*, *Aegle marmelos* (Bel), and leaves of *Azadirachta indica* (Neem), *Aegle marmelos* (Bel), and *Psidium guajava* (Guava). The average drying rate for *Phyllanthus emblica* (Anvla) is identified as 0.46 g/min, 0.44 g/min for Aloe Vera, and 0.39 g/min for *Aegle marmelos* (Bel). The overall results of the experiment reveal that solar drying is significantly superior to conventional ambient drying. In addition, a quicker drying rate is observed as 0.18 g/min for *Azadirachta indica* (Neem), 0.17 g/min for *Aegle marmelos* (Bel), and 0.14 g/min for *Psidium guajava* (Guava) leaves [17].

The nighttime hot temperature maintenance of the drying chamber is a significant bottleneck issue for the solar dryer. To address the issue, the solar dryer is coupled with solar thermal devices to ensure an uninterrupted drying process [18]. This article reviewed several dryer types, dried products, operation settings, and the elements utilized for latent and sensible heat storage [19]. Passive solar dryers with heat storage may decrease intermittent use while increasing efficiency. Sand

and rocks, solid-phase materials, are frequently employed as thermal power storage units and maintain the sensible heat for 3–4 h after the effective sunshine hours. The thermal effectiveness of the drier, collector efficiency, and capacity for storage of thermal energy storage materials have been estimated and identified as 45, 43, and 74.5%, respectively [20, 21].

Solanum trilobatum Linn, a member of the Solanaceae family, is a significant medicinal plant to treat various ailments. Here, the review focused on the drying process of *S. trilobatum* leaves using the Lewis and Page models. The procedure of air-drying *S. trilobatum* leaf correlated well with the computational model, which had an estimation factor of 0.9342. In *S. trilobatum* biscuits, proximate investigation of multigrain palm glucose identified 5.4% proteins, 19% fat, 0.9% fiber, and 52% carbohydrates [22].

The objective of this investigation was to see how various thermal methods of drying (conventional ambient drying, microwave drying, and hot air oven drying) affected the amount of total phenolic compounds (TPC), total anthocyanin content, and antioxidant capabilities of *Vitex negundo* (VN) tea. Hot air drying in the oven has a significant drop ($P < 0.05$) in antioxidant characteristics, indicating that it is not ideal for preserving antioxidant components in Vietnamese tea. Finally, microwave drying effectively preserves the TPC, anthocyanin content, and AEAC when drying VN tea samples [23].

The 40–75°C temperature range is appropriate for reducing moisture content in the sun dryer used, resulting in higher-quality products. Heat storage devices preserve heat energy during the day and use it later to maintain an ongoing hot temperature within the dryer's chamber. Biogas combustion makes a backup power source for the dryer on wet days. Chemical heat storage and heating pump-assisted solar heaters can provide constant energy without changes in hot air [24, 25].

An experiment on drying oregano leaves was conducted in solar drying and an electric oven to investigate moisture diffusivity, activation energy, and calorimetric analysis. It resulted in a reduced drying time of 210 min in an electric oven set to 65° C. The minimal sun drying time (250 min) was reached at a maximum drying rate of 0.025 g water/g dry matter per minute, and the minimum final moisture ranged from 0.28 to 0.13 g water/g dry matter with an indirect solar drier [26].

3 Design and Experimental Setups

3.1 Greenhouse Solar Dryer

The study "Value Addition of Medicinal Herbs Using Sustainable Solar Dryer: A Case Study of Sirumalai Hills" has been carried out in the Greenhouse Solar Dryer, which has been constructed in the "The Gandhigram Rural Institute—Deemed to be University" in Dindigul. The Greenhouse Solar Dryer is a direct natural convection system, as illustrated in Fig. 1. It is equipped with a high transmittance cover made of galvanized iron bars and a 6 mm polycarbonate sheet thickness, which can create a good

Fig. 1 Greenhouse type solar dryer

greenhouse effect inside the dryer. The polycarbonate sheet can prevent contamination of products, which specks of dust and other external causes can cause. The dryer had dimensions of 3704 mm × 2805 mm × 2104 mm (L × B × H) with a capacity of 21859970.88 m^3 and was oriented towards the north. Additionally, the dryer has a fixed door measuring 1708 mm × 706 mm (L × B) and four lower inlets, each with 18 holes with a diameter of 70 mm. A chimney of 1401 mm in height is provided with a width of 100 mm, and it is placed at a height of 1904 mm from the dryer to provide a passage for exhaust air; hence, natural convection takes place.

3.2 Integrated System

The integrated system was constructed using an evacuated tube collector water heater and solar dryer. The evacuated tube collector water heater has six evacuated tubes inclinedly attached to an insulated tank. The inlet cold water from the tank gets heated in evacuated tubes, then moves upward and is stored on top of the tank. The hot water circulates via a heat exchanger, which transfers it into hot air and sends it into the dryer. The dryer is constructed in a 3 × 3.5 ft. area with polycarbonate sheets and galvanized steel iron frames. The significant novelty of the fabricated integrated dryer unit is its round-the-clock drying process without compromising traditional or native drying technologies. In addition, a solar photovoltaic powered DC convective fan is operated to ensure the effective heat exchange from the hot water to the hot chamber.

Fig. 2 Solar dryer integrated with evacuated tube collector

A 100 W_P polycrystalline module coupled with a 12 V/40 Ah lead acid battery is used for a continuous drying process during the night. Here, a 12 V/8A PWM-based charge controller is employed for cost-effective design for rural development. The overall fabricated images of the proposed system are illustrated in Fig. 2.

4 Results and Discussion

The Experiment was carried out on March 2024 in The Laboratory of Center for Rural Energy, Gandhigram, Dindigul. Samples of 200 g of oregano, purple fruited plant, chaste tree, guava, and neem were taken and under the process of drying in conventional, GHS, and integrated solar drying. The parameters observed were discussed below:

4.1 Temperature

Temperature is a vital parameter for effective drying. Here, the conventional, solar, and integrated solar drying temperatures were compared and illustrated in Fig. 3. The maximum temperatures of 45°C in conventional drying, 56.4°C in the GHS dryer, and 58°C in the integrated dryer were attained, respectively. In the study, the maximum temperature attained in the fabricated integrated solar dryer system is 13°C compared to a conventional dryer and 2°C in a GHS dryer.

4.2 Humidity

The humidity is a critical factor to be monitored throughout the drying operation. Since the degree of the drying operation is significantly influenced by the humidity inside the drying chamber. The effective removal of the moisture and humidity from

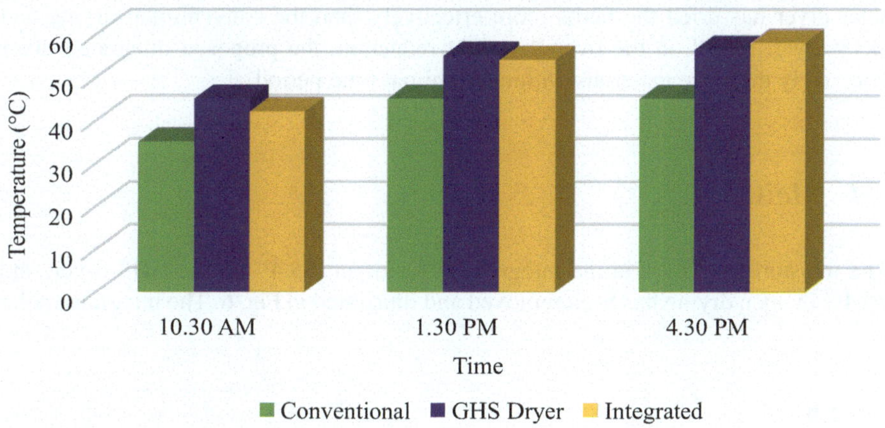

Fig. 3 Time versus temperature of dryer

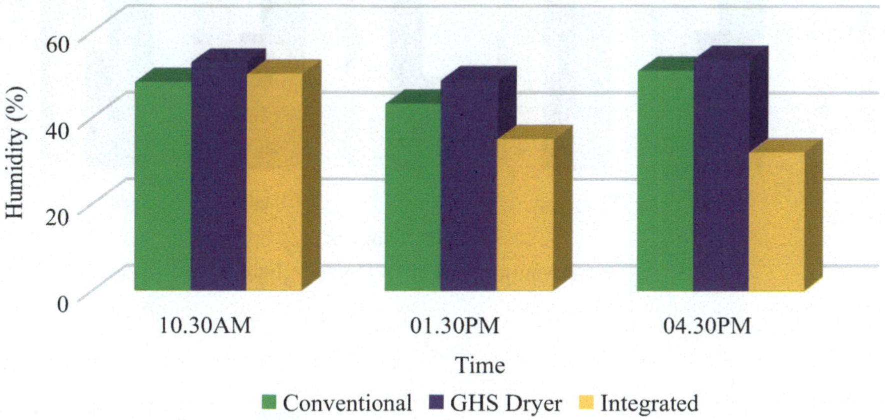

Fig. 4 Time versus humidity

the material ensures effective drying. The conventional drying and GHS dryer reached 8.3% and 12.5% humidity higher than the integrated solar dryer, as represented in Fig. 4. The minimum level of humidity helps to dry the herbs effectively. The integrated solar dryer performs more efficient drying than other methods of drying.

4.3 Weight Reduction

The drying quantity of herbs has been analyzed for weight reduction (g). In the integrated solar dryer, 35 g of herbs has been reduced to conventional drying and 33 g of herbs to GHS dryer, and the results are represented in Fig. 5. The integrated solar dryer has dried the herbs more effectively than the conventional drying and GHS dryer. Based on the overall weight reduction, the proposed integrated dryer effectively dries the materials within a minimal time period.

4.4 Moisture

The moisture of 60.4% in the integrated solar dryer, 55.4% in the GHS dryer, and 70.4% in open drying has been removed and illustrated in Fig. 6. The integrated solar

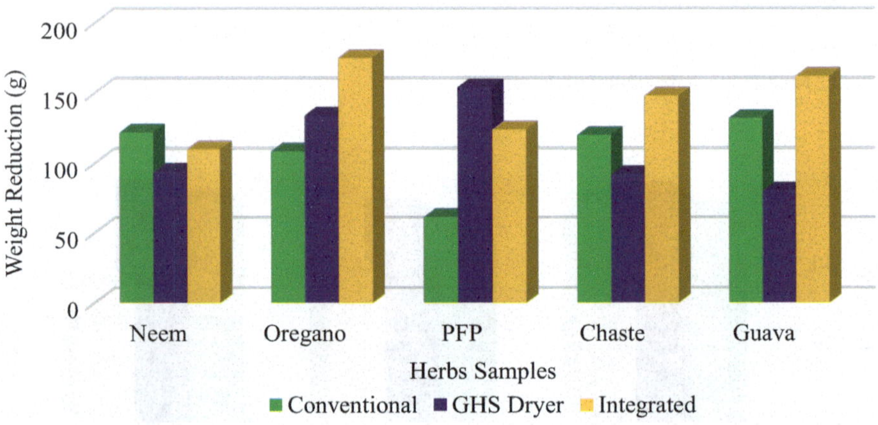

Fig. 5 Sample versus weight reduction

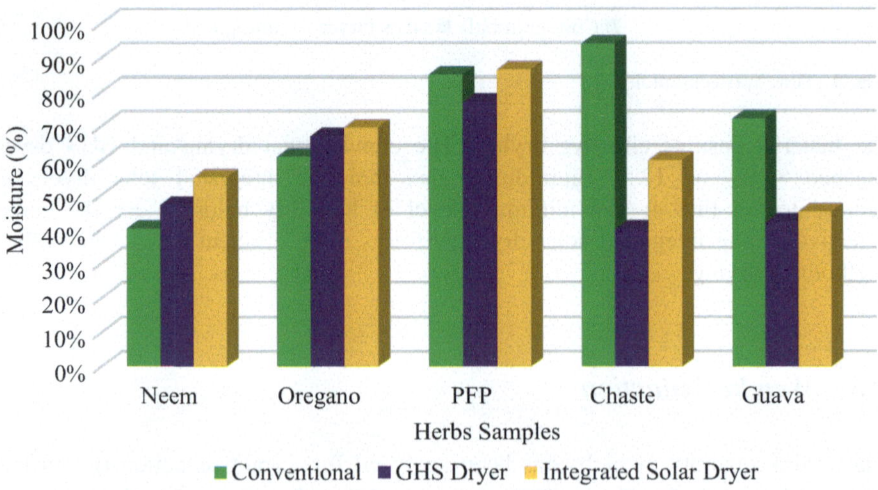

Fig. 6 Moisture versus herb samples

dryer effectively removed the moisture at 4.6% more than the GHS dryer and 10% more than the conventional drying method.

4.5 *Influence of Temperature on Moisture Removal*

Figure 7 illustrates the temperature versus weight reduction of the samples under investigation during solar dryers without an integrated solar water heating system. The effective weight reduction is achieved in the neem sample. The convective heat transfer ensures the average weight reduction of 154 g in the needed sample. Also, Fig. 8 illustrates the overall weight reduction of the samples under investigation by

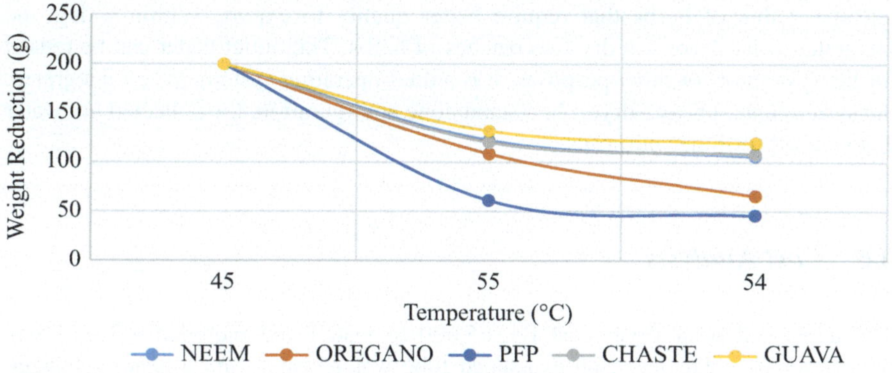

Fig. 7 Temperature versus weight reduction in solar dryer without integration

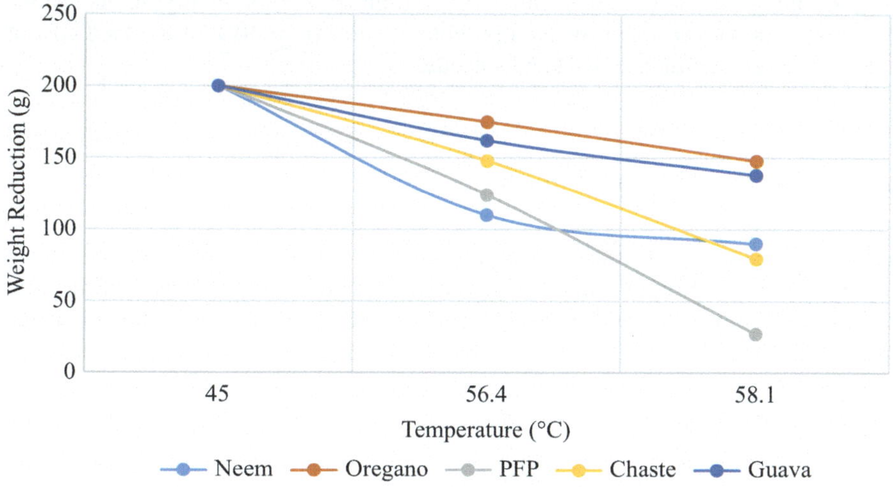

Fig. 8 Temperature versus weight reduction of herbs in solar dryer with integration

Table 1 Analysis of effective drying on herb samples

Solar dryer	Neem	Oregano	PFP	Chaste	Guava
Without integration	47%	67%	77%	46%	42%
With integration	55%	70%	86%	60%	45%

using an integrated solar drying process. The amount (or) quantum of significant weight reduction is achieved in the proposed and fabricated integrated solar drying unit.

From the effective drying analysis with and without integration, among the five dried herb samples, we analyzed the effective drying attained in the integrated solar dryer rather than the solar dryer without integration and tabulated the results in Table 1. Conventional drying and GHS dryer can perform efficiently only during sunshine, but the integrated solar dryer can perform 24×7 without degrading the nutritive value of herbs that require better quality to export. Within a day, the integrated solar dryer can dry two batches of herbs. The outlet water can be reused for the system or for other purposes. It is a dual-operating system. When integrated, the system operates as a dryer. In a standalone configuration, the collected hot water can be used for other purposes.

4.6 *Cost Analysis*

The integrated solar dryer cost Rs. 65,000 to install and included a 3 × 3.5 ft. polycarbonate solar dryer, an evacuated tube water heater, and a heat exchanger. The cost of dried leaves in the Khadi and Village Industries (KVIC) market is given below Table 2,

Selling dried herbs in the Khadi market can earn Rs. 650/day. The payback period for this integrated solar dryer can reach within 100 days of drying the herbs. Moreover, the Government of India provides a subsidy of 40% of the total cost of solar dryer installation as a backend subsidy.

4.7 *Effective Drying Process*

Figs. 9, 10 and 11 illustrate the drying characteristics of conventional ambient drying, greenhouse dryer without integration, and green house dryer with integration, respectively. The overall drying performance is significantly higher in the integrated dryer.

Table 2 Market value for the samples under investigation

Dried herbs	Cost (Rs/kg)
Neem leaves	65
Guava leaves	110
Oregano leaves	200
Chaste tree leaves	80
Purple fruited plant leaves	150

Herbs	Fresh	Dried
Neem	1	1'
Oregano	2	2'
PFP	3	3'
Chaste Tree	4	4'
Guava	5	5'

Fig. 9 Herbs of conventional drying

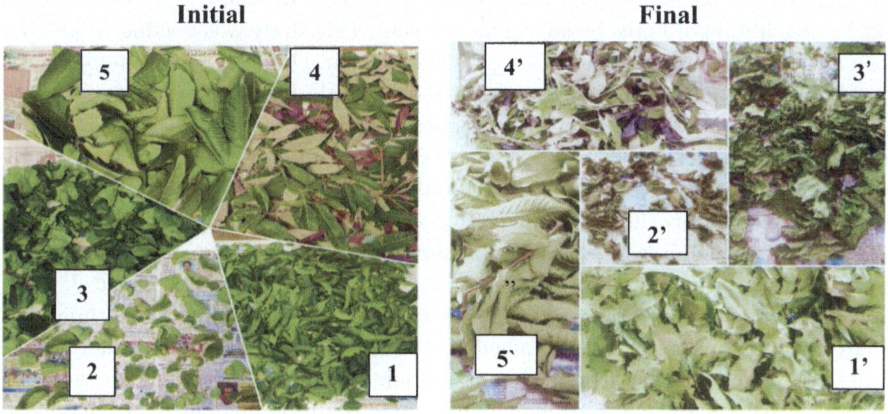

Fig. 10 Herbs of GHS dryer

Fig. 11 Herbs of integrated solar dryer

5 Conclusions

This study concentrated on evaluating the drying performance of conventional drying, GHS solar dryers, and integrated solar dryers to identify the best method for value addition of herbs by sustainable drying. This study was conducted in "The Laboratory of Center for Rural Energy, Gandhigram, Dindigul." The sample herbs taken for this study were *Solanum trilobatum* (purple fruited plant), *Azardirachta indica* (neem leaves), *Vintex negundo* (chaste tree), *Origanum vulgare* (oregano), *Psidium guajava* (guava leaves). Each herb was taken in 200 g to dry in every drying system. The parameters of moisture, humidity, weight reduction, and temperature were observed in this study. The integrated solar dryer attained a maximum temperature of 13°C, which is higher than conventional drying and 2°C than GHS dryer. From the weight reduction evaluation, the integrated solar dryer reduced 35 g of herbs to conventional drying and 32 g to GHS Dryer. According to the humidity analysis, conventional drying attained 8.3%, and the GHS dryer attained 12.5% humidity, which is higher than the integrated solar dryer. From moisture analysis, the integrated solar dryer removed the moisture by 4.6% more than the GHS dryer and 10% more than the conventional drying method. By observing the physical change of dried herbs, the samples in the integrated solar dryer achieved minimum changes in color, texture, and wrinkle compared to the other two drying methods. Based on the observations, we can conclude that the integrated solar dryer can dry herbs with optimal therapeutic properties and better quality than the conventional drying and GHS solar dryer, which supports the value addition of herbs in a sustainable method.

References

1. Karuppusamy S (2007) Medicinal plants used by Paliyan tribes of Sirumalai hills of southern India. Nat Prod Radiance 6:436–442
2. Khallaf AE-M, El-Sebaii A (2022) Review on drying of the medicinal plants (herbs) using solar energy applications. Heat Mass Transf 58:1411–1428. https://doi.org/10.1007/s00231-022-03191-5
3. Nadimuthu LPR, Victor K (2021) Optimization of energy-intensive process in Ayurvedic medicine manufacturing unit—a case study. Process Integration and Optimization for Sustainability. 5:975. https://doi.org/10.1007/s41660-021-00194-3
4. Al-Hamdani A, Jayasuriya H, Pathare PB, Al-Attabi Z (2022) Drying characteristics and quality analysis of medicinal herbs dried by an indirect solar dryer. Foods 11:4103. https://doi.org/10.3390/foods11244103
5. Bhaskara Rao TSS, Murugan S, Solar drying of medicinal herbs: A review (2021) Sol Energy 223:415–436. https://doi.org/10.1016/j.solener.2021.05.065
6. Chopade SS, Lalith Pankaj Raj GN, Prabakaran G (2018.): Special Issue 33) Bench marking of grid tied solar roof top photovoltaic system: a case comparison. Int J Eng Technol 7(2):33. https://doi.org/10.14419/ijet.v7i2.33.14832
7. Sai Kandukuri K, Kishor Sharma P, Kumar Arun R (2024) A comparative assessment of distributive mode active solar dryers: flat plate collector vs evacuated tube collector with thermal energy storage and perforated baffled trays. Sol Energy 271:112421. https://doi.org/10.1016/j.solener.2024.112421
8. Raj LP, N., Kirubakaran, V. (2021) Energy efficiency enhancement and climate change mitigations of SMEs through grid interactive solar photovoltaic system. Int J Photoenergy. 2021:1. https://doi.org/10.1155/2021/6651717
9. Yematawu S, Tamrat B, Tarekegn D, Mulugeta H (2024) Experimental testing on the performance of solar dryer equipped with evacuated tube collector, rock bed heat storage and reflectors. Energy Rep 12:453–471. https://doi.org/10.1016/j.egyr.2024.06.027
10. Nadimuthu LPR, Victor K (2021) Environmental friendly micro cold storage for last-mile Covid-19 vaccine logistics. Environ Sci Pollut Res 29:23767. https://doi.org/10.1007/s11356-021-17584-2
11. Singh P, Shrivastava V, Kumar A (2017) Recent developments in greenhouse solar drying : a review. Renew Sust Energ Rev 0–1:3250. https://doi.org/10.1016/j.rser.2017.10.020
12. Lalith Pankaj Raj Nadimuthu GN, Karthik VM, Mohanraj M, Kirubakaran V (2019) Fast thermal degradation of biomass using scrapped solar cell with special focus on photovoltaic (PV) waste disposal. In: Waste valorisation and recycling. Springer Singapore, pp 349–361. https://doi.org/10.1007/978-981-13-2784-1_33
13. Hamdi I, Kooli S, Elkhadraoui A, Azaizia Z, Abdelhamid F, Guizani A (2018) Experimental study and numerical modeling for drying grapes under solar greenhouse. Renew Energy 127: 936–946. https://doi.org/10.1016/j.renene.2018.05.027
14. Hegde VN, Hosur VS, Rathod SK, Harsoor PA, Narayana KB (2015) Design, fabrication and performance evaluation of solar dryer for banana. Energ Sustain Soc 5:23. https://doi.org/10.1186/s13705-015-0052-x
15. Zoukit A, El Ferouali H, Salhi I, Doubabi S, Abdenouri N (2019) Simulation, design and experimental performance evaluation of an innovative hybrid solar-gas dryer. Energy 189: 116279. https://doi.org/10.1016/j.energy.2019.116279
16. Tham TC, Ng MX, Gan SH, Chua LS, Aziz R, Chuah LA, Hii CL, Ong SP, Chin NL (2017) Effect of ambient conditions on drying of herbs in solar greenhouse dryer with integrated heat pump. Dry Technol 0:1–12. https://doi.org/10.1080/07373937.2016.1271984
17. Singh P, Kumar A, Tekasakul P (2015) Applications of software in solar drying systems : a review. Renew Sust Energ Rev 51:1326–1337. https://doi.org/10.1016/j.rser.2015.07.025

18. Lalith PRGN, Prabakaran G, Murugaiyan A, Kirubakaran V (2018) Hybrid photovoltaic-thermal systems: innovative CHP approach. In: 2018 4th International Conference on Electrical Energy Systems (ICEES), pp 726–730. https://doi.org/10.1109/ICEES.2018.8442352
19. Srinivasan G, Rabha DK, Muthukumar P (2021) A review on solar dryers integrated with thermal energy storage units for drying agricultural and food products. Sol Energy 229:22. https://doi.org/10.1016/j.solener.2021.07.075
20. Rulazi EL, Marwa J, Kichonge B, Kivevele T (2023) Development and performance evaluation of a novel solar dryer integrated with thermal energy storage system for drying of agricultural products. ACS Omega 8:43304. https://doi.org/10.1021/acsomega.3c07314
21. Selvaraj D, Nadimuthu L, Victor K (2021) Monitoring and simulation of parabolic trough collector powered vapor absorption refrigeration system for rural cold storage. Therm Sci 298–298:975. https://doi.org/10.2298/TSCI201110298S
22. Dasthageer N, Shanmugam P, Gnanamangai M (2020) Materials today: proceedings Analysing drying characteristics of Solanum trilobatum leaves and nutritive assessment of cookies. Material Today: Proc 48:316. https://doi.org/10.1016/j.matpr.2020.07.596
23. Division FT (2013) Effect of different drying methods on the antioxidant properties of Vitex negundo Linn tea. Int Food Res J 20:3171–3176
24. Kumar SP, Kumar G (2023) Iranian (Iranica). J Energ Environ 14:58–64. https://doi.org/10.5829/ijee.2023.14.01.08
25. Nadimuthu L, Selvaraj D, Victor K (2021) Simulation and experimental study on performance analysis of solar photovoltaic integrated thermoelectric cooler using MATLAB Simulink. Therm Sci 301–301:999. https://doi.org/10.2298/TSCI201211301N
26. Castillo-Téllez M, Martínez OS, Miranda-Mandujano E, Galvez GH, López DA, Castillo-Téllez B, Cetina-Quiñones AJ (2024) Solar drying of oregano leaves (Plectranthus amboinicus, Lour): analysis of experimental performance under a tropical climate. Energ Explor Exploit 42(5):1895–1921. https://doi.org/10.1177/01445987241234389

Evaluation of Machine Learning Algorithms for Predicting Compressive Strength of Geopolymer Concrete at High Temperatures

Aashi Gupta, Prachi Sarda, Faisal Mehraj Wani, and Jayaprakash Vemuri

Abstract In recent years, fly ash and slag, both industrial by-products, have been integral to sustainable geopolymer concrete, but their fire resistance lacks clarity. The complex interaction between its composition and elevated temperatures has a significant impact on its resulting mechanical properties, thereby suggesting the need for precise predictive models. This significant knowledge gap can be bridged by using advanced machine learning methods to accurately forecast the compressive strength of sustainable geopolymer concrete under high-temperature conditions. In this study, an experimental dataset consisting of over 200 data points is collected from literature. Several machine learning techniques, encompassing linear regression, decision trees, support vector machine, Gaussian process regression, ensemble methods, neural networks, and kernel-based algorithms, are employed to forecast the compressive strength. Statistical and performance metrics for all machine learning models are tabulated to examine their efficiency. It is observed that the trilayered neural network model outperformed other algorithms, achieving an R^2 value of 0.95 and MSE of 37.55. The study provides structural designers with a reliable computational tool for assessing the strength of green concrete exposed to elevated temperatures.

Keywords Concrete · High temperature · Compressive strength · Neural network · Performance metrics

1 Introduction

Concrete is the most widely used building material in the construction of infrastructures such as buildings, bridges, highways, dams, and many other facilities [1]. The properties and performance of concrete can be tailored to meet design requirements by varying the type and quantity of the mixture constituents (e.g., cement, water,

A. Gupta (✉) · P. Sarda · F. M. Wani · J. Vemuri
Mahindra University, Hyderabad, India

P. Pathak et al. (eds.), *Advances in Environmental Sustainability, Energy and Earth Science*, https://doi.org/10.1007/978-3-031-73820-3_12

aggregate, and admixtures) [2]. An alternative to making environmentally friendly concrete is the development of geopolymers [3, 4]. Geopolymer concrete is an emerging eco-friendly alternative to Portland cement [4]. The geopolymer is synthesized from the chemical reaction, geopolymerization, between the aluminosilicate materials (industrial by-product, such as fly ash and/or blast furnace slag) and alkaline activators [5]. Fly ash and slag are widely being used because they are considered cheap and widely available materials [6]. Reinforced concrete (RC) is increasingly being used in the construction of industrial buildings, where there is often a significant risk of fire [7]. Such RC buildings are also at risk for explosions during daily operations [8].

Machine learning (ML) is becoming increasingly popular in the scientific process, replacing traditional models and mechanical models in various commercial applications [9]. ML algorithms, which comprise various techniques such as classification and regression learned from data, are one of the most significant scientific topics of the current era [10, 11]. With the ability to tackle complex tasks autonomously, ML has demonstrated its transformative potential in concrete research [2]. With the recent increase in the use of geopolymer concrete made from fly ash and slag, there is a limited understanding of the fire resistance of this concrete. The complex interaction between sustainable geopolymer concrete composition and elevated temperatures has a significant impact on its resulting mechanical properties, thereby suggesting the need for precise predictive models. In this study, this knowledge gap is bridged by using ML algorithms for predictions.

2 Description of Dataset

The data points used to run the models via ML algorithms were obtained from the literature [12–20]. There were 207 data points in total. Table 1 shows the details of these parameters, and Fig. 1 shows a plot of the parameters versus the output response variable. Several ML methods were used, including both holdout validation

Table 1 Details of input parameters

S.N.	Parameter	Minimum value	Maximum value
1	Cement (kg/m^3)	250	786
2	Water (kg/m^3)	123	385
3	Fine aggregate (kg/m^3)	0	1345
4	Coarse aggregate (kg/m^3)	0	1681
5	Fly ash (kg/m^3)	0	150
6	Super plasticizer (kg/m^3)	0	25.9
7	Silica fume (kg/m^3)	0	150
8	Nano silica (kg/m^3)	0	22.5
9	Temperature (C)	20	1000
10	Compressive strength (MPa)	3	133.6

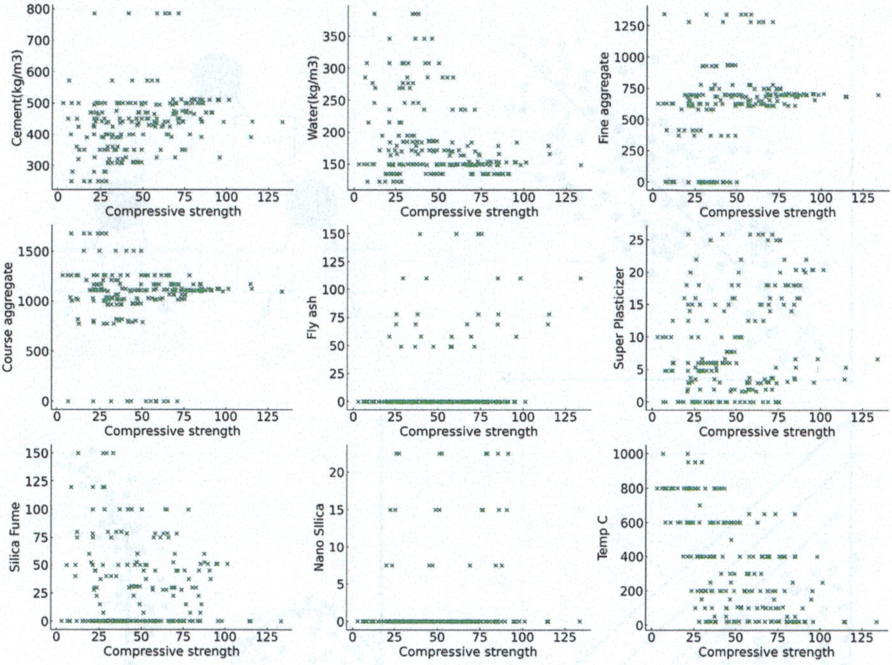

Fig. 1 Plots showing the variation of the input variables with the response variable

and cross-validation. The efficiency of the ML techniques is assessed using performance metrics.

3 ML Techniques

A brief discussion of the ML techniques is presented in this section. Figure 2 shows illustrations of the various ML techniques. Interactions linear regression captures how the dependency between the variables that are independent and dependent is influenced by another independent variable. Robust linear regression handles outliers and violations of assumptions, providing more reliable estimates. Stepwise linear regression automatically selects variables based on significance or criteria, simplifying the model. Decision trees started gaining traction in the late 1980s. They are employed in various fields due to their simplicity, and effectiveness in handling nonlinear relationships. Medium trees may have a moderate number of internal nodes and decision rules, resulting in a relatively interpretable model. Coarse trees aim to provide a higher-level overview of the data and emphasize broader patterns rather than specific nuances. Coarse trees typically have fewer internal nodes and decision rules, leading to a more straightforward and less interpretable model.

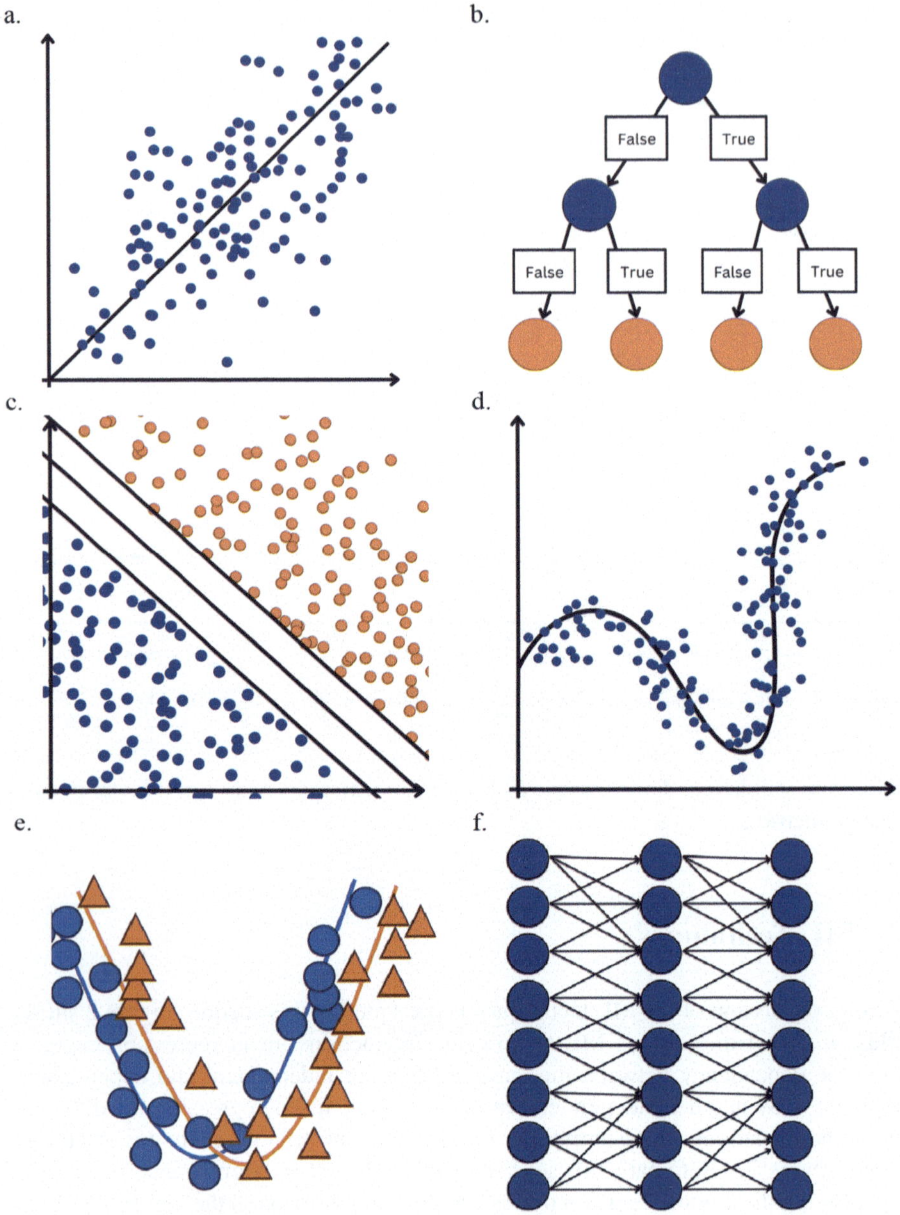

Fig. 2 Illustration of ML techniques used in the study. (**a**) Linear regression, (**b**) Decision trees, (**c**) Support vector machines, (**d**) Gaussian process regression, (**e**) Ensemble methods, (**f**) neural networks

In support vector machines, the different types of kernels have varying capabilities in capturing relationships between variables. The linear kernel handles linear relationships. It calculates the inner product between data points in the original feature space. Quadratic kernel captures quadratic relationships and can capture curved decision boundaries. Cubic kernel models cubic relationships. Fine Gaussian kernel calculates similarity based on the Gaussian distribution, assigns more weight to nearby data points, and is more sensitive to local details in the data. Medium Gaussian strikes a balance between local and global patterns in the data. Coarse Gaussian provides a smoother representation of the data by assigning more weight to a broader range of neighboring points, captures broader trends, and is less sensitive to local variations [21].

GPR, which is used for regression tasks, provides a flexible and powerful way to model complex relationships between variables and estimate uncertainty in predictions. It uses different kernels to model correlations between data points. Squared exponential (SE) assumes a smooth, infinitely differentiable interdependence of data. It is characterized by a parameter called the length scale, which determines the rate at which the correlation decreases as the distance between points increases. Matern 5/2 is characterized by two parameters—the length scale and the amplitude. It represents a compromise between capturing both long-range and short-range dependencies in the data. Exponential assumes an exponentially decreasing correlation between data points as the distance between them increases. It is characterized by the length scale parameter, which controls the rate of decay of the correlation. Rational quadratic handles various smoothness levels. It is characterized by three parameters: the length scale, the amplitude, and the alpha parameter, which controls the smoothness of the function. The rational quadratic kernel is robust enough to overfit and can handle both short-range and long-range dependencies [22].

Ensemble methods leverage diverse models to achieve robust predictions, particularly with noisy data. Boosted Trees combines multiple weak learners, correcting mistakes iteratively to create a strong model. Boosted Trees reduce bias and handle complex relationships. Bagged Trees train independent decision trees on different subsets of data, aggregating predictions through averaging or voting. Bagged Trees improve stability and handle high-dimensional data.

Narrow, medium, and wide neural networks refer to the number of neurons or units in a specific layer of a neural network. Narrow neural networks have a smaller number of neurons in a particular layer. Medium neural networks have a moderate number of neurons in a specific layer. They strike a balance between model complexity and computational efficiency. Wide neural networks have a larger number of neurons in each layer. Wide networks are advantageous when the dataset is large and complex, allowing for the potential to capture intricate patterns and relationships. In addition to the width of neural networks, layered networks can have different layers of arrangements; the types are shown below: bilayered networks have input, hidden, and output layers and are used for simple tasks. Trilayered networks have one or more hidden layers [23].

Kernel methods are used to transform data into higher-dimensional feature spaces, enabling linear algorithms to capture complex patterns and relationships. Least Squares Regression Kernel measures similarity based on Euclidean distance.

4 Results

The performance assessment of ML methods is shown in Table 2. These include the coefficient of correlation (R^2), mean absolute error (MAE), mean squared error (MSE), and root mean squared error (RMSE). These metrics are defined as follows:

Table 2 Performance metrics obtained from various ML techniques (holdout validation)

Preset	RMSE (validation)	MSE (validation)	R^2 (validation)	MAE (validation)
Linear				
Linear	14.05	197.49	0.73	10.94
Interactions linear	10.70	114.52	0.84	8.08
Robust linear	14.08	198.16	0.73	10.94
Stepwise linear	11.38	129.55	0.82	9.04
Trees				
Fine tree	13.03	169.82	0.77	9.13
Medium tree	16.33	266.65	0.63	10.74
Coarse tree	20.21	408.39	0.44	15.98
SVM				
Linear SVM	14.30	204.44	0.72	11.18
Quadratic SVM	8.13	66.16	0.91	5.94
Cubic SVM	9.51	90.53	0.88	6.26
Fine Gaussian SVM	19.82	392.74	0.46	13.67
Medium Gaussian SVM	9.63	92.76	0.87	7.12
Coarse Gaussian SVM	15.01	225.29	0.69	10.58
Ensemble				
Boosted trees	10.49	110.11	0.85	7.04
Bagged trees	12.12	146.98	0.80	8.05
GPR				
Squared exponential GPR	7.56	57.22	0.92	5.31
Matern 5/2 GPR	7.53	56.66	0.92	5.17
Exponential GPR	7.85	61.64	0.92	5.22
Rational quadratic GPR	7.53	56.64	0.92	5.13
Neural Networks				
Narrow neural network	13.02	169.49	0.77	9.79
Medium neural network	9.70	94.01	0.87	6.92
Wide neural network	9.53	90.74	0.88	6.53
Bilayered neural network	9.67	93.43	0.87	6.51
Trilayered neural network	6.13	37.55	0.95	4.67
Kernel				
SVM kernel	22.46	504.58	0.31	17.31
Least squares regression kernel	13.06	170.44	0.77	9.65

$$R^2 = 1 - \frac{\sum_{i=1}^{n} (y_i - \widehat{y}_i)^2}{\sum_{i=1}^{n} (y_i - \overline{y})^2} \tag{1}$$

$$MAE = \frac{1}{n} \sum_{i=1}^{n} | y_i - \widehat{y}_i | \tag{2}$$

$$MSE = \frac{1}{n} \sum_{i=1}^{n} (y_i - \widehat{y}_i)^2 \tag{3}$$

$$RMSE = \sqrt{\frac{1}{n} \sum_{i=1}^{n} (y_i - \widehat{y}_i)^2} \tag{4}$$

The analysis used 207 observations with 9 predictors and 1 response variable. The obtained results are tabulated in Table 2.

Figure 3 shows the performance metrics of various ML techniques. Figure 4 shows the plots of the predicted response versus the actual response. Upon analyzing the findings obtained through holdout validation and cross-validation techniques, significant insights into the performance of diverse models within the dataset have emerged. It appears that in the holdout validation, the neural network model stands out as the best performer among the evaluated models. It demonstrates impressive

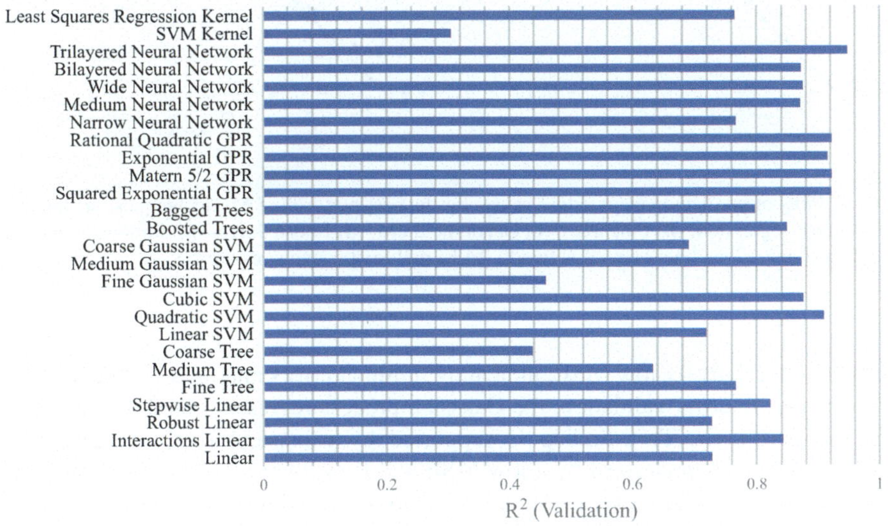

Fig. 3 Performance metrics of various ML techniques

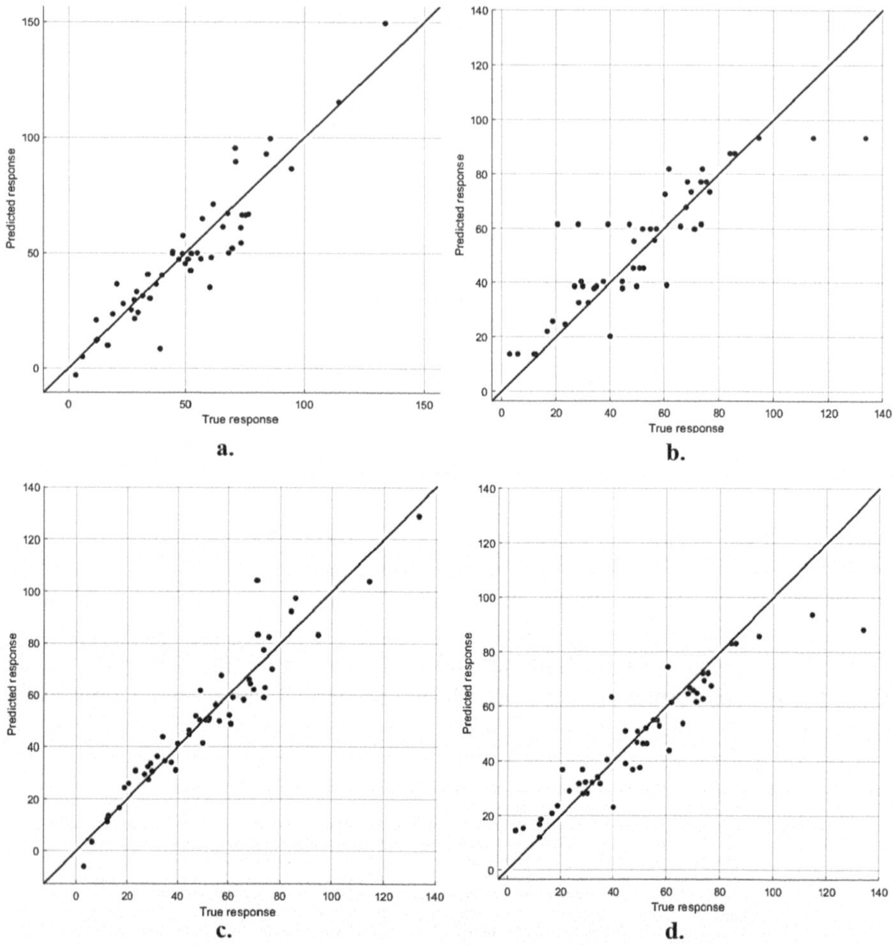

Fig. 4 Plots of predicted versus actual values for the best-performing ML techniques. (**a**) Interactions linear regression, (**b**) Fine trees, (**c**) Quadratic SVM, (**d**) Boosted trees, (**e**) Rational quadratic GPR, (**f**) Interactions linear neural network, (**g**) Kernel least squares regression

results with relatively low MSE, indicating minimal errors and an R^2 value of 0.95. Conversely, in cross-validation, the Gaussian process regression model emerges as the top performer. This finding is supported by consistent performance across multiple folds, characterized by a low MSE and R^2 value of 0.91. In summary, the comprehensive evaluation of ML models through holdout validation and cross-validation techniques illuminates the multifaceted nature of predictive modeling.

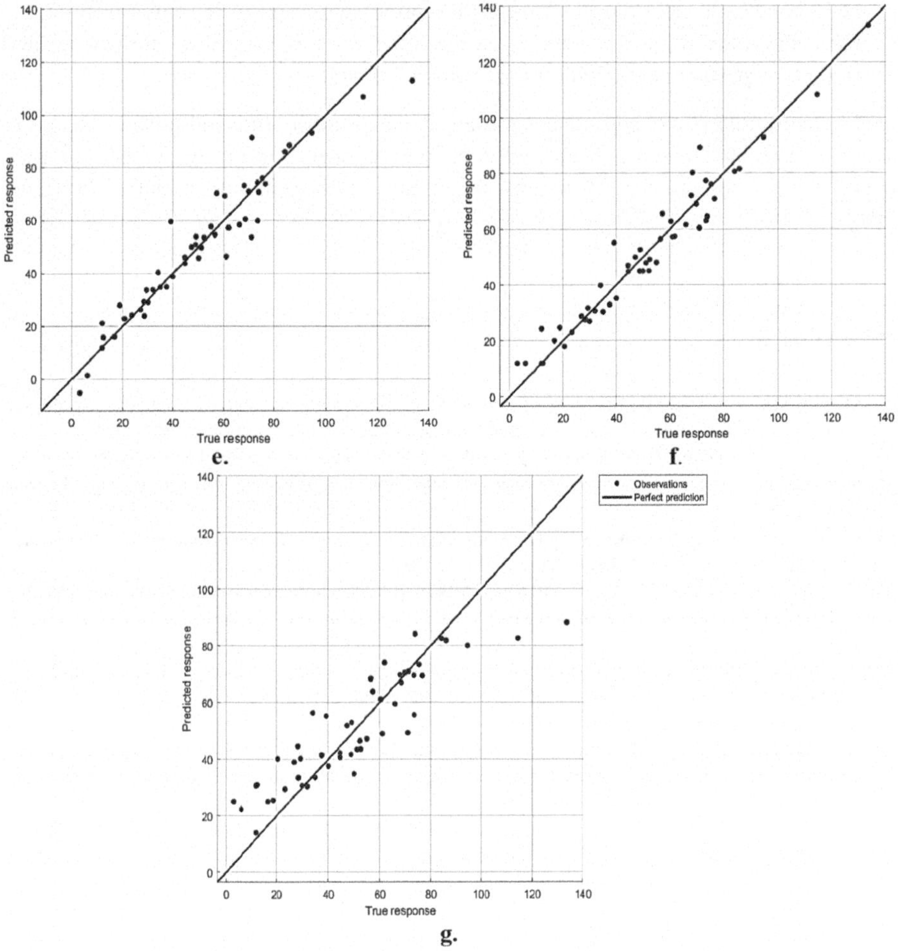

Fig. 4 (continued)

5 Conclusions

The efficiency of various ML algorithms in predicting the compressive strength of geopolymer concrete at elevated temperatures is examined in this study. The findings successfully bridge the knowledge gap regarding the complex relationship between elevated temperatures, geopolymer concrete composition, and the resulting mechanical properties. The ML algorithms were trained on an experimental dataset exceeding 200 data points. To assess the performance of the models, we employed the holdout validation technique, which holds a certain % of data for testing.

- The holdout validation identified the trilayered neural network (TNN) model as the top performer, achieving a high R^2 value of 0.95.

- The Gaussian process regression (GPR) model achieved an R^2 value of 0.92.
- The quadratic support vector machine (Q-SVM) model also showed strong predictive performance, with an R^2 value of 0.91.

The comprehensive approach adopted in the present study empowers the construction industry to make informed decisions regarding the usage of geopolymer concrete in fire-resistant structures, ultimately promoting sustainable building practices.

References

1. Shehab HK, Eisa AS, Wahba AM (2016) Mechanical properties of fly ash based geopolymer concrete with full and partial cement replacement. Constr Build Mater 126:560–565
2. Li Z, Yoon J, Zhang R, Rajabipour F, Srubar WV III, Dabo I, Radlińska A (2022) Machine learning in concrete science: applications, challenges, and best practices. npj Comp Material 8(1):127
3. Ma CK, Awang AZ, Omar W (2018) Structural and material performance of geopolymer concrete: a review. Constr Build Mater 186:90–102
4. Hadi MN, Al-Azzawi M, Yu T (2018) Effects of fly ash characteristics and alkaline activator components on compressive strength of fly ash-based geopolymer mortar. Constr Build Mater 175:41–54
5. Fernández-Jimenez A, De La Torre AG, Palomo A, López-Olmo G, Alonso MM, Aranda MAG (2006) Quantitative determination of phases in the alkali activation of fly ash. Part I. Potential ash reactivity. Fuel 85(5–6):625–634
6. Nikoloutsopoulos N, Sotiropoulou A, Kakali G, Tsivilis S (2021) Physical and mechanical properties of Fly ash based Geopolymer concrete compared to conventional concrete. Buildings 11(5):178. https://doi.org/10.3390/buildings11050178
7. Alaskar A, Alfalah G, Althoey F, Abuhussain MA, Javed MF, Deifalla AF, Ghamry NA (2023) Comparative study of genetic programming-based algorithms for predicting the compressive strength of concrete at elevated temperature. Case Stud Const Material 18:e02199
8. Peng S, Shen L, Du X, Wu B, Yu Z (2024) Dynamic mechanical properties and mechanisms of ordinary concrete after high temperature. J Mater Civ Eng 36(3):04023632
9. Vadyala SR, Betgeri SN, Matthews JC, Matthews E (2022) A review of physics-based machine learning in civil engineering. Results Eng 13:100316
10. Ray S (2019) A quick review of machine learning algorithms. In 2019 International conference on machine learning, big data, cloud and parallel computing (COMITCon), IEEE, pp. 35-39. https://doi.org/10.1109/COMITCon.2019.8862451
11. Wani F, Vemuri J, Chenna R (2023) Prediction of storey drift for reinforced concrete structures subjected to pulse-like ground motions using machine learning classification models. Int J Struct Integr 15:409. https://doi.org/10.1108/IJSI-06-2023-0054
12. Husem M (2006) The effects of high temperature on compressive and flexural strengths of ordinary and high-performance concrete. Fire Saf J 41:155–163
13. Ergün A, Kürklü G, Serhat BM, Mansour MY (2013) The effect of cement dosage on mechanical properties of concrete exposed to high temperatures. Fire Saf J 55:160–167
14. Cülfik MS, Özturan T (2010) Mechanical properties of normal and high strength concretes subjected to high temperatures and using image analysis to detect bond deteriorations. Constr Build Mater 24:1486–1493
15. Behnood A, Ziari H (2008) Effects of silica fume addition and water to cement ratio on the properties of highstrength concrete after exposure to high temperatures. Cem Concr Compos 30: 106–112

16. Bastami M, Baghbadrani M, Aslani F (2014) Performance of nano-silica modified high strength concrete at elevated temperatures. Constr Build Mater 68:402–408
17. Chen L, Fang Q, Jiang X, Ruan Z, Hong J (2015) Combined effects of high temperature and high strain rate on normal weight concrete. Int J Impact Eng 86:40–56
18. Xiong Y, Deng S, Wu D (2016) Experimental study on compressive strength recovery effect of fire-damaged high strength concrete after realkalisation treatment. Procedia Eng 135:476–481
19. Mousa MI (2017) Effect of elevated temperature on the properties of silica fume and recycled rubber-filled high strength concretes (RHSC). HBRC J 13:1–7
20. Fu YF, Wong YL, Poon CS, Tang CA (2005) Stress-strain behaviour of high-strength concrete at elevated temperatures. Mag Concr Res 57:535–544
21. Wani FM, Vemuri J (2024) Shapelet-informed machine learning classifiers: a path towards precise identification of pulse-like ground motions. J Earth Sys Sci 133(2)
22. Wani FM, Vemuri J, Reddy KK, Rajaram C (2024) Forecasting duration characteristics of near fault pulse-like ground motions using machine learning algorithms. Stoch Env Res Risk A 1-30: 98
23. Wani FM, Vemuri J, Rajaram C, Reddy KK (2023) Investigating the efficiency of machine learning algorithms in classifying pulse-like ground motions. J Seismol 27(5):875–899

Predictive Modeling for Compressive Strength in Sustainable Concrete Using Machine Learning Techniques

Shivatmika Bolla, Yashaswi Matla, Faisal Mehraj Wani, and Jayaprakash Vemuri

Abstract The large-scale fly ash and slag integration in concrete has not only aided in reducing the environmental impact of cement, but also contributed to economic development by reducing waste, cutting carbon emissions, conserving resources, and offering cost-effective, sustainable construction solutions. This aligns with the global shift towards green building practices. However, the optimal mix design of such green concrete continues to remain a significant challenge since it requires optimizing component proportions to achieve several desired attributes of concrete. This study focuses on optimizing concrete mix design by forecasting a crucial variable: 28-day compressive strength. It is analyzed using seven input factors: cement, slag, fly ash, water, superplasticizer (sp), coarse aggregate, and fine aggregate. A dataset comprising 103 data points from the literature is employed to investigate the chosen input and output variables. Various machine learning methods, such as linear regression, logistic regression, decision trees, random forests, support vector machines, k-nearest neighbors, k-means clustering, neural networks, and Gaussian process regression, are evaluated on the gathered dataset. The performance metrics from all machine learning models are compared to evaluate their efficiency. The results from the analysis indicate that Gaussian Process Regression (GPR) demonstrates exceptional accuracy in forecasting 28-day compressive strength, making it a robust choice for precision–critical applications in concrete mix design.

Keywords Fly ash · Slag · Sustainable concrete · Compressive strength · Prediction · Machine learning

S. Bolla (✉) · Y. Matla · F. M. Wani · J. Vemuri
Ecole Centrale School of Engineering, Mahindra University, Hyderabad, India
e-mail: se22ucse055@mahindrauniversity.edu.in

P. Pathak et al. (eds.), *Advances in Environmental Sustainability, Energy and Earth Science*, https://doi.org/10.1007/978-3-031-73820-3_13

1 Introduction

Concrete, valued for its robustness and longevity, primarily consists of cement, aggregates, and water, which chemically interact to form a hard, durable material extensively utilized in diverse construction projects, such as buildings, bridges, highways, and dams [1]. The evolution of concrete technology has played a crucial role in construction throughout history, significantly influencing building techniques across various eras. The origins of concrete trace back to ancient Greece and Rome, where engineers developed an innovative mixture that included lime as a binder, volcanic ash to enhance durability and enable underwater settings, and various aggregates, such as stones or gravel, to improve structural integrity. This early form of concrete facilitated the creation of numerous architectural masterpieces that remain standing today, a testament to its effectiveness and durability [2]. Advancements in concrete technology continued into the eighteenth century by discovering that heating a mixture of limestone and clay produced a more effective hydraulic lime capable of setting underwater [3]. This development expanded concrete's utility in marine construction and other water-related projects. The discovery of Portland cement was a breakthrough in technology during the nineteenth century to improve concrete. Produced by superheating a mixture of limestone and clay and then grinding it into a fine powder, Portland cement hardened into a robust, durable bond when mixed with water [4]. Today, concrete technology continues to evolve with innovations like high-performance concrete that offers superior durability and load-bearing capacities and self-compacting concrete that flows into molds without the need for mechanical vibration. These advancements underscore the ongoing development of concrete technology and highlight its critical importance in modern engineering and construction projects [5].

Concrete has many uses in the construction industry. However, concrete-related environmental issues are gaining importance because of its ubiquitous manufacturing processes around the world as well as towards sustainable development goals. Cement, the main ingredient of concrete and an essential component in construction materials, is well-established as a carbon dioxide (CO_2) intensive product [6]. The emission of CO_2, which is a greenhouse gas from the cement production process, adds greatly to one of its major environmental impacts. It occurs through two principal pathways: the calcination of limestone (calcium carbonate) driven chemically to CO_2 release and its transformation into lime (CaO), and combustion processes in order to achieve, among other rotary kilns, hot temperatures up to 1450°C for clinker production. These procedures transform the production of cement into one of most significant industrial sources' emissions globally. In addition to CO_2 emissions, the production of concrete demands substantial water resources, both in the processing of materials and in the mixing of concrete itself. This significant water use raises concerns in regions facing water scarcity or where the competition for water resources is intense [7]. Raw materials such as sand, gravel, and limestone required for concrete production are also extracted by mining which wreaks substantial havoc on local ecosystems. Efforts are underway in the industry to develop more sustainable practices, such as improving energy efficiency in cement production, capturing and isolating CO_2 emissions, and recycling concrete. Additionally, research into alternative materials like fly ash, slag, and natural pozzolans as partial replacements for cement in concrete mixtures is expanding [8]. This has the ability to outweigh its environmental cost gains; it can also add benefits in terms of increasing the strength and durability aspects of a construction material, which enables its usage more sustainably.

Incorporating fly ash and slag into concrete mixtures offers a robust opportunity to improve sustainability within the building sector [9]. Utilizing these materials as supplemental cementitious materials (SCMs) can yield significant positive environmental impacts, given their origins as by-products of coal combustion in power plants and iron production in blast furnaces [10]. Their integration into concrete offers significant environmental benefits:

Reducing Carbon Footprint: One of the main environmental advantages associated with fly ash and slag is a significant reduction in CO_2 emissions because of Portland cement manufacturing. This is why, aside from the environmental concerns for manufacturing them (the making of Portland cement requires great amounts of energy-consuming calcination in kilns and can release a lot of CO_2), natural cements are often recommended since they achieve high compressive strengths quickly. By partially replacing of Portland cement with unprocessed fly ash or slag, the emissions per unit volume of concrete produced can be significantly reduced [11].

Enhanced Material Properties: This allows fly ash and slag to react chemically with calcium hydroxide, which is also a by-product of cement hydration when water comes into contact. This reaction produces extra cementing materials, leading to even greater strength and durability of the concrete. Concrete mixtures incorporating these materials exhibit increased resistance to common degradation processes such as corrosion, sulfate attack, and alkali-silica reaction, thereby extending structure lifespan and reducing maintenance needs [12].

Waste Minimization: Incorporating fly ash and slag into concrete contributes to waste minimization by repurposing industrial by-products instead of disposing of them in landfills. This practice conserves landfill space and reduces the environmental impact associated with the production and disposal of conventional cementitious materials, aligning with the principles of the circular economy [11].

Improved Workability: Adding fly ash and slag to concrete improves workability, which can result in more efficient construction methods, as less energy is required to achieve the placement of concrete. Furthermore, fewer additives are needed to achieve optimal flow characteristics, contributing to overall resource efficiency in construction processes [12].

Concrete formulations involving fly ash and slag represent a significant step towards more environmentally friendly practices in the building sector. This approach not only mitigates environmental impacts associated with traditional cementitious materials but also improves the functionality of the structures. It exemplifies the practical implementation of circular economy principles, transforming industrial waste into valuable resources for sustainable construction. Estimating the compressive strength of concretes with such new formulations remains a challenge due to the inherent complexity involved in the mix design and strength gain processes.

2 Methodology

2.1 Optimal Mix Design with Machine Learning

Designing the optimal mix for green concrete, incorporating additional cementitious ingredients such as fly ash and slag, presents several challenges due to the need to balance various requirements and qualities. Achieving the desired properties in both freshly poured and hardened concrete involves considerations of workability, strength, durability, and environmental impact. Ensuring the correct water-to-binder ratio is crucial for concrete's strength and workability. Balancing low water content with adequate workability to enhance strength is a key aspect of mix design [13]. The inclusion of slag and fly ash may sometimes result in slower setting times for concrete. Striking a balance between achieving early strength and allowing for reasonable setting times presents a challenge in practical mix design. This focus is on improving the concrete mix, addressing the complexities involved in optimizing the blend while considering several factors such as workability, setting time, strength, and environmental impact.

In this study, ML models are used to predict the output variable, i.e., the 28-day compressive strength. The work was performed on a dataset of 103 samples [14]. By employing machine learning techniques, this study aims to accurately estimate compressive strength, addressing the challenges inherent in designing green concrete mixes [15]. Through the exploration of various predictive models and the examination of different input factors, it strives to provide insights into optimizing concrete compositions incorporating fly ash and slag while effectively managing considerations of workability, setting time, and strength requirements. Table 1 shows details of input variables.

Table 1 Details of input parameters

S.N.	Parameter	Minimum value	Maximum value
1	Cement (kg/m^3)	137	374
2	Slag (kg/m^3)	0	193
3	Fly ash (kg/m^3)	0	260
4	Water (kg/m^3)	160	240
5	Super plasticizer (kg/m^3)	4.4	19
6	Coarse aggregate (kg/m^3)	708	1049.9
7	Fine aggregate (kg/m^3)	640.6	902

2.2 Significance of 28-Day Compressive Strength in Concrete Performance Assessment

Compressive strength development in concrete is a complex process; the exact prediction of strength depends on various interdependent parameters [16]. It has been suggested that a detailed investigation of the combined effect of various supplementary cementitious materials on the prediction of the compressive strength of concrete is needed [17]. The 28-day Compressive Strength is a crucial parameter in assessing concrete quality and performance within the construction sector, especially in structural applications [18]. It signifies the maximum load or stress a concrete sample can endure under compression after undergoing a 28-day curing period [19]. The 28-day period is widely utilized because it allows for essential hydration and curing processes crucial for the development of the durability and strength of concrete. During this standardized period, concrete undergoes significant chemical reactions, resulting in the formation of hydrated cementitious compounds. These reactions are instrumental in enhancing the concrete's resistance and structural integrity. Testing concrete for compressive strength at the 28-day mark enables a reliable assessment of its suitability for real-world structural applications. The safety, dependability, and longevity of constructed infrastructure are contingent upon a comprehensive understanding and precise evaluation of this parameter. This also underscores the critical imperative for the construction industry to adhere rigorously to stringent testing protocols and implement robust quality control measures [20]. Such proactive measures are essential for preserving structural integrity and ensuring full compliance with regulatory standards, thereby safeguarding the integrity and longevity of infrastructure projects.

3 Machine Learning Models

3.1 Neural Networks

Neural networks draw inspiration from the intricate processing mechanisms of neurons in the human brain. They are made up of interconnected nodes called artificial neurons that are laid in layers. That is, input signals enter each neuron, and it processes them before outputting one; the latter may serve as an input to other neurons in subsequent layers. Neural networks consist of layers such as, Input Layer: It is the beginning point, and this layer accepts input data and passes it to the next layer. The layer name Hidden Layer: Computes intermediate representation of input data. The presence of multiple hidden layers in deep neural networks facilitates the acquisition of intricate data representations. Output Layer: Culminating the process, this layer furnishes the final output of the network [21]. Each neuron within a neural network applies a transformative operation to its input, conventionally utilizing an activation function. Popular activation functions [22] encompass sigmoid, tanh, ReLU (Rectified Linear Unit), and SoftMax, instilling nonlinearity within the network, thereby facilitating the assimilation of intricate data relationships. Neural networks absorb insights from data via backpropagation, a mechanism wherein the network forecasts outcomes, gauges the discrepancy (loss) between predicted and actual outputs, and adjusts its parameters (weights and biases) to minimize this disparity [23]. Gradient descent optimization techniques, including stochastic gradient descent (SGD) and Adam, are typically employed for parameter updates [24].

There are distinct types of neural networks, namely, *Feedforward Neural Networks (FNN):* Characterized by unidirectional data flow devoid of feedback loops, FNNs find application in classification and regression tasks [25]; *Recurrent Neural Networks (RNN):* With cyclic connections enabling data sequence processing, RNNs are apt for tasks like language modeling and time series prediction; *Convolutional Neural Networks (CNN):* Tailored for grid-like data, particularly images, CNNs leverage convolutional layers to autonomously discern features from input data; *Generative Adversarial Networks (GAN):* GANs, comprising a generator and a discriminator network, are trained conjointly to generate realistic data samples. The applications of neural networks range from image classification, object detection, and machine translation to sentiment analysis, medical diagnosis, and autonomous driving [26].

3.2 Support Vector Machines

Support vector machines (SVMs) are used for classifying data points in two classes. They are state-of-the-art in high-dimensional spaces used for pattern recognition, image classification, text categorization, and bioinformatics. The goal of SVMs is to find the hyperplane that separates data points corresponding to two different classes in feature space [27]. SVMs can handle both linearly and nonlinearly separable data using appropriate kernel functions [28]. The kernels enable SVMs to implicitly compute the dot product in higher-dimensional space, circumventing the computational burden of explicit data transformation. SVC (Classification variant): This form observes to determine the optimal hyperplane that separates data points into different classes with maximum margin. Binary and Multiclass classifications with strategies such as one-vs-one or one-vs-all are managed by SVC. The regression variant SVR seeks a hyperplane that best fits the data points (line in case of 1D, plane for all others) and decreases margin violations. SVR is beneficial for tasks where linear regression may not suffice, especially when dealing with nonlinear relationships between variables [29].

3.3 Linear Regression

Linear regression links a dependent variable with an independent variable or line and implies a linear connection between the independent and dependent variables [30]. The models are evaluated using metrics like Mean Squared Error (MSE) and

Root Mean Squared Error (RMSE) and R squared (R^2)—the fraction of the variation in the dependent variable that can be predicted using the independent factors.

3.4 Gaussian Process Regression

A Gaussian process regression (GPR) is an advanced probabilistic method used in machine learning and statistical modeling. The GPR model expresses the relationship between input data X and output variables Y as a distribution of function. It is assumed that any finite collection of function values has a joint Gaussian distribution [31]. GPR gives both predictions and uncertainty estimates, making it especially valuable for Bayesian optimization and uncertainty quantification. GPR relies heavily on the kernel function used. Commonly used kernels include the squared exponential kernel (radial basis function). There are three types of kernels: matérn, rational quadratic, and periodic. GPR is useful for regression and interpolation tasks in machine learning and statistics [32].

4 Results and Discussion

For neural networks, we have R^2 values of 0.84 in the training dataset and 0.95 in the test dataset, respectively; this model demonstrates temperate performance in predicting compressive strength as shown in Fig. 1a, b. The RMSE values of 2.93 for training and 1.83 for the test, respectively, suggest some variability, as shown in Table 2. In the case of SVM, the R-squared values suggest an average fit of the

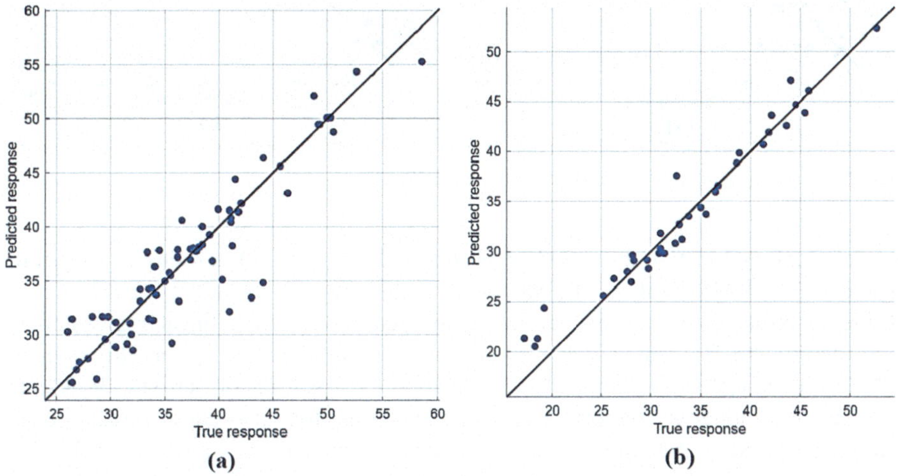

Fig. 1 (**a**) Training phase, NN; (**b**) testing phase, NN

Table 2 Comparison of models using evaluation metrics-R-Squared, RMSE, and MSE

Model	Hyperparameters	R^2		RMSE		MSE	
		Train	Test	Train	Test	Train	Test
Neural networks	Preset: trilayered neural network; Number of fully connected layers: 3; First layer size: 10; Second layer size:10; Third layer size:10	0.84	0.95	2.93	1.83	8.59	3.36
Support vector machine	Preset: Quadratic SVM; Kernel function: quadratic	0.93	0.98	1.98	1.21	3.91	1.46
Linear regression	Preset: interactions linear	0.96	0.98	1.52	1.19	2.31	1.43
Gaussian process regression	Preset: squared; Exponential GPR; Basis function: constant; Kernel function: squared exponential	0.97	0.99	1.25	0.71	1.55	0.51

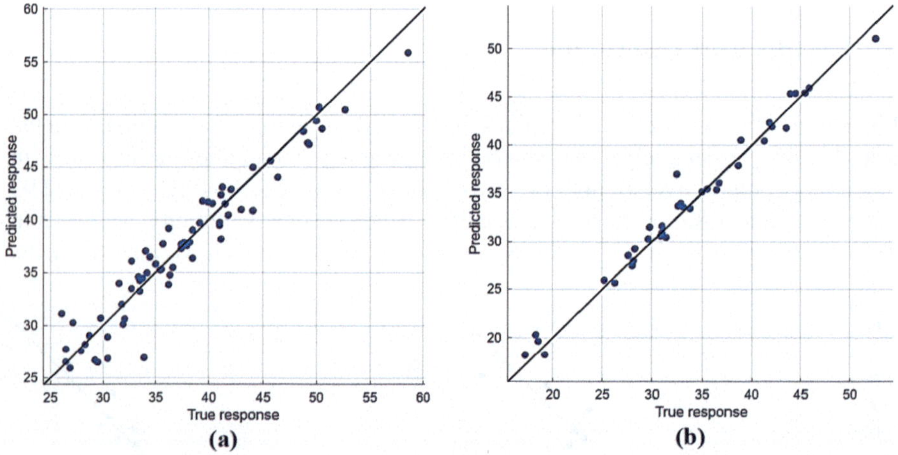

Fig. 2 (**a**) Training phase, SVM; (**b**) testing phase, SVM

model as plotted in Fig. 2a, b with 0.93 train value and 0.98 test value as shown in Table 2. The RMSE values of 1.98 for the train and 1.21 for the test, as shown in Table 2, indicate the models' accuracy in prediction. When linear regression is considered, the R-squared value for the train is 0.96, while a very high value (98%) for the test suggests that this mode is an excellent fit for our data. It depicts how much a model will be off, on average, when predicting strength values for the train set: RMSE, 1.52 (train). The test RMSE is even lower (1.19; see Table 2), meaning better prediction of new data. This indicates a better fit of the model to the data as plotted in Fig. 3a, b. For Gaussian process regression, exceptional

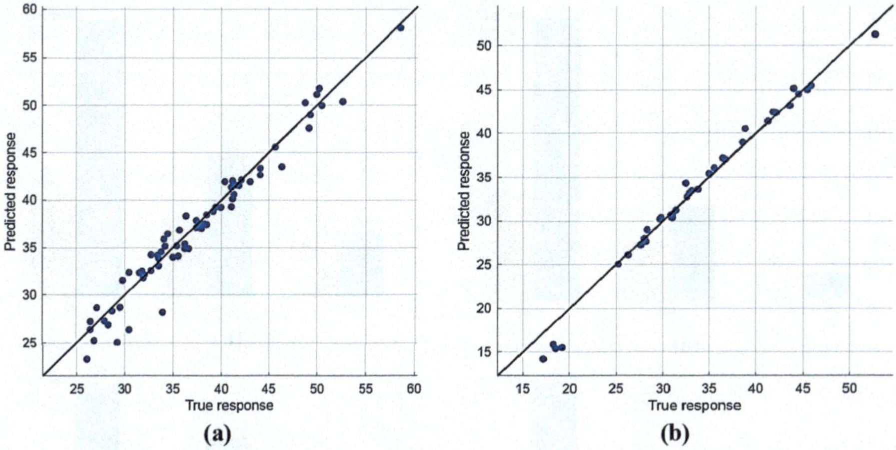

Fig. 3 (**a**) Training phase, LR; (**b**) testing phase, LR

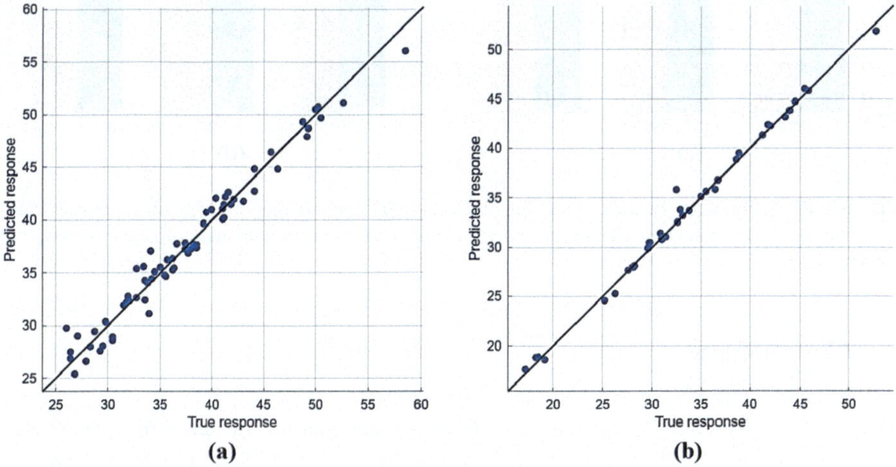

Fig. 4 (**a**) Training phase, GPR; (**b**) testing phase, GPR

performance is evident in the R-Squared values, reaching 0.97 in the train set and 0.99 for the test set. Low RMSE values of 1.25 for train and 0.71 for testing emphasize the precision of the model, as shown in Table 2. This model stands out as a robust choice for applications where precision in prediction is paramount, as plotted in Fig. 4a, b. The performance of machine learning models is compared in Table 2. Further, Fig. 5a–d shows the performance metrics for the four machine learning methods.

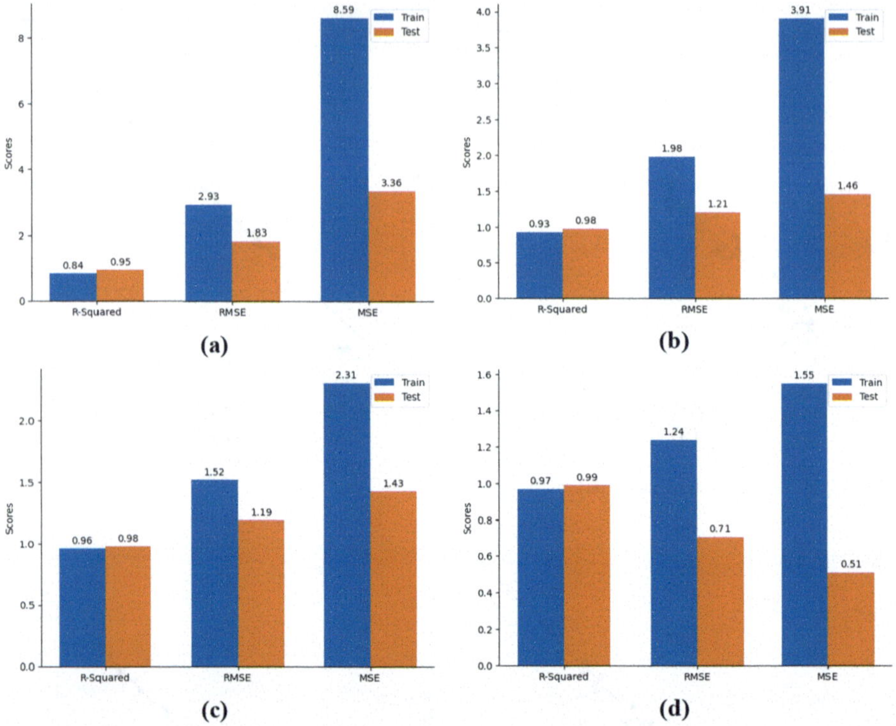

Fig. 5 (**a**) Performance metrics of neural networks; (**b**) performance metrics of support vector machine; (**c**) performance metrics of linear regression; (**d**) performance metrics of gaussian process regression

5 Conclusions

This study delves into optimizing concrete mix designs for sustainability, focusing on incorporating fly ash and slag while predicting the 28-day compressive strength. By employing various machine learning models, this study provides insights into accurately forecasting compressive strength, which is essential for precision–critical applications in the concrete mix design.

- Gaussian process regression outperformed all other accessible machine learning models with an impressive R-Squared score of 0.97 (train) and 0.98 (test). GPR excels in detecting complicated correlations in data, particularly when the underlying patterns are not simply linear or predictable. Its capacity to represent uncertainty and adapt to data features adds to its accuracy.
- The performance of GPR is followed by linear regression, support vector machines, and neural networks. Linear regression is less flexible than GPR. However, it performed well, exhibiting an R-squared score of 0.93 (train) and 0.98 (test).

SVMs may not have performed as well in this scenario due to the selection of kernel and hyperparameters that were not optimal for the dataset. Overfitting might be the cause of the neural networks' sub-par performance, particularly when dealing with limited datasets.

References

1. Wani FM, Vemuri J, Chenna R (2023) Evaluation of ground motion parameters and seismic response of reinforced concrete buildings from the mw 6.9, 2011 Sikkim earthquake. Geohazard Mechanics 1(2):162–178
2. Palley R (2010) Concrete: a seven-thousand-year history, 1st edn. The Quantuck Lane Press
3. Idorn GM (1991) Marine concrete technology. J Coast Res 7(4):1043–1056
4. Ryan JF (1929) The story of Portland cement. J Chem Educ 6(11):1854. https://doi.org/10.1021/ed006p1854
5. Sabouni AR (2023) Advances in reinforced concrete integrity and failure. In: Advances in structural integrity and failure. IntechOpen
6. Kaptan K, Cunha S, Aguiar J (2024) A review: construction and demolition waste as a novel source for CO2 reduction in Portland cement production for concrete. Sustainability 16(2):585
7. Scrivener KL, John VM, Gartner EM (2018) Eco-efficient cements: potential economically viable solutions for a low-CO_2 cement-based materials industry. Cem. Concr. Res. 114:2–26. https://doi.org/10.1016/j.cemconres.2018.03.015
8. de Brito J, Saikia N (2012) Recycled aggregate in concrete: use of industrial, construction and demolition waste. Springer Science & Business Media
9. Panarese WC, Allen GE, Cumming S (1991) In: Kosmatka SH (ed) Design and control of concrete mixtures. Canadian Portland Cement Association [CPCA]
10. Mindess S, Young JF, Darwin D (2003) Concrete, 2nd edn. Prentice Hall
11. Mohammadi A, Ramezanianpour AM (2023) Investigating the environmental and economic impacts of using supplementary cementitious materials (SCMs) using the life cycle approach. J Build Eng 79:107934
12. Fode TA, Jande YAC, Kivevele T (2023) Effects of different supplementary cementitious materials on durability and mechanical properties of cement composite–comprehensive review. Heliyon 9(7):E17924. https://doi.org/10.1016/j.heliyon.2023.e17924
13. Alaneme GU, Olonade KA, Esenogho E, Lawan MM (2024) Proposed simplified methodological approach for designing geopolymer concrete mixtures. Sci Rep 14(1):15191
14. Yeh I-C (2009) Concrete slump test. UCI Machine Learning Repository. https://doi.org/10.24432/C5FG7D
15. Li D, Tang Z, Kang Q, Zhang X, Li Y (2023) Machine learning-based method for predicting compressive strength of concrete. PRO 11(2):390. https://doi.org/10.3390/pr11020390
16. Kamath MV, Prashanth S, Kumar M, Tantri A (2024) Machine-learning-algorithm to predict the high-performance concrete compressive strength using multiple data. J Eng Design Technol 22(2):532–560
17. Karim R, Islam MH, Datta SD, Kashem A (2024) Synergistic effects of supplementary cementitious materials and compressive strength prediction of concrete using machine learning algorithms with SHAP and PDP analyses. Case Stud Constr Material 20:e02828
18. Wani FM, Khan MA, Vemuri J (2022) 2D nonlinear finite element analysis of reinforced concrete beams using total strain crack model. Materials Today: Proc 64:1305–1313
19. Neville AM (2011) Properties of concrete, 5th edn. Pearson
20. Mehta PK, Monteiro PJM (2014) Concrete: microstructure, properties, and materials, 4th edn. McGraw-Hill Education

21. LeCun Y, Bengio Y, Hinton G (2015) Deep learning. Nature 521(7553):436–444. https://doi.org/10.1038/nature14539
22. Goodfellow I, Bengio Y, Courville A (2016) Deep learning. MIT Press
23. Orr GB, Müller K-R (1998) Introduction. In: Neural networks: tricks of the trade. Springer, pp 1–5. https://doi.org/10.1007/3-540-49430-8_1
24. Lechevallier Y, Saporta G (2010) Proceedings of COMPSTAT'2010: 19th International Conference on Computational StatisticsParis France, August 22–27, 2010 Keynote, Invited and Contributed Papers. Springer Science & Business Media
25. Bishop CM (1995) Neural networks for pattern recognition. Oxford University Press
26. Wani FM, Vemuri J, Rajaram C, Reddy KK (2023) Investigating the efficiency of machine learning algorithms in classifying pulse-like ground motions. J Seismol 27(5):875–899
27. Wani FM, Vemuri J (2024) Shapelet-informed machine learning classifiers: a path towards precise identification of pulse-like ground motions. J Earth Syst Sci 133(2):96
28. Cortes C, Vapnik V (1995) Support-vector networks. Mach Learn 20(3):273–297. https://doi.org/10.1007/BF00994018
29. Cristianini N, Shawe-Taylor J (2000) An introduction to support vector machines and other kernel-based learning methods. Cambridge University Press
30. Montgomery DC, Peck EA, Vining GG (2015) Introduction to linear regression analysis. John Wiley & Sons
31. Rasmussen CE, Williams CKI (2005) Gaussian processes for machine learning. MIT Press
32. Wani FM, Vemuri J, Reddy KK, Rajaram C (2024) Forecasting duration characteristics of near fault pulse-like ground motions using machine learning algorithms. Stoch Env Res Risk A:1–30

Solid Waste Disposal Site Selection Using Geospatial Technology: A Comprehensive Study

Ch. Manasa Reddy and S. S. Asadi

Abstract Solid waste management stands as a significant municipal service vital for urban cleanliness worldwide. Amidst the conveniences of modern civilization, there emerges a pressing issue: the proliferation of solid waste, necessitating meticulous disposal efforts. The efficacy of waste disposal hinges upon judicious site selection, influenced by various factors. Recent years have seen notable strides in employing advanced techniques like remote sensing and Geographic Information Systems (GIS) to address regional waste management challenges. These technologies expedite the acquisition, processing, and dissemination of pertinent data crucial for effective waste management strategies. This paper presents a comprehensive examination of leveraging remote sensing and GIS methodologies integrated with the Analytical Hierarchy Process (AHP) to tackle environmental concerns associated with waste disposal. The study underscores the potential of remote sensing and GIS techniques in optimizing waste management systems. A geospatial multi-criteria analysis, employing weighted overlay technique, integrates diverse parameters encompassing physical, social, and demographic dimensions of urban landscapes to identify suitable solid waste disposal sites. The Land Suitability Index (LSI) aids in quantifying site suitability based on environmental considerations. Employing an equal interval classification method, the resultant index model categorizes sites into four distinct suitability tiers: "unsuitable," "less suitable," "moderately suitable," and "suitable."

Keywords AHP · Land suitability index · GIS · Site selection · Solid waste management

---

C. M. Reddy · S. S. Asadi (✉)
Department of Civil Engineering, Vignan's Foundation for Science, Technology & Research (Deemed to be University), Guntur, Andhra Pradesh, India

P. Pathak et al. (eds.), *Advances in Environmental Sustainability, Energy and Earth Science*, https://doi.org/10.1007/978-3-031-73820-3_14

1 Introduction

Solid waste encompasses all undesired or discarded substances that are neither in a liquid state nor in a gaseous state. This comprises household refuse, industrial byproducts, construction remnants, and agricultural detritus [1]. "Waste and solid waste" were not "problems when everything was used to the end and nature took care of the leftover materials that were left behind. Historically, early societies prioritized the removal of solid waste from their generation areas. This was likely due to a low human population and biodegradable solid waste production. Solid waste management has become a serious issue due to population growth, changing lifestyles, rising disposable income, technological advancements, and increased consumption of new products [2]. For instance, it was perfectly legal for Hooker Electrochemical Company to dump hundreds of tons of chemical sludge into Love Canal in the 1940s and early 1950s. Since infectious diseases were associated with unsanitary settings, waste management became centered around health and sanitation. As the 1970s saw the start of the Industrial Era, worries about waste production, pollution, and the discharge of hazardous substances increased. The problems with climate change associated with methane leaks from landfills came to light after 2000. In addition to contributing to climate change, the massive trash generation also suggested strategies for entrepreneurship, income development, and the eradication of poverty. The "Municipal Solid Waste (Management and Handling) Rule" notification in September 2000 evolved to create scientific methods for disposing of municipal waste in a safe manner. The Ministry of Urban Development released a Guidance Manual on Municipal Waste Management contemporaneously in 2000. The waste-management hierarchy comprising trash minimization, recycling, processing, transformation, and disposal was emphasized in the handbook [3, 4]. An estimate of 2.24 billion (10^9) tonnes of solid waste will be spawned globally within 2020, or 790 g/person/day. An annual rise in garbage production might attain 3.88 billion tonnes by 2050 owing to rapid urbanization and population explosion [5]. Effective management of solid waste is crucial for both environmental conservation and safeguarding public health. Methods of managing solid waste include recycling, composting, landfill disposal, and waste-to-energy technologies [6, 7]. Solid waste disposal poses environmental challenges including land as well as water contamination, green-house gas emissions [8]. Economically developing countries face waste management challenges due to their focus on industrial development, which might lead to environmental and public health issues in the future [9]. An appropriate disposal site must meet environmental safety guidelines and segregate waste to minimize harm to people and the environment during operation. Site selection criteria will consider both natural physical traits and economical, ecological, and land-use variables. Traditionally, garbage disposal has been done by dumping it on any available free area in the town [10]. In developing countries, waste reduction, recycling, and reuse are common methods for managing solid waste. To eliminate remaining materials, the most cost-effective and environmentally benign solution is to use landfilling, which may be done consistently and scientifically [9]. Conventional methods like landfills and incineration exacerbate pollution and health risks.

1.1 Importance of Efficient Site Selection

Efficient site selection is paramount in various industries and projects on the basis of its profound repercussions on socio-economic, environmental factors. It involves meticulous evaluation and consideration of multiple criteria such as accessibility, proximity to resources, environmental sensitivities, and community concerned guidelines as listed by the Ministry of Environment, Forest and Climate Change (MoEFCC) of India and government firms likely CPCB (Central Pollution Control Board) [11]. Optimal site selection can enhance operational efficiency, minimize transportation costs, reduce environmental impacts, and mitigate risks associated with land-use conflicts or natural disasters [12]. Furthermore, it can foster positive community relations and promote sustainable development by ensuring projects are compatible with local socio-economic conditions and regulatory requirements. Overall, efficient site selection is fundamental for achieving project success while balancing economic viability, environmental stewardship, and social responsibility [13–16].

2 Solid Waste Management—Challenges

2.1 Improper Disposal of Waste—Adverse Impact on Biotic-Life Health and Environment

Improper disposal of waste has profound environmental and health ramifications, encompassing contamination of vital resources like soil, water, and air, exacerbating ecosystem degradation and biodiversity loss [17, 18]. The release of harmful chemicals and pathogens through inadequate disposal practices contributes to widespread air and water pollution, precipitating various health issues in humans and wildlife [19]. Urgent implementation of robust waste minimization strategies is indispensable to alleviate detrimental consequences resulting in safeguarding of environmental and public health.

2.2 Regulatory Frameworks and Standards for Waste Management

Owing to rapid urbanization, Municipal Solid waste (MSW) management in towns and cities is becoming a grave task. Despite MSW rules 2016, site allocation and

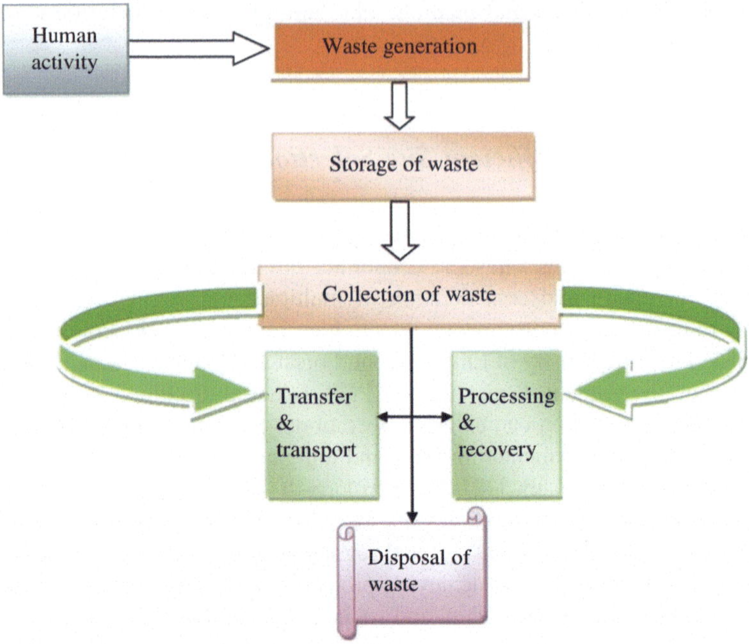

Fig. 1 Schematic view of waste management system

waste monitoring have become immediate issues to concerned regulatory bodies. To address these issues and meet the requirements of Swachh Survekshan, the part of Swachh Abhiyan, monitory bodies can employ RS&GIS to MSW for effective site allocation as well as waste segregation monitoring at this site. A schematic presentation on waste management is depicted in Fig. 1.

3 GIS and AHP Application for Solid Waste Management

3.1 GIS in Spatial Database Management for Waste Management

GIS plays a crucial role in managing spatial data for waste management, facilitating sophisticated analysis, planning, and decision-making [20]. It enables the efficient handling and examination of spatial information related to waste generation, collection, transportation, treatment, and disposal. By integrating diverse datasets, including geographical, environmental, and demographic data, GIS enables the creation of spatially referenced databases [21, 22]. These databases empower stakeholders to identify optimal locations for waste facilities, streamline collection routes, evaluate

Table 1 List of software and their applications

S. No	Software	Availability	Benefits
1	QGIS	Open source	• Utilizing geoprocessing tools to calculate distances, buffer zones, and other spatial metrics to aid in decision-making. • It aids planning by visualizing, analyzing, and modeling spatial data. • It also provides open street map data. • QGIS also provides different plug-in tools in network analysis.
2	SAGA GIS	Open source	• Terrain analysis for evaluating terrain suitability and constraints. • Hydrological analysis for assessing drainage patterns and watershed characteristics. • Geostatistics for analyzing spatial distributions of data and interpolating values.
3	ENVI	Proprietary	• Analysis of satellite and aerial imagery for identifying potential landfill sites and assessing environmental factors. • Image classification to distinguish land cover types relevant to site selection criteria. • Change detection to monitor land-use changes over time and their impact on potential sites.
4	ArcGIS	Proprietary	• Utilizing for spatial mapping. • Spatial analysis to identify suitable locations. • Integration with other data sources and models for comprehensive analysis.

environmental impacts, and ensure compliance with regulations. Furthermore, GIS facilitates the visualization of spatial patterns and trends, empowering decision-makers to make well-informed choices regarding waste management strategies and resource allocation [23, 24]. Overall, the utilization of GIS in spatial data management significantly enhances the efficiency, effectiveness, and sustainability of waste management endeavors. Some of the software products which are prominently designed to solid waste disposal site selection and analysis are mentioned below in Table 1.

3.2 *Analytical Hierarchy Process (AHP)*

The Analytical Hierarchy Process (AHP) methodology is a structured decision-making approach widely utilized across various disciplines [24, 25]. Developed by Thomas Saaty, AHP provides a systematic framework for evaluating complex decision problems involving multiple criteria and alternatives [4, 26, 27]. AHP involves breaking down decision criteria into hierarchical structures, allowing decision-makers to compare and prioritize criteria based on pairwise comparisons [28]. Through mathematical computations, AHP synthesizes these comparisons to derive priority weights for each criterion, enabling quantitative analysis and rational decision-making [29]. The flexibility and scalability of AHP make it a valuable tool for addressing diverse decision problems [30].

3.3 GIS and AHP Implications in Selection of Optimal Site for Waste Disposal

GIS along with AHP synergizes spatial analysis with structured decision-making in site selection [16]. GIS enables mapping and analysis of spatial factors [17], while AHP provides a method to weigh and compare criteria [18]. This integration optimizes ultimate decision-making giving consideration to both quantitative and qualitative aspects, aiding various disciplines in selecting the most suitable sites for projects [19]. The studies are presented in Table 2.

4 Methodological Frameworks for Waste Site Selection Using GIS-AHP

The waste site selection framework combines GIS with the Analytical Hierarchy Process (AHP), merging spatial analysis with systematic decision-making [29]. It organizes criteria hierarchically, integrating factors like environmental sensitivity and accessibility as shown in Fig. 2 [26–28]. GIS integrates spatial data, while AHP prioritizes criteria through pairwise comparisons, facilitating objective evaluation of options [30]. This synergy leads to optimal selections, minimizing environmental impact and complying with regulations.

5 Results & Discussion

Studies have shown that combining GIS and AHP techniques guarantees the selection of landfill sites with optimal precision, thereby helping management authorities identify the most appropriate locations for waste disposal [13, 22, 26, 31]. This integrated approach aids decision-makers in tackling waste management challenges while minimizing pollution and public health risks with lower economy [11]. GIS is essential in the site selection process for solid waste disposal, providing critical spatial analysis and visualization tools [22, 23]. While [32] emphasized that GIS integrates diverse datasets, including topography, land use/land cover, population density and transportation maps and hydrology, to generate detailed maps and spatial models. These maps identify suitable locations by overlaying criteria and sub-criteria, such as social factors, environmental impact zones, and economic

Table 2 Global scenario of remote sensing and GIS applications in solid waste management

S.No.	Title	Outline of the work	Reference
1	GIS approach for optimal urban solid waste disposal sites in Khulna City, Bangladesh	**Modalities:** Geographical information system (GIS) **Scope of study:** Choosing appropriate disposal locations. • As an information tool, GIS technology has aided in the acquisition of current land usage in information studies meant to address environmental issues. • The study's analysis has proven the effectiveness of GIS application to conventional one and effectiveness vary with accurate utilization of GIS tool. • The following results were derived using the GIS tool llikely land use, road, and slope. • Such mechanism would diminish the soaring improper municipal trash management in rural, urban areas.	[2]
2	GIS and remote sensing for urban waste disposal site selection in Kottayam, India	**Modalities:** Geographical information system and remote sensing **Scope of study:** Identifying suitable site to disposal. • GIS is a tool that contributes to reducing the time and cost involved in choosing a site and creates a digital archive for prospective site management monitoring projects. • GIS is utilized to evaluation of the suitability of landfill sites across the region, based on specific criteria categorized into both social and physical factors. These parameters, which were chosen in accordance with regional features consist of lithology, geomorphology, slope, drainage, population density, closeness to important thoroughfares and streams, and separation from drainage systems.	[6]
3	Selecting solid waste disposal sites in Madanapalle using remote sensing, GIS, and MCDA	**Modalities:** Remote sensing, GIS, and Multi Criteria Decision Analysis (MCDA) method **Scope of study:** Identifying a dumping site for solid waste disposal. • With the use of the arcGIS 10.1 program, the site appropriateness has been recorded in raster format using the "Weighted Overlay" approach. • Assessment of site suitability for solid waste disposal involves multiple parameters like geomorphology, lithology, soil, slope, land use, and road networks are given the complexity of their interaction, the MCDA strategy is used to make appropriate decisions.	[7]
4	Solid waste safe disposal site selection, Thiruvananthapuram Corporation.	**Modalities:** GIS, AHP, remote sensing method **Scope of study:** Selecting a site for safe disposal of solid waste. • GIS is a pivotal tool for identifying the best places to put disposal sites and for figuring out the most economical travel routes for disposing of and	[13]

(continued)

Table 2 (continued)

S.No.	Title	Outline of the work	Reference
		transporting waste, and AHP used for decision-making. • The land use, contour, drainage, slope, road network, geomorphology, and base maps are considered for research.	
5	Selecting solid waste management sites in Lucknow using remote sensing, GIS, and AHP	**Modalities:** Remote sensing, GIS, and AHP methods **Scope of study:** To select most suitable disposal site for solid waste. • This research illustrates the potential of remote sensing data in creating different thematic maps and conducting analysis within a GIS framework. • The thematic layers under consideration comprise land use/cover, soil, geomorphology, maps of transportation, slope, habitat, surface water, and groundwater table. The analytical hierarchy process (AHP) model is shown to be effective in assigning weightage, performing pairwise comparisons for the finding, and choosing the best dump locations in the research region.	[8]
6	Waste disposal site suitability in Kano Metropolis using multi-criteria evaluation, AHP, and GIS	**Modalities:** AHP and GIS **Scope of study:** Waste disposal suitability map. • The land use, slope, soil, road, population components are considered for present study. • A methodology for choosing suitable locations for municipal solid waste management is provided by the research. Waste landfills using fuzzy function standardization, AHP logic, and multi-criteria evaluation in a GIS. • The best land parcel was one that was level, had no settlements nearby, and was away from surface waterways. For moving waste, the mode of transportation is ideal. • Additional research should take into account the urban master plan, groundwater level, community preferences, and construction costs.	[9]
7	GIS based modeling for optimal landfill site selection	**Modalities:** AHP and GIS **Scope of study**: Selection of appropriate landfill sites. • From two studies, this study offered a viable methodology. Initially, the sites were sorted by suitability using the weights assigned to the sustainability criteria through a qualitative study. As a result, the AHP combined with a GIS framework model produced a sustainable structure with three levels of assessment for the locations of landfills. • In the future, recycling solid waste might be taken into consideration to lower the total acreage required for landfills.	[10]

(continued)

Table 2 (continued)

S.No.	Title	Outline of the work	Reference
8	Identification of suitable sites for solid waste disposal using GIS multi-criteria analysis in Peshawar, Pakistan	**Modalities:** ArcGIS 9.3, EPA parameters **Scope of study:** Selection of suitable site. • The following layers are considered for research such as patterns of land use, valuation of land, slope of the study area, road network, hydrological network, water table, airport. • Local planning authorities often lack the resources and knowledge necessary to carry out a sustainable siting process. By using GIS, time and resources can be saved by evaluating possible dumping sites.	[11]
9	Selecting landfill sites in Mohammedia, Morocco	**Modalities:** Multi-criterion decision analysis and GIS **Scope of study:** Potential sites to establishment of the landfill. • The approach used in this work is based on techniques like remote sensing as well as GIS. GIS is considered as robust software for spatial analysis. • To find the best suitable site, the following thematic maps are considered likely surface level water, land-use patterns, map of slopes, road network, geology. The rehabilitation is done for current landfill site of the study area.	[12]
10	Utilizing GIS and remote sensing for solid waste management: An analysis of Mellur Municipality, India	**Modalities:** GIS, weighted overlay analysis **Scope of study:** Selection of acceptable site of solid waste disposal. • There was a direct correlation between the number of data layers taken into consideration for analysis and the accuracy of the suggested locations. • Prior to making any decisions, adequate weights had to be assigned to each data layer in order to evaluate its impact on the community, the environment, and society. Weighted overlay analysis in GIS was utilized to identify the suitable locations.	[14]
11	Landfill site selection in Behbahan, Iran, using GIS and AHP	**Modalities:** GIS, AHP, Simple additive weighting (SAW) method **Scope of study:** Land suitability • Information regarding groundwater, surface level water, ecologically fragile regions, terrain coverage, proximity to urban and rural zones, land usage, distance from roads, slope, soil composition, and distance from waste generation sites was gathered and prepared within a Geographic Information System (GIS) platform for this particular investigation.	[15]

(continued)

Table 2 (continued)

S.No.	Title	Outline of the work	Reference
		• Subsequently, the SAW approach was utilized to evaluate land suitability, while the AHP method was employed to determine the hierarchical significance of the criteria in relation to each other.	
12	Landfill site selection in Kolkata, India, using AHP-MCDM	**Modalities:** Analytical hierarchy process and economic viability index (EVI) **Scope of study:** Siting sanitary landfill. • To find suitable disposal site accessibility, receptor, environment, public acceptance, and geology criterions are considered. • The relatively significant weights of which were determined using the AHP method. Compared to other widely used approaches, time is saved during data collection and computation when there are fewer evaluation criteria. • EVI aids in decision-making by emphasizing the financial costs associated with landfilling.	[16]

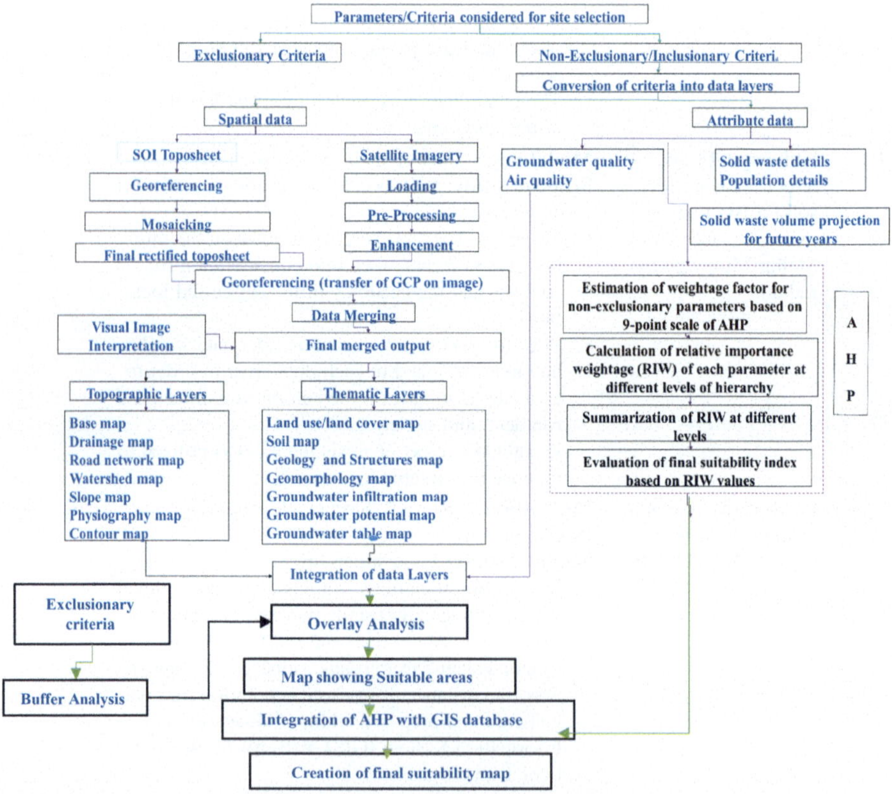

Fig. 2 Flowchart showing overall methodology of solid waste site selection

costs values [14, 33]. Using spatial analysis, GIS identifies areas that meet predefined thresholds for each criterion, ensuring that the selected site minimizes lower cost value, environmental impacts and achieves social acceptance as well [34, 35]. This geospatial approach improves the decision-making process by providing a clear, visual representation of potential sites and their evaluations [12]. In parallel, the Analytic Hierarchy Process (AHP) is used as advanced multi-criteria decision-making (MCDM) tool to determine best sites for solid waste disposal [29]. AHP breaks the decision problem down into a hierarchical framework that includes the aim, criteria, sub-criteria, and options [17]. Pairwise comparisons are conducted to assess the relative noteworthiness of each basis and sub-criterion, taken after by normalization of comparison lattices to determine criteria weights [17, 23, 29, 36]. Eigenvalue computation and consistency proportion (CR) investigation guarantee the vigor and unwavering quality of the comparisons [10, 12, 13, 18, 25, 26, 29, 30, 32]. Alternatives are evaluated against each sub-criterion, and local priority vectors are produced and combined to produce global priority vectors [15]. The option with the highest global priority vector is chosen as the most appropriate site, allowing for a methodical, translucent, and level headed decision-making process backed up by quantitative rigor.

6 Conclusions

This review paper summarizes the fusion of advanced satellite data utilization in geospatial technology with the Analytical Hierarchy Process (AHP) method for selecting sites for solid waste disposal presents a robust and systematic approach. This comprehensive study underscores the importance of advanced spatial analysis in evaluating diverse criteria, ensuring transparency, stakeholder engagement, and environmental sustainability. By optimizing decision-making processes for Government officials and operational efficiency, this approach contributes to the long-term viability and effectiveness of waste management practices with lower economy. Moving forward, further research is needed to incorporate emerging technologies like AI and Internet of Things (IoT) with refine methodologies like collection of data, prior processing, spatial analysis in GIS, and suitability analysis followed by a detailed evaluation by Analytical Hierarchy Process (AHP) method and address evolving challenges in the field. Overall, this review highlights the pivotal role of geospatial technology and AHP in advancing strategies for sustainable solid waste management without any damage of environment. Continued advancements in technology, methodology, and interdisciplinary collaboration offer opportunities to enhance the accuracy, efficiency, and sustainability of waste management strategies. Addressing emerging challenges such as dynamic spatial modeling, community engagement, and policy implications will be critical to ensure effective decision-making and equitable outcomes for long term. By embracing these future research directions, stakeholders can work toward building resilient waste management systems that protect from air and water borne diseases for public health, preserve environmental integrity, and promote social equity for generations to come.

References

1. George Tchobanoglous (2009) Solid waste management. In: Environmental engineering: environmental health and safety for municipal infrastructure, land use and planning, and industry, 6th edn. John Wiley & Sons, New Jersey, pp 177–307
2. Rahman MM, Sultana KR, Hoque MA (2008) Suitable sites for urban solid waste disposal using GIS approach in Khulna city, Bangladesh. Proc Pak Acad Sci 45(1):11–22
3. Asadi SS, Rao BVTV, Raju MV, Padmaja V (2011) An integrated multistage and multicriteria analytical hierarchy process GIS model for landfill siting: a model. Int J Mod Eng Res (IJMER) 1(1):40–51
4. Saaty TL (1980) The Analytic Hierarchy Process. McGraw-Hill, New York
5. The World Bank, The World Bank. https://www.worldbank.org/en/topic/urbandevelopment/brief/solid-waste-management. Accessed 07 Apr 2024
6. Vijith H, Nishanth T, Prakash MN (2010) Suitable site determination for urban solid waste disposal using GIS and remote sensing techniques in Kottayam municipality, India. Int J Geo Geosci 1(2):197
7. Shubhasmita Sahani VR, Prasad HS (2014) Location of solid waste disposal site using remote sensing, GIS and MCDA techniques in Madanapalle municipality, Andhra Pradesh. Institutes of Town Planners, India J 11(2):65–74
8. Virendra Kumar VR, Yadav K (2014) Selection of suitable site for solid waste Management in Part of Lucknow City, Uttar Pradesh using remote sensing, GIS and A.H.P. Method. Int J Eng Res Technol (IJERT) 3(3):1461
9. Usman MK, Aminu A (2015) Site suitability analysis for waste disposal in Kano Metropolis, Nigeria (using multi-criteria evaluation AHP, and GIS techniques). Int J Sci Eng Res 6(11): 1072–1081
10. Alsulaili AD, Alrukaibi DS (2017) GIS-based modeling for appropriate selection of landfill sites. J Eng Res 5(2):87–109
11. Nawaz S, Ali SA, Khan Z (2016) Identification of suitable sites for solid waste disposal using GIS multi-criteria analysis in Peshawar, Pakistan. Proc Pak Acad Sci: Pak Acad Sci B. Life Environ Sci 53(4):309–321
12. El Maguiri A, Kissi B, Idrissi L, Souabi S (2016) Landfill site selection using GIS, remote sensing and multicriteria decision analysis: case of the city of Mohammedia, Morocco. Bull. Eng. Geol. Environ. 75(3):1301–1309. https://doi.org/10.1007/s10064-016-0889-z
13. Asha Poorna VP (2016) Solid waste disposal site selection by data analysis using GIS and remote sensing tools: a case study in Thiruvananthapuram corporation area. Int J Geo Geosci 6(4):1734–1747
14. Rathinam VK, Sivasankar S (2017) Application of remote sensing and GIS in solid waste management: a case study of Mellur municipality, India. Int J Innov Res Sci Eng. Technol. 6(8): 174–180
15. Rahmat ZG et al (2017) Landfill site selection using GIS and AHP: a case study: Behbahan, Iran. KSCE J. Civ. Eng. 21(1):111–118. https://doi.org/10.1007/s12205-016-0296-9
16. Majumdar A, Hazra T, Dutta A (2017) Landfill site selection by AHP based multi-criteria decision making tool: a case study in Kolkata, India. J Inst Eng (India): Series A 98(3):277–283. https://doi.org/10.1007/s40030-017-0211-5
17. Zehra S, Afsar S, Anis Y (2019) Municipal solid waste disposal site selection in Jacobabad using RS/GIS techniques. J Geosci Environ Protect 07(10):13–22. https://doi.org/10.4236/gep.2019.710002

18. Kamdar I, Ali S, Bennui A, Techato K, Jutidamrongphan W (2019) Municipal solid waste landfill siting using an integrated GIS-AHP approach: a case study from Songkhla, Thailand. Resour Conserv Recycl 149:220–235. https://doi.org/10.1016/j.resconrec.2019.05.027
19. Bhavana YNE (2019) Site selection for urban solid waste disposal for Warangal city using geospatial techniques. Int J Adv Acad Stud 1(2):09–12
20. Nyoni T, Mandizvidza T, Mutongi C, Nyoni SP (2020) Application of GIS In landfill site selection for solid waste management in an urban environment. Int J Adv Res Innov Ideas Educ 6(12):1359–1368
21. Sk MM, Ali SA, Ahmad A (2020) Optimal sanitary landfill site selection for solid waste disposal in Durgapur City using geographic information system and multi-criteria evaluation technique. KN J Cartogr Geogr Inf 70(4):163–180. https://doi.org/10.1007/s42489-020-00052-1
22. Balew A, Alemu M, Leul Y, Feye T (2022) Suitable landfill site selection using GIS-based multi- criteria decision analysis and evaluation in robe town, Ethiopia. GeoJournal 87(2): 895–920. https://doi.org/10.1007/s10708-020-10284-3
23. Mussa A, Suryabhagavan KV (2021) Solid waste dumping site selection using GIS-based multi- criteria spatial modeling: a case study in logia town, Afar region, Ethiopia. Geol Ecol Landsc 5(3):186–198. https://doi.org/10.1080/24749508.2019.1703311
24. Majid M, Mir BA (2021) Landfill site selection using GIS based multi criteria evaluation technique. A case study of Srinagar city, India. Environ Chall 3:100031. https://doi.org/10.1016/j.envc.2021.100031
25. Aishwarya V, Salma S, Dodamani BM (2022) Identifying municipal solid waste dumping site location using AHP and GIS techniques: a case study of Coimbatore District, India. J Indian Soc Remote Sens 50(12):2337–2357. https://doi.org/10.1007/s12524-022-01605-9
26. Helal AH (2022) Multicriteria evaluation-GIS integration framework for landfill site selection in limited space regions: a case study in the West Bank. Adv Civil Eng 2022:1–13. https://doi.org/10.1155/2022/9367256
27. Tadese B, Wagari M, Tamiru H (2022) MCA and geospatial analysis-based suitable dumping site selection for urban environmental protection: a case study of Shambu, Oromia Regional State, Ethiopia. Heliyon 8(7):e09858. https://doi.org/10.1016/j.heliyon.2022.e09858
28. Karabulut Aİ, Yazici-Karabulut B, Derin P, Yesilnacar MI, Cullu MA (2022) Landfill siting for municipal solid waste using remote sensing and geographic information system integrated analytic hierarchy process and simple additive weighting methods from the point of view of a fast-growing metropolitan area in GAP area of Turkey. Environ Sci Pollut Res 29(3): 4044–4061. https://doi.org/10.1007/s11356-021-15951-7
29. Mihiretie A (2022) Suitable solid waste disposal site selection using geographical information system: a case of Debre Markos town, Ethiopia. Int J Environ Geoinformatics 9(2):1–8. https://doi.org/10.30897/ijegeo.784087
30. Ramu P, Sai Santosh B, Praveen S (2023) An integrated GIS-AHP approach for municipal solid waste landfill siting in Srikakulam District, Andhra Pradesh. Nat Environ Pollut Technol 22(1): 139–148. https://doi.org/10.46488/NEPT.2023.v22i01.012
31. Arabeyyat OS, Shatnawi N, Shbool MA, Al Shraah A (2024) Landfill site selection for sustainable solid waste management using multiple-criteria decision-making. Case study: Al-Balqa governorate in Jordan. Methods 12:102591. https://doi.org/10.1016/j.mex.2024.102591
32. Kebede HA, Ayenew WA (2023) Solid waste disposal site selection analysis using geospatial technology in Dessie city Ethiopia. Heliyon 9(8):e18865. https://doi.org/10.1016/j.heliyon.2023.e18865
33. Dande J, Mallikarjuna V (2023) Site selection for solid waste management in Krishna district, Andhra Pradesh using GIS. E3S Web Conf 399:02001. https://doi.org/10.1051/e3sconf/202339902001

34. Desta MD, Tesseme T, Yigezu TT, Nigussie AB (2023) Assessment of landfill site suitability using GIS, remote sensing, and the multi-criteria decision-making (AHP) approach, Ethiopia. Geol Ecol Landsc:1–14. https://doi.org/10.1080/24749508.2023.2256549
35. Saketa Y, Tamene N, Melknew M (2023) Municipal solid waste disposal site suitability analysis using multi-criteria evaluation in Assosa, Ethiopia. Int J Environ Sci Technol 20(4): 3815–3830. https://doi.org/10.1007/s13762-022-04221-9
36. Radhakrishnan T, Manimekalan A, Ghosh D, Prasanna R (2024) Identifying high-vulnerable garbage accumulation areas in Coimbatore City, India: an AHP-GIS approach for effective waste management. Environ Sci Pollut Res 31(14):21797–21810. https://doi.org/10.1007/s11356-024-32486-9

Exploring Spatial Technology for In-Depth Natural Resource Management: A Comprehensive Study

Mariya Hasnat, Ruqaiya Bano, and Farhana Khatoon

Abstract The integration of Remote Sensing and Geographical Information System (GIS) provides extensive opportunities for monitoring and overseeing natural resources with capabilities spanning multi-temporal, multi-spectral, and multi-spatial resolutions. There is an immediate imperative to comprehend the specialized functionalities of a continuously expanding range of image sources and analytical techniques, particularly for those involved in natural resource management. This comprehensive study compiles diverse applications of remote sensing and GIS tools relevant to the management of natural resources, including agriculture, water, forests, soil, and natural hazards. The major applications of remote sensing and GIS in natural resource management are resource management, change detection, impact assessment, suitability analysis, biodiversity, and forest management, etc. The collaborative effort is essential for the effective use and application of remote sensing techniques for monitoring and assessment of natural resource management. Remote Sensing (RS) proves to be an efficient instrument for monitoring spatial and temporal changes in the landscape, by facilitating land use and land cover mapping. When integrated with on-site ecosystem assessments and other relevant biophysical and socio-economic data, the resulting land cover and land use maps offer crucial insights. By integrating and modifying spatial data in a Geographic Information System (GIS), it is possible to carry out analyses that improve understanding of the dynamics and processes associated with land use changes.

Keywords Remote sensing · GIS · Natural resource · Change detection · Biodiversity

---

M. Hasnat (✉)
Department of Environmental Science, Integral University, Lucknow, UP, India
e-mail: hmariya@iul.ac.in

R. Bano
Department of Agriculture, Integral Institute of Agricultural Science and Technology, Integral University, Lucknow, UP, India

F. Khatoon
Department of Geography, University of Allahabad, Prayagraj, UP, India

P. Pathak et al. (eds.), *Advances in Environmental Sustainability, Energy and Earth Science*, https://doi.org/10.1007/978-3-031-73820-3_15

1 Introduction

Recently, remote sensing data has been extensively utilized across diverse fields of natural resource management. The availability of data from numerous sensors on different platforms, offering varying spatio-temporal, radiometric, and spectral resolutions, has established remote sensing as a primary data source for extensive applications and research. These datasets play a crucial role in modeling ecological processes [1]. Integrating remote sensing data with GPS and GIS will facilitate the monitoring and assessment of natural resource management plans for government agencies, policymakers, conservation organizations, and industry. This allows to study changes in land cover, forest density, coastal morphology, and biodiversity even in remote locations. The integration of modern geospatial technologies such as Remote Sensing (RS), Geographic Information Systems (GIS), and Global Positioning Systems (GPS) is a powerful way to survey, identify, classify, map monitor, characterize, and track various forms of change provided. It includes both renewable and non-renewable resources, living and non-living entities. Remote sensing plays an important role in providing spatial geographic information and contributes to defining, improving, and monitoring the overall potential of the Earth. Satellite observations of land, oceans, and atmosphere, especially during natural and human-induced hazards, are vital for environmental protection, disaster risk reduction, and sustainable development. RS using space-borne sensors offers repetitive (from minutes to days) and synoptic (global coverage) observations, supporting applications such as crop inventory, drought and flood damage assessment, and land use monitoring.

India has emerged as a major provider of Earth observation data worldwide, offering various spatial, spectral, and temporal resolutions to meet diverse national development needs [2]. GPS provides globally accurate coordinates, aiding in establishing geographic location and context. It enables cost-effective acquisition of recent in situ data, crucial for repetitive monitoring of Earth features and phenomena. GIS facilitates the visualization of geospatial data, enhancing communication and understanding of complex information such as natural resources. It also enables data generation, editing, storage, and analysis, crucial for planning and decision-making. GIS allows the investigation of natural resource and environmental problems in a geographic context, enabling the modeling of phenomena and analysis of cause-and-effect relationships at local to global scales (Fig. 1).

Remote Sensing (RS) is valuable for assessing rates and trends, while GIS excels in analyzing causes and consequences. However, integrating spatial technologies with other analytical approaches enhances the generation of more comprehensive information, thereby improving our understanding of effective natural resource management. The emerging field of geographic information science encompasses modern tools such as RS, GIS, Global Positioning Systems (GPS), and simulation models. By combining these technologies, a cost-effective method is established for acquiring high-resolution, real-time data through remote sensing, managing and analyzing data using GIS, georeferencing ground data with GPS, and integrating all data into information systems tailored for specific purposes. Geoinformatics

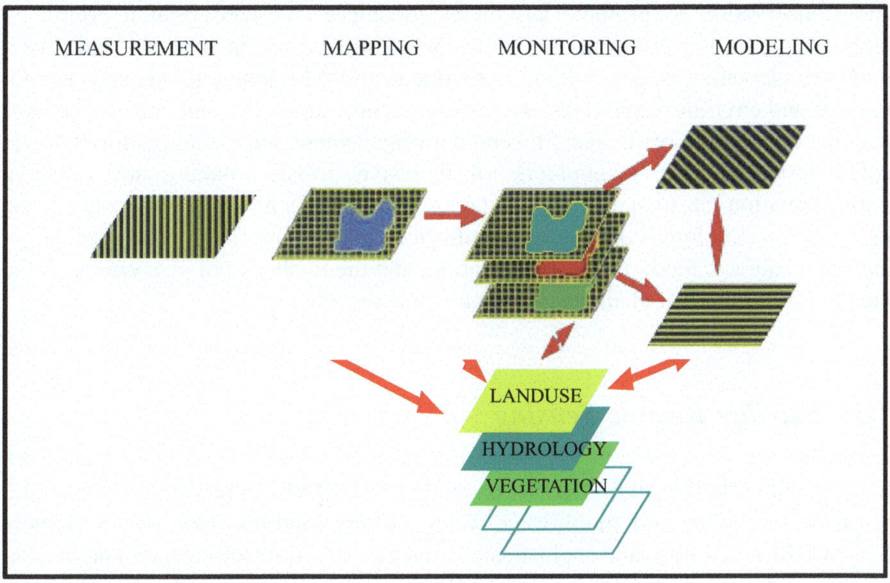

Fig. 1 GIS process (source: Lillisand and Kiefer)

integrates elements from various disciplines dealing with geographic data by encoding input data into three-dimensional space and associating it with specific locations on Earth's surface. Consequently, geoinformatics applies information technology to the study and management of Earth resources.

Remote Sensing involves the sensing, identification, delineation, and measurement of surface features and processes from a distance without physical contact. It captures Electromagnetic Energy (EME) from the sun or emitted by sensors, which is reflected, scattered, or emitted by Earth's surface features. Passive sensors detect natural radiation, while active sensors emit electromagnetic radiation. Remote sensing provides multi-spectral, multi-sensor, and multi-temporal data for generating accurate, timely, and cost-effective information on natural resources.

Optical remote sensing detects solar radiation across various wavelength regions, producing images resembling photographs, while aerial photography remains widely utilized for stereoscopic analysis, offering essential three-dimensional perspectives.

The primary aim of this paper is to explore the application of geospatial technology in natural resource management, enhancing the understanding, monitoring, and sustainable utilization of natural resources through spatial data analysis, modeling, and decision support systems. This overarching objective can be subdivided into several specific goals: Mapping and Monitoring to create precise maps and geographic databases of natural resources, track changes over time, assess current conditions, and identify areas of concern; Modeling and Prediction to replicate natural processes, forecast future events, and evaluate management impacts; Decision Support Systems (DSS) to assist stakeholders in making informed decisions by integrating spatial data, statistical tools, and visualization approaches; Sustainability and Conservation to promote sustainable practices and conservation efforts by identifying ecologically important areas, assessing ecosystem services, and prioritizing conservation actions; Risk Assessment and Mitigation to identify natural hazards and environmental risks, develop mitigation strategies, and enhance ecosystem and community resilience; Stakeholder Engagement and Collaboration to foster collaboration among stakeholders for inclusive decision-making and effective implementation of management strategies. Through achieving these objectives, geospatial technology contributes significantly to the sustainable management of natural resources, biodiversity conservation, and the resilience of ecosystems in the face of environmental challenges and human activities.

1.1 Satellite Remote Sensing

Starting with satellites like TIROS-1 and later LANDSAT has revolutionized natural resource surveying and monitoring. With various satellites like Indian Remote Sensing (IRS) satellites and sophisticated image processing facilities, remote sensing has become vital in meteorology, agriculture, geology, hydrology, urban development, ecology, and pollution monitoring (Fig. 2). Satellite data, including microwave data, offers advantages such as all-weather capability and day-night observation [3]. GIS is a database management system that stores, retrieves,

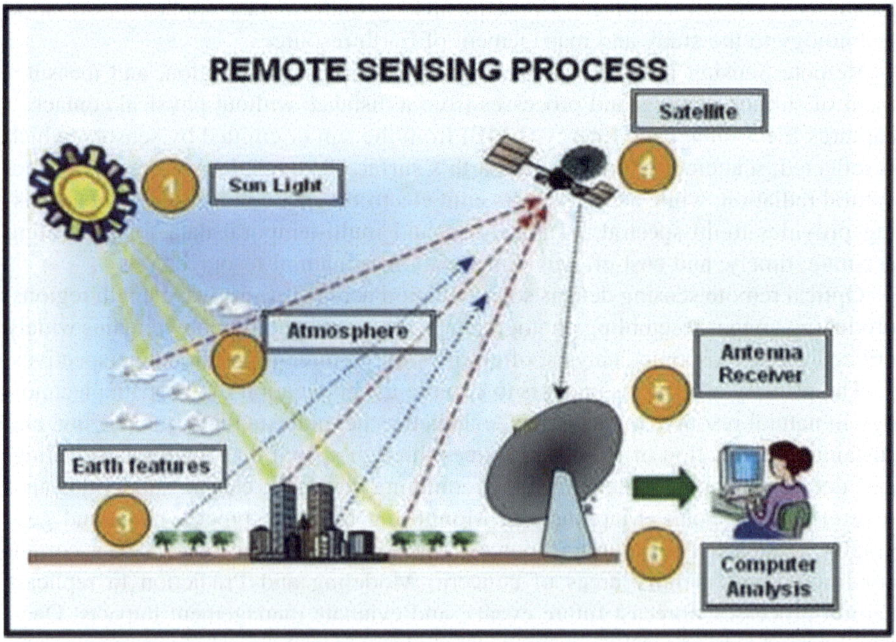

Fig. 2 Remote sensing process (source: Lillisand & Kiefer)

manipulates, analyzes, and displays spatial information from diverse sources. It handles large volumes of spatial data from field surveys, aerial surveys, and remote sensing, integrating them into a common platform for effective analysis and decision-making. GIS components include data inputting, encoding, management, analysis, and presentation. GIS enables the querying of data sets, answering complex questions, and solving a broad range of problems related to location, condition, pattern, routing, and modeling [4]. It has been used for integrating spatial data on various resource themes and generating alternate development plans showing site-specific primary production activities, linking land parameters to primary production activities through rule-based decision capabilities.

1.2 Global Positioning System (GPS)

The Global Positioning System (GPS) has transformed positioning methodologies, originally designed as a navigation tool. The Navigation system with Satellite Timing and Ranging (NAV- Navigation system, STAR- Satellite Timing and Ranging) is a satellite-based radio navigation system that delivers precise three-dimensional position, navigation, and time data to appropriately equipped users. GPS receivers passively receive signals without transmitting any themselves. At least 24 operational GPS satellites are maintained continuously by the US Air Force, each with an orbital period of 12 h. Ground stations meticulously track each satellite's orbit, and every GPS satellite is equipped with an atomic clock and transmits data indicating its location and time. Synchronization among GPS satellites ensures that signals are transmitted simultaneously. The receiver calculates its position in three dimensions by estimating the time it takes for signals from at least four GPS satellites to reach it [5]. Typically, hand-held GPS units offer locational accuracy within 10–20 meters. To achieve greater accuracy, Differential GPS (DGPS) is employed, requiring an additional receiver positioned at a known location nearby. Observations made by the stationary receiver (Base station) are utilized to correct the position recorded by covering units, resulting in accuracy surpassing one meter. In the past, the US Government deliberately introduced errors in orbit data and clock accuracy for L1 signals to limit civilian user accuracy; this practice, known as "Selective Availability," was discontinued in May 2000 due to the widespread civilian applications of GPS. GPS, with its attributes of all-weather operation, real-time availability, continuous operation, cost-effectiveness, and high precision, finds application across various fields due to its ability to accurately position points separated by hundreds to thousands of kilometers. Key applications include:

I. Geodetic control surveys, such as locating precise control points.
II. Geodynamics surveys for determining recent crustal movements.
III. Establishing local networks for engineering project control.
IV. Monitoring mass movements like ground subsidence and landslides.
V. Conducting geophysical and cadastral surveys.
VI. Facilitating precise navigation, marine geodesy, and hydrography.

2 Methodology Flowchart

Formula-based methodology as shown in Fig. 3 is discussed here

1. Data Collection:

$$(\mathbf{Dc} = \mathbf{sum\_\{i = 1\}^\wedge\{n\}Di})$$

Where Dc is the collected data and Di represents individual data sources.

2. Data Preprocessing

$$(\mathbf{Pd} = \mathbf{F}(\mathbf{P\{r\}}, \mathbf{P\{c\}}, \mathbf{P\{g\}}))$$

Where Pd represents preprocessed data.

(P{r}), (P{c}), and (P{g}) denote radiometric correction, geometric correction, and data cleaning, respectively.

3. Spatial Analysis

$$(\mathbf{As} = \mathbf{G}(\mathbf{Ao}, \mathbf{Ts}))$$

Where (As) is the spatial analysis output. (Ao) denotes original spatial data. (Ts) represents the spatial analysis technique.

4. Modeling and Simulation

$$(\mathbf{Mm} = \mathbf{S}(\mathbf{Mo}, \mathbf{Vm}))$$

Where (Mm) represents modeled output. (Mo) denotes original data. (Vm) represents variables used in the model.

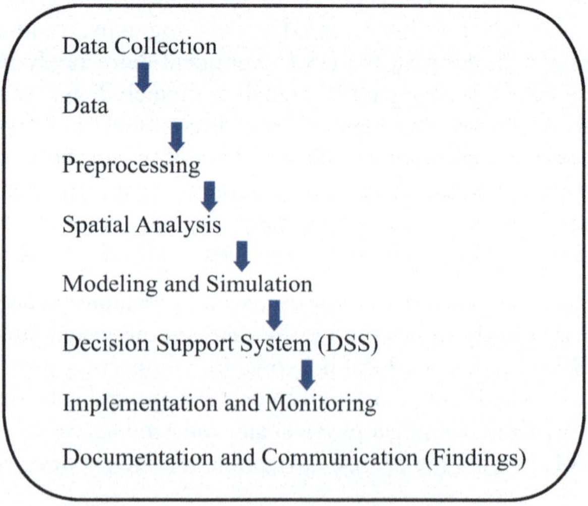

Fig. 3 A flowchart for the methodology

5. Decision Support Systems (DSS)

$$(\mathbf{DSS} = \mathbf{C}(\mathbf{Ms}, \mathbf{V}\{\mathbf{v}\}, \mathbf{T}\{\mathbf{p}\}))$$

Where (DSS) represents the decision support system. (Ms) denotes model outputs.

(V{v}) represents visualization tools.

(T{p}) denotes scenario planning techniques.

6. Implementation and Monitoring

$$(\mathbf{Im} = \mathbf{M}(\mathbf{Is}, \mathbf{Rm}))$$

Where (Im) represents implemented management strategies. (Is) denotes model simulations.

(Rm) represents monitoring results.

7. Documentation and Communication

$$(\mathbf{Dc} = \mathbf{C}(\mathbf{Dm}, \mathbf{Rc}))$$

Where (Dc) represents documented data.

(Dm) denotes methodology documentation. (Rc) represents results communication.

This formula-based approach provides a structured framework for implementing geospatial technology in natural resource management.

3 Remote Sensing and GIS Application Areas

Some of the broad natural resource management application areas wherein considerable progress has been made in the past years:

3.1 Geospatial in Agriculture

In the realm of agriculture, there is a growing emphasis on using remote sensing platforms for real-time assessments of agricultural landscapes. Precision agriculture, which tailors management practices within fields according to site conditions, relies on modern technologies such as GPS, GIS, yield monitoring devices, soil and pest sensors, and remote sensing [6]. Satellite remote sensing, coupled with GIS, proves to be a powerful tool for detecting land use and land cover changes, providing cost-effective multi-spectral and multi-temporal data for understanding and monitoring land development patterns [4].

Table 1 shows the application of RS GIS in natural resource management.

Table 1 Geospatial Methodologies for Natural Resource Assessment and Management

Application	Description	Data source	Methodology	Key metrics	Results	Author
Assessment of water quality	Monitoring and assessing water quality in lakes	Satellite imagery and field data	GIS analysis and remote sensing	Water clarity and pollutant levels	Improved detection of pollutants and water quality trends	Philipson & Lindell (2003)
Monitoring of crop stress	Identifying and managing crop stress	Multi-spectral satellite data	Crop models and remote sensing	Normalized Difference Vegetation Index (NDVI) and crop yield	Enhanced nitrogen management and increased crop yield	Baret, Houles & Guérif (2007)
Analysis of land use/land cover change	Detecting changes in land use and land cover	Satellite imagery and historical maps	GIS analysis and remote sensing	Land cover types and change rate	Accurate mapping of urbanization and deforestation rates	Ghosh et al. (1996)
Estimation of soil moisture	Estimating soil moisture levels	Satellite observations	Remote sensing and hydrological	Soil moisture	Improved irrigation practices and drought	Ford, Harris, & Quiring
		(e.g., Soil Moisture and Ocean Salinity (SMOS))	modeling	content	monitoring	(2014)
Groundwater prospecting zones	Identifying potential groundwater zones	Remote sensing data and geological surveys	GIS modeling and fuzzy analytical hierarchy process	Groundwater potential zones	Enhanced groundwater resource management and new water sources	Singh, Hasnat, Rao, & Singh (2021)
Mapping of impermeable surface	Mapping and monitoring impermeable surfaces	Aerial photograph and satellite data	Remote sensing and image characterization	Surface area and change detection	Better urban planning and flood risk assessment	Melesse & Wang (2007)
Disaster management	Assessing and managing natural disasters	Satellite imagery and field surveys	GIS analysis and remote sensing	Disaster extent and impact zones	Improved disaster response and risk mitigation strategies	Madhavan Nair (2004)
Planning of basin development	Providing land and soil information for basin development	Satellite imagery and field surveys	Remote sensing and GIS analysis	Soil types and land suitability	Optimized land use planning and sustainable development	Das (2001)
Agricultural productivity enhancement	Enhancing productivity through crop models	Satellite imagery and field surveys	Crop modeling and remote sensing	Crop yield and productivity variance	Increased agricultural productivity and optimized resource use	Batchelor, Basso, & Paz (2002)
Environmental impact assessment	Evaluating environmental impacts of land use changes	Satellite imagery and field surveys	Geospatial analysis	Environment al impact scores	Informed decision-making and sustainable land management	Samant & Subramanyam (1998)

3.2 Geospatial in Soil Science

Soil properties exhibit spatial variability, necessitating their estimation as continuous variables for higher accuracy and broader applications. Traditional soil analysis methods are laborious and time-consuming, leading to increased expenses. Kriging and its variants have emerged as vital spatial interpolation techniques in land resource inventories [4]. With advancements in GIS and remote sensing, predictive soil mapping techniques have been developed, correlating in situ point measurements of soil quality with satellite-derived indices to create spatial maps ideal for spatially distributed models [7, 8].

3.3 Geospatial in Crop-Irrigation Demand Monitoring

Remote sensing technology plays a crucial role in monitoring irrigated lands, aiding in optimizing irrigation water use, irrigation system performance, and crop water deficit management [9]. By monitoring plant water status, measuring evapotranspiration rates, and estimating crop coefficients, remote sensing helps determine optimal irrigation timing and water application amounts [10]. This is of particular interest to irrigation water policymakers for effective water resource management.

3.4 Geospatial in Crop Modeling/Agricultural Monitoring and Precision Agriculture

Remote sensing data can be used to monitor crop health, estimate crop yields, detect pests and diseases, and optimize irrigation practices. GIS can be used to create precision agriculture maps, delineate management zones, and plan optimal planting patterns based on soil and environmental conditions. Combining crop models with remote sensing data allows for assessing yield variables and transferring results from field to regional scales. Various methods, such as estimating leaf area index values and quantifying stress through assimilation approaches, have been employed to reduce uncertainties in crop models using remote sensing data [11, 12]. Additionally, remote sensing facilitates the classification of agricultural fields and crop types, providing essential data for accurate crop modeling and assessment of methane emissions and global food production impacts [13, 14].

3.5 Geospatial in Water Resource Management

Water, a fundamental resource for human survival, faces a concerning trend of decreasing availability amid a growing global population. The imperative to monitor and understand water usage has never been more urgent, as it lays the foundation for effective water management strategies, particularly in regions where water scarcity is a pressing issue [15]. Achieving a comprehensive grasp of the intricate water system demands an interdisciplinary approach, integrating concepts from various fields for sustainable water resource management [16]. While field-scale studies offer initial insights into the complexities of the water cycle, policy decisions are typically made at regional or national levels. Therefore, it is crucial to upscale these studies effectively. Hydrological models often grapple with data scarcity or inadequate input data quality, highlighting the potential of remote sensing technologies to complement such models by providing continuous input data, especially in data-scarce regions. The launch of advanced Earth Observation (EO) satellites has revolutionized global hydrological monitoring by bridging data gaps and offering continuous measurements of various hydrological components [16, 17]. Remote sensing can aid in the assessment of water resources by monitoring surface water bodies, mapping wetlands, estimating water quality parameters, and monitoring changes in water levels over time. GIS facilitates the analysis and visualization of water-related data, helping in effective water resource planning and management.

3.6 Geospatial in Water Quality Monitoring

Satellite remote sensing, in conjunction with GIS, plays a pivotal role in monitoring hydrological components such as rainfall, soil moisture, evapotranspiration, and groundwater levels. These technologies facilitate the 3D mapping of water bodies and the generation of spatial water availability maps, empowering authorities to identify and protect vulnerable regions. Continuous monitoring of water quality is essential for management and improvement [18, 19]. While traditional methods like in situ measurements are accurate, they lack the spatial and temporal coverage necessary for a comprehensive assessment. Remote sensing techniques provide both spatial and temporal information, enabling the monitoring of water quality parameters such as turbidity, chlorophyll, and temperature [20, 21].

3.7 Geospatial in Forest Management

Forests, crucial components of ecosystems, face rapid depletion globally. Remote sensing data coupled with GIS techniques offers invaluable insights into forest cover, types, human encroachment, and habitat suitability for wildlife

[17, 22]. This information is indispensable for formulating effective forest management plans and decision-making processes aimed at sustainable resource utilization. Remote sensing can provide valuable information for forest inventory, monitoring forest health, and detecting forest disturbances such as wildfires, insect infestations, and illegal logging activities. GIS allows for the integration of remote sensing data with other spatial data layers for comprehensive forest management planning.

3.8 Geospatial in Natural Disaster Management

Remote sensing plays a crucial role in disaster monitoring, early warning systems, damage assessment, and post-disaster recovery efforts. GIS enables the integration of remote sensing data with infrastructure, population, and environmental data to assess vulnerability, plan evacuation routes, and coordinate emergency response activities [20]. In disaster management, satellite remote sensing serves as an indispensable tool for providing timely information over large areas during various phases, from prevention to reconstruction. Alongside, GIS techniques play a vital role in handling vast spatial datasets and enhancing disaster management capabilities [23, 24].

3.9 Geospatial in Land Use/Cover Mapping

Remote sensing data, especially satellite imagery, can be used to classify and map different land use and land cover types. This information is vital for assessing changes in land use over time and monitoring urban sprawl, deforestation, agricultural expansion, and habitat loss [10]. Remote sensing technology has revolutionized the acquisition of vast amounts of information across a broad spectrum of applications. Earth observation satellites serve as indispensable tools for various tasks such as mapping urbanization, identifying potential sites for tourism development and assessing their impact on the environment, monitoring mining activities, and analyzing transportation networks [16]. Moreover, remote sensing techniques enable the study of complex land use and land cover systems in challenging terrains, facilitating the preparation of comprehensive management plans [25]. In mountainous regions where terrain is challenging and access is limited, obtaining accurate information for natural resource management is difficult [26, 27].

3.10 Geospatial in Urban and Regional Planning

Infrastructure development is crucial for improving the quality of life, particularly in developing countries like India. The integration of high spatial resolution remote sensing data with advanced image analysis techniques, GIS, and GPS has facilitated various infrastructure projects. These include urban development plans, road alignment, rural connectivity projects, micro-hydel site ranking for power generation, and detailed facility planning [17]. In addition, remote sensing technologies are applied in urban flood modeling, environmental impact assessment, automated feature extraction for roads, urban drainage planning, three-dimensional (3D) city modeling for disaster preparedness, peri-urban area mapping and monitoring, solid waste management, identification of archeological and heritage sites, and utility GIS.

3.11 Geospatial in Wasteland Mapping

The development of degraded lands, including erosion-affected and waterlogged soils, necessitates up-to-date information on their geographical location and spatial distribution. Remote sensing has revolutionized the acquisition of such information, moving away from statistical compilation from village records toward more precise spatial assessments. Management of watersheds involves implementing development programs aimed at maximizing the utilization of natural resources within specific hydrological entities [10]. A watershed, typically defined as a region where all rainfall runoff converges to a particular channel, is crucial for planning watershed management. Remote sensing data can be utilized to map and monitor various parameters of watersheds such as stream network, physiography, land use, vegetation/forest cover, and snow cover [7]. Strategic planning for watershed development includes delineating and categorizing watersheds, prioritizing them, conducting detailed soil inventories, implementing treatments, and evaluating the impact of these treatments. This strategic approach is essential for conserving and managing water and land resources to enhance productivity. Remote sensing techniques can be applied for delineating watershed areas, characterizing watersheds, assessing priorities, identifying erosion-prone areas, mapping soil and land use/land cover, mapping drainage patterns, devising water conservation strategies, selecting sites for check dams/reservoirs and rainwater harvesting structures, and evaluating treatment impacts [10].

3.12 Geospatial in Floods/Cyclones

Regarding floods and cyclones, satellite data are currently used for near real-time flood inundation and damage assessment, post-flood river configuration analysis, preliminary flood hazard risk mapping, and flood forecasting. However, comprehensive flood monitoring requires integrating ground measurements and aerial/satellite remote sensing data into a Geographic Information System (GIS), particularly focusing on floodplain characteristics and topography [28]. For landslide and seismic hazard assessment, remote sensing data aid in preparing hazard zonation maps by incorporating information on lithology, geological structures, slope, vegetation, and land use [10]. High-resolution data are crucial for micro-seismic hazard zonation.

3.13 Geospatial in Marine Fisheries

In marine fisheries, space technology assists in identifying potential fishing areas, suitable sites for sea ranching, and monitoring harmful algal blooms and industrial effluent discharge patterns to mitigate adverse effects on vulnerable fish stocks [20].

3.14 Geospatial in Wildlife Conservation and Management

Remote sensing can assist in wildlife habitat mapping, monitoring wildlife populations, tracking animal movements, and identifying critical habitats for conservation efforts. GIS can be used to analyze spatial relationships between wildlife habitats, human activities, and protected areas, guiding conservation planning and management decisions [17].

3.15 Geospatial in Mineral Exploration and Mining

Remote sensing techniques such as hyperspectral imaging and geological mapping can aid in mineral exploration by identifying mineral deposits, mapping geological structures, and assessing terrain suitability for mining activities [16]. GIS can help in the integration and analysis of geological, geochemical, and geophysical data for effective mineral resource management [17].

3.16 Geospatial in Geomorphological Research

Geomorphology, intertwined with disciplines such as soil science, hydrology, geology, and environmental science, delves into the evolutionary processes shaping the Earth's relief forms, closely examining landforms. Its application extends to various sectors like planning, mining, and hydrology. Remote sensing and GIS represent powerful technologies for acquiring information about terrestrial objects by measuring electromagnetic radiation [23, 29]. Since the advent of early Landsat data, remotely sensed imagery has been extensively employed in geomorphology, serving the investigative needs of geomorphologists. While geomorphology focuses on describing/classifying landforms and characterizing processes, remote sensing provides crucial insights into landform distribution, surface/subsurface composition, and elevation [2]. The contemporary application of remote sensing in geomorphology is particularly influenced by advancements like the widespread availability of digital elevation models (DEMs) and innovations such as hyperspectral imaging, radiometrics, and electromagnetics. Additionally, remote sensing now offers close-range (<200 m) techniques for high-resolution imaging [30]. Remote sensing is reshaping geomorphology into a more global science, impacting environmental policy concerning geomorphological issues. For instance, the evolution of remote sensing applications in terrain analysis in British Columbia over the past 25 years serves as an instructive example. The paper further illustrates how geomorphological applications contribute to land management, resource development planning, land use planning, project planning, and policies addressing natural hazards.

3.17 Geospatial in Coastal Zone Management

Coastal areas face considerable threats from both climate change and human intervention. Remote sensing plays a pivotal role in monitoring coastal erosion, fluctuations in sea levels, and alterations in land cover [31, 32]. This data serves as a crucial foundation for implementing sustainable strategies in coastal zone management, effectively safeguarding communities and ecosystems from detrimental consequences [33, 34].

4 Future Perspectives

Looking ahead, despite the routine use of data from various earth observation systems in natural resource management, several gaps remain to be addressed. These include enhancing data accuracy and resolution for identifying short-duration and marginal crops, refining yield models, detecting crop stress factors, accessing information on soil subsurface horizons, quantifying soil erosion, estimating reservoir water depth and groundwater quality, and improving contour accuracy in micro-level watershed development plans. The future will witness an increased emphasis on integrating data from diverse sources such as satellite imagery, aerial surveys, ground-based sensors, and unmanned aerial vehicles (UAVs). This integrated approach will provide more comprehensive and current information crucial for decision-making in natural resource management. Ongoing advancements in remote sensing technologies, including higher spatial and spectral resolution sensors, extended temporal coverage, and enhanced data processing techniques, will enable more precise and detailed monitoring of natural resources and environmental changes. Integration of machine learning and artificial intelligence algorithms with remote sensing and GIS data will automate feature extraction, classification, and pattern recognition, revolutionizing data analysis capabilities at a finer scale than previously feasible. GIS software will continue evolving with more sophisticated spatial analysis tools, enabling complex analyses, dynamic process modeling, and innovative data visualization to support decision-making in natural resource management. The adoption of cloud computing platforms and big data analytics will further enhance efficiency and cost-effectiveness in storing, processing, and analyzing large volumes of remote sensing and GIS data, democratizing access to advanced analytical tools globally.

Remote sensing will expand into new domains such as biodiversity conservation, ecosystem services assessment, climate change adaptation, and urban resilience planning, necessitating interdisciplinary collaboration across sectors. Engaging citizens and local communities through citizen science initiatives and participatory mapping efforts will broaden data collection coverage, enhancing the relevance and legitimacy of management decisions. Remote sensing and GIS will play pivotal roles in developing evidence-based policies and governance frameworks for sustainable natural resource management. Decision-makers will increasingly rely on spatially explicit information to evaluate trade-offs, prioritize interventions, and monitor policy effectiveness.

5 Conclusions

As human population growth exerts increasing pressure on natural resources, remote sensing and GIS emerge as indispensable tools for effective resource management. Geospatial information aids in identifying and analyzing factors influencing resource utilization, facilitating informed decisions to ensure the sustainable use of natural resources for present and future generations. Remote sensing technology has brought about a significant transformation in the management of natural resources across a variety of sectors. Its applications range from preserving biodiversity to addressing natural calamities and promoting sustainable development, yielding diverse and meaningful impacts. Through the utilization of remote sensing capabilities, we are empowered to make well-informed decisions that foster the responsible guardianship of our planet's invaluable resources.

References

1. Matthews RB, Wamann R, Arah J (2000) Model development. Nutr Cycling Agric Ecosyst 58: 141–159
2. Sharma PK (2004) New technologies in remote sensing and GIS for spatial information generation. Second Dr. R.R. Agarwal Memorial Lecture
3. Hasnat M, Singh P (2018) A review on groundwater investigations using remote sensing in India. Disaster Adv 11(3):34–51

4. Hengl T, Heuvelink GBM, Stein A (2004) A generic framework for spatial prediction of soil variables based on regression-kriging. Geoderma 120(1–2):75–93. International Science Council (2002)
5. Burrough P (1993) Principles of geographical information systems for land resources assessment. Oxford University Press
6. Seelan SK, Laguette S, Casady GM, Seielstad GA (2003) Agriculture: a social approach to education. Journal of the Andaman Science Association, 20(1), 2015–2016. Remote Sens Environ 88(1):157–169
7. Das SN (2001) Land and soil information for basin development. All India land and land use survey technical bulletin-1/20018
8. Krishan G, Kushwaha SPS, Velmurugan A (2009) Traditional ecological knowledge of the Sahariya tribe in Central India. Indian J Tradit Knowl 8(1):49–59
9. Michaelidis A, Matas K, Zoramani I, Karamouzis D, Navalgunde RR (2002) Prerequisites for agricultural cultivation. In: Hindu survey of Indian agriculture, pp 203–208
10. Ghosh S, Sen KK, Rana U, Rao KS, Saxena KG (1996) Application of GIS in land use/land cover change analysis in mountainous areas. J Indian Soc Remote Sens 24(3):193–202
11. Batchelor WD, Basso B, Paz JO (2002) Strategies to distinguish average variances and enhance productivity over time using crop models. Eur J Agron European J Agron 18:141–158
12. Panigrahy S, Chakraborty M, Palihar JS (1996) Agricultural Research – an approach using remote sensing data. Promotion. RSAM/SAC/CAPE II/TR/61/96. ISRO Space Applications Centre, Ahmedabad, p 27
13. Feng G, Shengyue J, Gang H (2002) GIS@kev 6(11):25–26
14. Baret F, Houles V, Guérif M (2007) Measuring plant stress through remote sensing and crop models: the argument for nitrogen management. J Exp Bot 58(4):869–880
15. Bastiaanssen WGM, Menenti M, Feddes RA, Holtslag AAM (1998) A remote sensing surface energy balance algorithm for land (SEBAL): 1. Formulation. J Hydrol 212–213:198–212
16. Singh P, Hasnat M, Rao MN, Singh P (2021) Fuzzy analytical hierarchy process based GIS modelling for groundwater prospective zones in Prayagraj, India. Groundw Sustain Dev 12: 100530
17. Van Westen CJ (2000) The role of GIS in landslide hazard assessment: a review. Nat Hazards 22(2–3):137–152
18. Chakraborty M, Panigrahy S (2000) ISPRS J Photogramm Remote Sens 55(2):119–128
19. Ritchie JC, Cooper CM (1991) Remote sensing techniques for determining water quality: applications to the conservation of water resources
20. Ritchie JC, Siebe FR, Cooper CM, Ritchie JC, Siebe FR, Cooper CM, Harrington JA Jr (1994) Geostatistical analysis of soil properties in the Palouse region of Washington. Soil Sci Soc Am J 58(5):1376–1383. Harrington, Jr., J.A. (1994)
21. Philipson P, Lindell T (2003) Using remote sensing to assess water quality. Ambio 32:586–593
22. Priya S, Shibasaki R. (Year) Ekolojik Modelleme 136(2):113–129
23. Anderson JE (1999) Photogramm Eng Remote Sens 65(10):1118–1123
24. Madhavan Nair G (2004) Disaster management field: Indian perspective. 55th International Astronautical Congress, 4–8 October. Canada Vancouver
25. Melesse A, Wang X (2007) Remote sensing of impermeable surfaces, vol 19. CRC Press/Taylor and Francis, pp 369–384
26. Panigrahy S, Ray SS, Sharma PK, Sood AK, Patel LB, Parihar JS (2003) RSAM/SAC/CS/SR/04/2003. Space Applications Centre, Ahmedabad and Punjab Remote Sensing Centre, Ludhiana
27. Samant HP, Subramanyam V (1998) Land use/land cover changes and their impact on rivers and canals in Mumbai-Navi Mumbai city – a study using GIS. J Indian Soc Remote Sens 26 (1 and 2):1–6
28. Panigrahy S, Sharma SA (1997) Facility agreement utilizes several days of IRS digital data. Int Photogramm Remote Sens Assoc Photogramm J 52:85–91

29. Panigrahy S, Sharma SA (1995) Proceedings of the National Workshop on remote environment with special reference to green revolution, Ludhiana, eleven months 22–24, pp 35–39
30. Delecolle R, Maas SJ, Guérif M, Baret F (1992) ISPRS J Photogramm Remote Sens 47(2–3):145–161
31. Liu Z, Ostrenga D, Teng W, Kempler S (2012) Tropical Rainfall Measuring Mission (TRMM) precipitation data and services for research and applications. Bull Am Meterol Soc 93:1317–1325. https://doi.org/10.1175/BAMS-D-11-00152.1
32. Ford TW, Harris E, Quiring SM (2014) Estimation of basal area soil moisture using near-surface observations from SMOS. J Hydrol Earth Syst Sci 18(1):139–154. https://doi.org/10.5194/hess-18-139-2014
33. Navalgunde RR (2006) Indian earth observing system. Asian J Geogr Inf 6
34. Navalgund RR, Jayaraman V, Roy PS (2007) Curr Res 93(12):1747–1766

Carbon Sequestration for Net-Zero Achievement in Africa and Asia: A Comprehensive Explanation

John Majok Manyok Thuch and Abhishek Gupta

Abstract The escalating threat of climate change demands urgent action to achieve net-zero greenhouse gas emissions. This paper explores the critical role of carbon sequestration in this global endeavor. We are specifically evaluating the potential of afforestation vs. Engineered techniques, analyzing the economic feasibility of large-scale implementation, or examining the social and environmental impacts of various methods. Through comparative analysis of existing data, and case studies of specific projects especially from Africa and Asia, we uncover that more effort is still needed in these continents since both continents have growing economies and populations, leading to rising carbon dioxide emissions, they are particularly vulnerable to climate change impacts and investing in carbon sequestration can offer economic opportunities (e.g., new jobs, green technologies), support sustainable development, and protect vulnerable communities. Our findings illuminate identifying promising pathways, informing policy decisions, and suggesting areas for further research and development. By examining the potential and challenges of carbon sequestration, this paper enhances the understanding of this vital strategy in combating climate change.

Keywords Carbon sequestration · Sustainable development · Climate resilience and mitigation · Net-zero emissions

1 Introduction

Achieving net-zero emissions is imperative within global endeavor to address issue associated with climate change and secure a sustainable future [1]. The 2015 Paris Agreement has established the objective of capping global warming at below 2 degrees Celsius, with the aim of constraining it to 1.5 degrees Celsius. Net-zero emissions play a vital role in balancing greenhouse gas (GHG) levels, mitigating climate change, fostering sustainability, driving economic transformation, safeguarding public health,

J. M. M. Thuch · A. Gupta (✉)
Department of Chemical Engineering, Marwadi University, Rajkot, Gujarat, India
e-mail: abhishek.gupta@marwadieducation.edu.in

P. Pathak et al. (eds.), *Advances in Environmental Sustainability, Energy and Earth Science*, https://doi.org/10.1007/978-3-031-73820-3_16

Table 1 Challenges faced by Africa and Asia in reaching net-zero [4]

Challenges and barriers	Africa	Asia
1. Land use and land rights	• Land tenure issues, competing land uses for agriculture and forestry • Lack of clear land ownership structures	• Intensive land use for agriculture and urbanization • Land degradation, conflicts over land rights and land-use change
2. Financial resources	• Limited access to funding for carbon sequestration projects • Lack of financial incentives for landowners and communities	• Unequal distribution of financial resources • Lack of investment in sustainable land management practices • Limited access to climate finance
3. Socio-economic factors	• Limited awareness and capacity among local communities • Cultural barriers to adopting new practices • Lack of community participation in decision-making processes	• Rapid urbanization and industrialization leading to increased emissions • Social inequalities impacting access to resources and benefits from carbon sequestration projects • Need for inclusive and equitable development approaches

and preserving biodiversity [2]. These objectives correspond with sustainable development targets and advocate for the integration of renewable energy sources [3]. Several nations have committed to achieving net-zero emissions as part of their obligations under the Paris Agreement, underscoring the importance of international collaboration, governmental strategies, corporate investments, and individual efforts to achieve this ambitious target (Table 1).

To assess the carbon sequestration potential in Africa and Asia, various methods like afforestation, reforestation, soil carbon storage, and sustainable land management practices are crucial [4, 5]. Understanding the regional challenges and opportunities in these continents is essential to identify barriers and opportunities for scaling up carbon sequestration efforts, considering socio-economic, environmental, and policy contexts. Enhanced carbon sequestration can play a crucial role in lowering greenhouse gas emissions and attaining net-zero targets in Africa and Asia., emphasizing the importance of impact assessment [6]. Evaluating co-benefits and trade-offs associated with carbon sequestration initiatives is vital, including impacts on biodiversity conservation, water resources, food security, and rural livelihoods. Analyzing existing policies, regulatory frameworks, and institutional capacities related to carbon sequestration in African and Asian countries is crucial for enhancing policy coherence and implementation, highlighting the significance of policy and institutional analysis. Community engagement and stakeholder consultation are vital for ensuring social inclusiveness and equity in carbon sequestration projects, highlighting the need to involve local communities, indigenous peoples, and civil society organizations in decision-making processes.

The research highlights the potential for carbon sequestration through agricultural lands, emphasizing the need for high-resolution soil maps and field data to improve estimates [7]. Additionally, it underscores the significance of land-use changes, climate conditions, and nutrient availability in determining the carbon sequestration potential of different land uses, emphasizing the importance of sustainable land management practices for effective carbon storage. The study also emphasizes geographical, temporal, methodological, sectoral, and cross-cutting themes in analyzing carbon sequestration across Africa and Asia, promoting collaboration with various stakeholders.

2 Contextualizing Global Climate Change: Africa and Asia's Role

Global climate change is a multifaceted issue impacting ecosystems, weather patterns, and natural disasters worldwide, with developing regions like Africa and Asia bearing a disproportionate burden due to their reliance on climate-sensitive sectors such as agriculture, fisheries, and forestry. The interplay of environmental, social, economic, and political factors heightens the vulnerability of these regions to climate change impacts, such as more frequent and severe heatwaves, storms, droughts, floods, and disruptions to agriculture and water resources [8]. The lack of resources and infrastructure in Africa and Asia hinders their ability to adapt, leading to increased socio-economic disparities. Addressing the unique challenges faced by these regions is crucial to mitigating the adverse effects of climate change and promoting resilience in the face of environmental shifts [8–10].

Africa and Asia are pivotal in the global efforts to address climate change through various mitigation and adaptation strategies. These continents are increasingly embracing renewable energy technologies and enacting climate-friendly policies such as carbon sequestration to reduce greenhouse gas emissions and transition toward low-carbon economies [11]. They are also investing in climate-resilient infrastructure, sustainable land management practices, and early warning systems to enhance resilience to climate change impacts [12]. Moreover, African and Asian nations are actively involved in international climate negotiations, advocating for fair solutions, and seeking financial and technological assistance from developed countries to combat climate challenges [13]. Leveraging their rich biodiversity and natural resources, Africa and Asia are prioritizing conservation, sustainable agriculture, and ecosystem restoration to mitigate impacts due to climate change [14, 15].

Global efforts to mitigate climate change hinge crucially on commitments to achieve net-zero emissions. Research underscores the urgent need for rapid greenhouse gas reductions to reach net-zero status. This goal necessitates a profound transformation across socio-technical-political systems, highlighting the complex interdependencies involved in achieving sustainable climate action [16]. Achieving net-zero emissions is anticipated to stabilize global temperatures, yet regions may still face warming and extreme events post-net-zero, requiring adaptation strategies [17]. Transitioning to net-zero emissions poses challenges, especially for countries with limited domestic mitigation options like Switzerland, which require significant investments in low-carbon technologies and renewable energy sources. The transition to net-zero emissions will impact traditional chemical feedstocks and catalytic elements, creating opportunities for emerging feedstocks and altering supply dynamics of critical metals [18]. International agreements and collaborative efforts are essential to ensure a smooth transition toward a net-zero future. Some of the key commitments and agreements include Paris Agreement, Race to Zero campaign, G7 and G20 Commitments, International Climate Summits, and Net Zero Alliance and Coalitions.

Failing to achieve net-zero goals poses significant implications for the environment, society, and economy. From a technical perspective, transitioning to net-zero carbon emissions requires substantial investments in low-carbon technologies, energy efficiency, and renewable sources [19]. However, the economic impacts of such a transition can be substantial, with potential GDP losses of up to 8.5% in 2050 for countries like Thailand [20]. Moreover, the health sector, a major greenhouse gas emitter, must also decarbonize to mitigate climate risks and reduce its environmental impact [21]. Collaborative efforts between researchers, local governments, and industries are crucial for successful implementation of sustainable energy systems to meet net-zero targets. Overall, failing to achieve net-zero goals not only threatens environmental sustainability but also poses economic challenges and health risks, demanding immediate attention.

3 Significance and Strategies for Achieving Net-Zero in Africa and Asia

Achieving net-zero emissions in these two continents is of paramount significance due to several key factors

a. Global Climate Impact: Africa and Asia are highly vulnerable regions to the impacts of climate change, including extreme weather events, sea-level rise, and disruptions to ecosystems and livelihoods. Achieving net-zero emissions is essential for mitigating the adverse effects of climate change and enhancing the resilience of communities and ecosystems.
b. Population and Economic Growth: Africa and Asia are home to a significant portion of the world's population, with rapidly growing economies and energy demand. Transitioning to net-zero emissions is critical for managing the environmental footprint of population growth, urbanization, industrialization, and economic development while ensuring sustainable and inclusive growth.
c. Biodiversity Conservation: The rich biodiversity and ecosystems of Africa and Asia are under threat from deforestation, habitat loss, and degradation. Carbon sequestration initiatives play a vital role in conserving forests, protecting biodiversity hotspots, and restoring degraded ecosystems, thereby safeguarding vital habitats and preserving biological diversity.

d. Public Health and Air Quality: Addressing air pollution and improving public health are urgent priorities in many countries in Africa and Asia. Transitioning to clean and renewable energy sources, reducing fossil fuel combustion, and promoting sustainable transportation can mitigate air pollution-related health risks and improve the well-being of communities [21].
e. Sustainable Development Goals (SDGs): Achieving net-zero emissions aligns with several Sustainable Development Goals (SDGs), including Goal 13 (Climate Action), Goal 7 (Affordable and Clean Energy), Goal 11 (Sustainable Cities and Communities), and Goal 15 (Life on Land). Net-zero strategies not only support poverty reduction and enhance food security but also promote access to clean energy and the development of resilient infrastructure. By advancing these objectives, net-zero initiatives play a pivotal role in accelerating progress toward the broader SDGs, ensuring sustainable development and environmental stewardship for future generations.
f. International Commitments: Africa and Asia are signatories to international agreements like the Paris Agreement, which seeks to cap global warming below 2 °C and strive for 1.5 °C above pre-industrial levels. Achieving net-zero emissions is critical for meeting these obligations and mitigating the severe consequences of climate change. By committing to net-zero targets, these regions aim to uphold their international pledges, safeguarding against the worst effects of climate change and promoting global sustainability efforts.

In order to achieve net-zero goals, a multifaceted approach is essential. Utilities must focus on decarbonizing electricity production, transitioning away from fossil fuels, and investing in other operational areas to reduce emissions [22]. Countries should set specific carbon reduction targets based on their capabilities and developmental stages, utilizing methods like inverse Data Envelopment Analysis to allocate these targets efficiently. Additionally, nature-based carbon sequestration through oceans, soils, and forests presents a viable solution to extract and store carbon, with remote sensing playing a crucial role in monitoring these carbon pools [23]. Forest management strategies with high sequestration capacity can reduce the need for costly negative emission technologies, emphasizing the importance of establishing effective forest management before investing in bioenergy with carbon capture and storage. Ultimately, converging human carbon emissions toward net zero is crucial for stabilizing the global climate, with potential for ocean-based carbon dioxide removal technologies for large-scale carbon sequestration [24].

4 Carbon Sequestration as a Solution

Carbon sequestration involves capturing and storing carbon to mitigate climate change. Various methods like agricultural practices, flue gas capture, and geologic sequestration are employed for this purpose. Factors influencing carbon sequestration include tree species, human disturbance intensity, and environmental conditions. The sequestered carbon can be stored in soil, vegetation, or engineered bio landfills. Carbon sequestration helps in reducing GHG emissions along with this it also offers socio-economic benefits and aids in climate change mitigation. Overall, understanding carbon sequestration is essential for formulating effective strategies to address climate change and advance sustainable development.

It plays a crucial role in mitigating climate change by extracting CO_2 from the air and storing it in various carbon sinks like forests, soil, and oceans for extended periods [25]. Forests, as highlighted in by Rodrigues [26] and Lin [27], act as significant carbon sinks, with their aboveground carbon sequestration rate being influenced by climate change and forest management practices. Soil, as emphasized in [28], is the largest terrestrial carbon store, with key factors affecting its carbon trapping capacity. Additionally, innovative approaches like Agro-Sequestration, as proposed by Helferty [29] and Yablonovitch [30], offer cost-effective and scalable solutions for long-term carbon sequestration by burying salted biomass in dry environments. These diverse strategies collectively contribute to reducing atmospheric carbon dioxide levels, thereby helping mitigate global warming and its adverse effects [31].

Processes like photosynthesis, ocean uptake, and soil carbon storage naturally remove CO_2 from the air and trap the same in terrestrial and aquatic ecosystems [32, 33]. Trees and other vegetation absorb CO_2 during photosynthesis, storing carbon in their biomass and soils. The oceans also act as a carbon sink, absorbing carbon dioxide from the atmosphere and transferring it to deep ocean layers. Artificial or anthropogenic carbon sequestration involves deliberate human interventions to enhance this process, using various techniques such as afforestation, reforestation, soil carbon sequestration, and carbon capture and storage (CCS) technologies [32, 34].

Different approaches for carbon sequestration include

- **Afforestation and Reforestation:** Afforestation refers to the process of planting trees on land that was not previously forested. This approach aims to create new forests where none existed before, thereby increasing the overall carbon uptake capacity of the area. Reforestation, on the other hand, involves restoring forests on land that has been deforested or degraded. By replanting trees in these areas, the goal is to rebuild forest ecosystems and enhance their ability to absorb carbon dioxide from the atmosphere, contributing to efforts aimed at mitigating climate change [32].
- **Soil Carbon Sequestration:** Soil carbon sequestration involves enhancing carbon storage in soils through agricultural methods like conservation tillage, cover cropping, crop rotation, and organic soil amendments. This process is crucial for mitigating climate change by reducing atmospheric CO_2 levels and enhancing soil fertility [35–37]. Various agricultural practices, including tillage, erosion control, nutrient management, and agroforestry, contribute to carbon sequestration by stabilizing carbon in soil through biotic and abiotic mechanisms [38]. Studies have shown that implementing practices like growing cover crops, conservation tillage, and proper nutrient management can significantly increase

soil carbon levels, thereby improving crop yields and soil health. Additionally, research indicates that the interaction between nematodes and bacterial communities plays a vital role in regulating soil carbon accumulation, highlighting the importance of understanding these relationships for effective carbon sequestration.

- **Carbon Capture and Storage (CCS):** CCS technologies are used in capturing CO_2 emissions from various point sources like industrial processes and power plants [39]. These technologies aim to prevent the release of CO_2, thereby aiding to global warming mitigation efforts. Different approaches such as advanced CO_2 capture technologies, electrochemical processes, and integrated systems like bioenergy with CCS are being explored to enhance the efficiency and versatility of CCS systems. Media analyses have shown evolving societal perceptions of CCS, highlighting both support and opposition based on factors like cost, technology maturity, and availability of alternative solutions. Overall, the development and deployment of CCS technologies are essential for achieving emission reduction targets and combating climate change effectively.
- **Ocean Afforestation:** Ocean afforestation, also known as marine afforestation or ocean reforestation, involves restoring and enhancing marine ecosystems such as mangroves, kelp forests, and seagrass meadows. These coastal and marine habitats act as blue carbon sinks, capturing carbon from the air and storing it in biomass and sediments.
- **Enhanced Weathering**: Enhanced weathering involves accelerating natural weathering processes, such as the dissolution of silicate rocks, to remove carbon dioxide from the atmosphere. This method increases the rate of carbon dioxide uptake by minerals, converting atmospheric carbon dioxide into stable carbonates that are stored in soils and sediments.

5 Technological Advancements in CCS

Technological advancements in the field of CCS have significantly progressed, offering diverse solutions for carbon mitigation. Recent studies have highlighted the effectiveness of molecular catalysts like cobalt (II) phthalocyanine in electrochemical reduction of CO_2 to valuable products like carbon monoxide (CO) [40]. Additionally, direct air capture of CO_2 with storage (DACCS) has emerged as a novel CCS technology, enhancing feasibility and fairness in achieving climate targets [41, 42]. These developments, coupled with ongoing research on CO_2 capture, utilization, and geological storage technologies, underscore the crucial role of CCS in achieving net-zero emissions [43].

Here are some notable advancements:

- Advanced Capture Technologies

Pre-Combustion Capture: These technologies involve capturing carbon dioxide from fuel gases before combustion. Integrated gasification combined cycle (IGCC) plants and chemical looping combustion (CLC) systems are examples of advanced pre-combustion capture processes that offer higher efficiency and lower emissions compared to traditional methods.

Post-Combustion Capture: Traditional post-combustion capture technologies involve capturing carbon dioxide from flue gases after combustion. Advanced solvent-based systems and solid adsorbents have been developed to enhance carbon dioxide capture efficiency and reduce energy consumption and operating costs.

- Membrane Separation Technologies

Membrane-based carbon dioxide separation technologies offer a promising alternative to traditional solvent-based capture methods. Advancements in membrane materials, such as polymer, ceramic, and mixed matrix membranes, have led to improved selectivity, permeability, and durability, making membrane separation more cost-effective and energy-efficient for carbon dioxide capture applications.

- Direct Air Capture (DAC)

DAC technologies directly capture carbon dioxide from ambient air using chemical processes or sorbent materials. Innovations in DAC systems, including novel sorbents, enhanced process designs, and renewable energy integration, have made DAC more viable for large-scale carbon dioxide removal and utilization applications, such as carbon removal and negative emissions technologies.

- Carbon Capture, Utilization, and Storage (CCUS)

CCUS technologies integrate carbon capture with utilization and storage options, such as enhanced oil recovery (EOR), mineralization, and carbon utilization in products and materials. Advances in carbon dioxide utilization pathways, catalyst development, and process optimization have expanded the range of economically viable CCUS applications, enabling carbon-negative solutions and value-added products from captured carbon dioxide.

- Carbon Storage Monitoring and Verification

Advanced monitoring, measurement, and verification (MMV) technologies are essential for ensuring the safe and secure storage of captured CO_2 in geological formations. Innovations in seismic imaging, remote sensing, geochemical analysis, and modeling techniques have improved our ability to accurately monitor carbon dioxide plumes, detect leakage risks, and assess long-term storage integrity.

- Cost Reduction and Scale-Up

Continuous research, development, and demonstration (RD&D) efforts, coupled with policy support and private sector investments, have led to significant cost reductions and scale-up of CCS technologies. Technological learning, standardization, modularization, and deployment incentives have helped drive down capital and operating costs, making CCS more economically competitive and commercially viable for diverse industrial applications.

6 Initiatives for Carbon Sequestration in Africa and Asia

Current carbon sequestration initiatives in Africa and Asia focus on diverse strategies to reduce GHG emissions and enhance carbon storage. In Africa, efforts include ecological restoration projects in karst regions, leading to increased vegetation cover and carbon stocks [44]. Additionally, agricultural land use systems play a crucial role in carbon sequestration, with potential strategies like erosion control, conservation tillage, and biofuel production contributing to enhancing soil organic carbon pools. In Asia, Integrated Crop-Livestock-Forestry (ICLF) systems are being explored to sequester carbon through practices like intercropping and tree biomass utilization, although challenges like technical complexity and high initial investments exist [45]. These initiatives underscore the importance of holistic approaches to carbon sequestration in terrestrial ecosystems across Africa and Asia.

Here is an overview of some current initiatives in these regions:

- **Community-Based Initiatives:** Community-based carbon sequestration initiatives, such as community forestry programs, participatory watershed management, and community-managed conservation areas are empowering local communities to take ownership of natural resource management and carbon sequestration efforts. These initiatives promote sustainable livelihoods, biodiversity conservation, and climate resilience at the grassroots level.
- **International Partnerships and Financing:** International organizations, development agencies, and donor institutions are providing technical assistance, capacity building, and financing support to scale up carbon sequestration initiatives in Africa and Asia. Partnerships such as the Africa Climate Change Fund, the Asia-Pacific Network for Global Change Research, and the Global Environment Facility are facilitating knowledge exchange and collaboration to address climate change challenges in the region.

7 Assessing Carbon Sequestration Capacity in Africa and Asia

Assessing the natural capacity for carbon sequestration in Africa and Asia involves considering multiple factors. Forest management, afforestation, and ecological restoration projects significantly impact carbon sequestration in vegetation [46, 47]. The vegetation carbon sequestration capacity varies across regions due to spatial heterogeneity, with forest tending showing higher effectiveness in certain areas [48]. Additionally, the impact of land-use changes associated with ecological restoration projects on net ecosystem productivity is substantial, contributing significantly to carbon sequestration trends in regions like southwest China. Understanding these factors is crucial for evaluating the natural capacity for carbon sequestration in Africa and Asia, guiding targeted land management and policy decisions to enhance carbon storage in terrestrial ecosystems.

Identifying key ecosystems and land use practices for effective carbon sequestration in Africa and Asia is crucial for mitigating global warming and enhancing sustainability [49]. Several strategies can be utilized, including conservation and reduced tillage, afforestation, peatland restoration, water conservation, and urban forest development. Additionally, the protection of fine roots and active nitrogen input have been highlighted as essential processes for soil carbon and nitrogen sequestration following land-use changes. It is important to note that soil holds twice as much carbon as the atmosphere and vegetation combined, making it a significant terrestrial carbon sink. By implementing comprehensive approaches that include biochar application, crop rotation, no-tillage, and reforestation, the potential for effective carbon sequestration in diverse landscapes can be maximized.

8 Challenges and Solutions

Addressing challenges and barriers to effective carbon sequestration in Africa and Asia involves considering various factors, refer Table 2. Socio-economic challenges include the need to enhance public and private finance for initiatives like REDD+ [50], as well as the significance of local community participation in forest management [51]. Political challenges encompass the necessity for robust policies and governance frameworks to support forest management planning and ecosystem functions [52]. Environmental challenges involve methodological, socio-economic, and implementation barriers in forest carbon offset projects, emphasizing the importance of addressing issues like additionality and leakage [53]. Overall, a comprehensive approach integrating socio-economic, political, and environmental considerations is crucial for successful carbon sequestration efforts in Africa and Asia [54].

To address challenges in carbon sequestration projects, strategies such as rigorous risk assessment of well integrity, use of mixed matrix membranes (MMMs) for enhanced gas separation, and learning from forest carbon offset (FCO) projects can

Table 2 Obstacles to Carbon sequestration in Africa and Asia [52, 54]

Types of challenges	Challenges		
Socio-economic challenges	Poverty and livelihoods	Access to resources and technology	Land tenure and property rights
Political challenges	Weak governance and institutional capacity	Policy inconsistencies and fragmentation	Lack of political will and commitment
Environmental challenges	Deforestation and land degradation	Biodiversity loss	Climate change impacts
Other challenges	Public acceptance and stakeholder engagement	Energy requirements and efficiency	Limited access to carbon markets

be adopted. The risk-management methodology for well integrity evaluation [55], development of MMMs with inorganic fillers for improved permeability and selectivity, and understanding challenges in FCO projects like additionality, permanence, and leakage is crucial [56]. Additionally, implementing decision-making models for forest species to maximize carbon sequestration without environmental harm can provide valuable insights for sustainable carbon sequestration practices. By integrating these strategies, carbon sequestration projects can enhance their efficiency and long-term viability [7, 57].

These identified challenges can be overcome through applying following strategies:

- Stakeholder Collaboration and Partnerships

Foster collaboration and partnerships among diverse stakeholders, including government agencies, NGOs, academia, private sector entities, and local communities. Create multi-stakeholder platforms for dialogue, coordination, and knowledge exchange to leverage resources, expertise, and networks for effective project implementation.

- Innovative Financing Mechanisms

Explore innovative financing mechanisms to mobilize resources and investment for carbon sequestration projects. This could include public–private partnerships, carbon finance mechanisms, green bonds, impact investing, and crowdfunding platforms. Develop financial incentives, subsidies, and grants to support community-led initiatives and small-scale projects.

- Policy and Regulatory Support

Advocate for supportive policy frameworks, incentives, and regulations that facilitate the implementation of carbon sequestration projects. Ensure alignment with national development priorities, climate action plans, and international commitments. Address legal and institutional barriers, streamline permitting processes, and provide clarity on land tenure rights and resource access.

- Technology Transfer and Innovation

Facilitate technology transfer and innovation to improve the efficiency, effectiveness, and scalability of carbon sequestration techniques. Promote the adoption of climate-smart technologies, agroforestry practices, and sustainable land management approaches. Invest in research and development to advance carbon capture and storage (CCS) technologies and nature-based solutions.

- Social Inclusion and Equity

Prioritize social inclusion and equity in project design and implementation to address the needs and interests of marginalized groups, indigenous peoples, and vulnerable populations. Ensure gender mainstreaming, cultural sensitivity, and respect for traditional knowledge and practices.

- Risk Management and Resilience

Identify and mitigate potential risks and uncertainties associated with carbon sequestration projects, including climate variability, market fluctuations, and socio-political instability. Develop risk management strategies, contingency plans, and insurance mechanisms to enhance project resilience and adaptive capacity.

9 Carbon Sequestration Projects in Africa and Asia: Case Studies

a) Great Green Wall Initiative (Africa)

Launched in 2007 by the African Union, the Great Green Wall initiative aims to rehabilitate degraded landscapes across 22 African countries in the Sahel region. This ambitious project focuses on transforming lives by revitalizing communities and is supported by over USD 8 billion in funding. The initiative aligns with the principles of green growth, emphasizing environmental productivity and policy responses to address climate change challenges. Green aid flows to developing countries have shown to significantly reduce carbon dioxide emissions, especially when targeted toward environmental objectives, emphasizing the importance of environmental aid in sustainability efforts. The initiative's focus on green infrastructure, such as green façade systems, can contribute to urban biodiversity and microclimatic improvements, showcasing the potential benefits of integrating nature into urban spaces [58].

b) Miyawaki Forest Restoration Projects (Asia)

Miyawaki afforestation, a method introduced by Professor Akira Miyawaki in the 1980s, focuses on restoring indigenous ecosystems and maintaining global environments through the use of native plants. This approach aids in disaster prevention and carbon dioxide mitigation [59]. Additionally, agroforestry practices integrate trees with arable crops, livestock, or fodder, enhancing soil fertility and providing ecological services like erosion control and biodiversity promotion [60]. Furthermore, the establishment of forest-medicine systems can improve soil quality and enhance microbial communities, benefiting the overall forest environment [61]. The development of forest measurement systems using mobile phones allows for accurate spatial structural assessments, aiding in forest management and optimization [60, 62]. These combined approaches underscore the importance of sustainable forest management for ecosystem restoration and environmental conservation [63].

Lessons learned from carbon sequestration initiatives highlight crucial factors influencing project outcomes. Successful projects emphasize the importance of tailored carbon-farming programs based on local conditions [64]. In contrast, unsuccessful initiatives underscore the challenges associated with large uncertainty estimates, particularly related to methane and nitrous oxide emissions, and the need for in situ sampling for accurate measurements [65]. Furthermore, lessons from corporate learning stress the necessity of integrating Lessons Learned processes into daily operations to drive continuous improvement and innovation, ensuring knowledge transfer and employee empowerment. Technical insights from Carbon Capture and Storage projects in Asia Pacific emphasize the complexities of CO_2 storage, including careful engineering to maintain desired phases and considerations like geothermal gradients for site selection and dynamic modeling for accurate capacity estimation [66].

10 Role of Policy and Governance

Africa and Asia have made significant strides in developing climate and environmental policies to address the challenges of climate change, biodiversity loss, deforestation, and environmental degradation.

Here is an overview of existing policies in these regions

- African Union Agenda 2063

The African Union Agenda 2063 provides a framework for achieving inclusive and sustainable development across the continent. It includes aspirations related to climate resilience, sustainable land use, renewable energy, and environmental conservation.

- African Ministerial Conference on the Environment (AMCEN)

AMCEN facilitates regional cooperation and policy dialogue on environmental issues, including climate change adaptation and mitigation, biodiversity conservation, desertification, and pollution control. It promotes the implementation of regional and international environmental agreements.

- African Renewable Energy Initiative (AREI)

AREI aims to promote renewable energy development and access in Africa, with a focus on scaling up investments in clean energy technologies, enhancing energy security, and reducing greenhouse gas emissions thereby aligning with the objectives of the Paris Agreement and the SDGs.

- African Forest Landscape Restoration Initiative (AFR100)

AFR100 is a regional initiative with the goal of restoring 100 million hectares of degraded land in Africa by 2030. It mobilizes governments, civil society, and private sector partners to undertake projects for restoring forest landscapes, preserving biodiversity, and enhancing livelihoods. The initiative focuses on collaborative efforts to combat land degradation through sustainable practices. AFR100 aims not only to restore ecosystems but also to promote resilience against climate change impacts, fostering socio-economic benefits for local communities. By integrating diverse stakeholders, it seeks to achieve long-term environmental sustainability and improve the quality of life in affected regions across the continent.

- National Climate Change Policies and Strategies

Many African countries have developed national climate change policies, strategies, and action plans to address mitigation, adaptation, and resilience-building efforts. These policies outline national priorities, targets, and measures for climate action, including renewable energy development, climate-smart agriculture, and ecosystem-based adaptation.

- Asia-Pacific Climate Action Partnership (APCAP)

APCAP fosters regional collaboration and facilitates knowledge exchange among Asia-Pacific countries on climate change mitigation and adaptation. It promotes collective initiatives, builds capacity, and facilitates policy dialogue to advance the objectives of the Paris Agreement and the 2030 Agenda for Sustainable Development. By supporting joint efforts and enhancing resilience to climate impacts, APCAP aims to achieve sustainable development goals across the region. Through partnerships with governments, civil society, and the private sector, it strives to foster innovation and build adaptive capacity, ensuring a coordinated approach to address climate challenges and promote environmental stewardship in Asia-Pacific nations.

- ASEAN Agreement on Transboundary Haze Pollution

This agreement addresses the cross-border impacts of land and forest fires in Southeast Asia. It seeks to prevent and alleviate transboundary haze pollution by fostering regional collaboration, sharing information, and undertaking collective actions to tackle the root causes of fires. The agreement emphasizes joint efforts among member states to enhance fire prevention measures, promote sustainable land management practices, and strengthen early warning systems. By addressing the environmental and socio-economic impacts of haze, it aims to safeguard public health, biodiversity, and regional stability, reinforcing a cooperative approach to environmental governance across ASEAN nations.

- Forest Conservation and Restoration Initiatives

Several countries in Asia have launched forest conservation and restoration initiatives to address deforestation, forest degradation, and biodiversity loss. These initiatives include afforestation/reforestation programs, protected area management, community forestry projects, and sustainable land use planning to conserve and restore forest ecosystems.

- National Climate Adaptation Plans

Asian countries are developing national climate adaptation plans and strategies to address the impacts of climate change on vulnerable communities, ecosystems, and economic sectors. These plans integrate climate risk assessments, early warning systems, resilient infrastructure development, and community-based adaptation measures to enhance adaptive capacity and reduce climate vulnerability.

11 Policy Recommendations for Governments and Stakeholders

Building a roadmap for achieving net-zero emissions in Africa and Asia requires a comprehensive and integrated approach that addresses key sectors, challenges, and opportunities. Here is a structured roadmap outlining the steps and strategies

- Assess Current Emissions and Trends
  Conduct a comprehensive assessment of current greenhouse gas emissions across key sectors, including energy, industry, transportation, agriculture, forestry, and waste. Analyze emission trends, drivers, and sources to identify priority areas for mitigation and sequestration efforts.
- Set Ambitious Targets and Timelines
  Establish ambitious and science-based targets for achieving net-zero emissions in Africa and Asia, taking into account national circumstances, development priorities, and global climate goals. Set clear timelines and milestones for emissions reductions, carbon neutrality, and transition to renewable energy sources.
- Develop Sectoral Action Plans
  Develop sector-specific action plans outlining mitigation and adaptation measures for each sector. Prioritize low-carbon technologies, energy efficiency improvements, sustainable land management practices, and nature-based solutions that contribute to emissions reductions and carbon sequestration.
- Promote Sustainable Land Use and Forest Restoration
  Promote sustainable land use practices, such as afforestation, reforestation, agroforestry, and sustainable agriculture, to enhance carbon sequestration and biodiversity conservation. Restore degraded ecosystems, conserve natural habitats, and protect biodiversity hotspots to maximize carbon sinks and ecosystem resilience.
- Transition to Low-Carbon Transportation
  Transition to low-carbon transportation systems by promoting public transit, electrification of vehicles, fuel efficiency standards, and alternative fuels. Invest in clean and efficient transport infrastructure, promote active transportation modes, and encourage behavioral changes to reduce emissions from the transport sector.
- Strengthen Climate Resilience and Adaptation

Enhance climate resilience and adaptation efforts to mitigate climate change impacts on vulnerable communities, ecosystems, and infrastructure. Implement strategies based on nature, develop resilient infrastructure, and improve early warning systems and disaster preparedness mechanisms.

- Foster Innovation and Technology Transfer

Foster innovation, research, and development of clean technologies and climate solutions suited to the African and Asian context. Facilitate technology transfer, capacity building, and South-South cooperation to accelerate the adoption of low-carbon technologies and practices.

- Mobilize Financing and Investment
  Mobilize domestic and international financing and investment for climate action, including public funding, private sector investment, and climate finance mechanisms. Develop innovative financing instruments, green bonds, and carbon pricing mechanisms to incentivize low-carbon investments and scale up climate finance flows.
- Strengthen Governance and Institutional Capacity
  Strengthen governance structures, institutions, and policy frameworks to support the implementation of climate action plans. Enhance coordination, collaboration, and stakeholder engagement at national, regional, and local levels. Build institutional capacity, establish regulatory frameworks, and mainstream climate considerations into decision-making processes.
- Monitor, Evaluate, and Review Progress
  Establish robust monitoring, evaluation, and reporting mechanisms to track progress toward net-zero emissions goals. Regularly review and update climate action plans based on evolving scientific knowledge, technological advancements, and changing socio-economic conditions. Foster transparency, accountability, and learning to ensure effective implementation and continuous improvement.

A call to action for achieving net-zero emissions in Africa and Asia necessitates collaborative efforts from various stakeholders, including governments, policymakers, businesses, civil society, and individuals, focusing on key sectors. These sectors include transitioning the energy sector away from coal and other fossil fuels, promoting low-carbon mobility and logistics, and leveraging international trade and investment to facilitate the transition to a low-carbon future [67]. The urgency for substantial and sustained reductions in greenhouse gas emissions is emphasized by the IPCC, with global temperatures projected to exceed 1.5 °C by 2041 unless significant measures are taken. To achieve carbon neutrality by mid-century, countries are developing long-term strategies, such as the Net-Zero-2050 System, to guide their efforts systematically and comprehensively [68]. Agriculture, a significant contributor to global emissions, can adopt technologies and innovations to reduce emissions by up to 45%, requiring additional measures like carbon dioxide removal technologies to achieve net-zero status [69, 70].

12 Conclusions

Achieving net-zero emissions in Africa and Asia requires urgent action to mitigate greenhouse gas emissions and enhance resilience. These regions offer significant potential for carbon sequestration through natural and anthropogenic processes like afforestation and renewable energy deployment. Engaging local communities in carbon sequestration projects is crucial for project acceptance and sustainability. Effective policy frameworks and governance structures are essential for driving climate action and transitioning to net-zero emissions. Despite facing challenges like socio-economic and technological barriers, collaboration among governments, businesses, and civil society is vital to overcome obstacles and achieve net-zero emissions. Prioritizing equity, innovation, and collaboration can accelerate the transition to net-zero emissions, contributing to global climate efforts and sustainable development in Africa and Asia.

References

1. Rosa L, Gabrielli P (2023) Achieving net-zero emissions in agriculture: a review. Environ Res Lett 18(6):063002
2. Gabrielli P, Rosa L, Gazzani M, Meys R, Bardow A, Mazzotti M, Sansavini G (2023) Net-zero emissions chemical industry in a world of limited resources. One Earth 6(6):682–704
3. Hasan L, Lizarralde G, Lachapelle E (2023, June) The problem with net zero building policy: reflecting on the making, and future, of a global discourse. In: IOP conference series: earth and environmental science, vol 1196, No. 1. IOP Publishing, p 012002
4. Saeed S, Makhdum MSA, Anwar S, Yaseen MR (2023) Climate change vulnerability, adaptation, and feedback hypothesis: a comparison of lower-middle, upper-middle, and high-income countries. Sustain For 15(5):4145
5. Lv Y, Zhang L, Li P, He H, Ren X, Zhang M (2023) Ecological restoration projects enhanced terrestrial carbon sequestration in the karst region of Southwest China. Front Ecol Evol 11: 1179608
6. Ewing PM, Tu X, Runck BC, Nord A, Chikowo R, Snapp SS (2023) Smallholder farms have and can store more carbon than previously estimated. Glob Chang Biol 29(6):1471–1483
7. Amissah E, Adjei-Gyapong T, Antwi-Agyei P, Asamoah E, Abaidoo RC, Jeppesen E, Andersen MN, Baidoo E (2023) Implications of changes in land use on soil and biomass carbon sequestration: a case study from the Owabi reservoir catchment in Ghana. Carbon Manag 14(1): 1–10
8. Lamptey B, Kolli RK, Amo-Boateng M (2023) Climate change information for regional impact and risk assessment. Front Environ Sci 11:1207320
9. Khalid AM, Okitasari M (2023) Accelerating climate action in Africa: insights from the 2022 voluntary National Reviews
10. Atwoli L, Erhabor GE, Gbakima AA, Haileamlak A, Ntumba JMK, Kigera J, Laybourn-Langton L, Mash R, Muhia J, Mulaudzi FM, Zielinski C (2022) COP27 climate change conference: urgent action needed for Africa and the world. Lancet Oncol 23(12):1486–1488
11. Elsen PR, Oakes LE, Cross MS, DeGemmis A, Watson JE, Cooke HA, Darling ES, Jones KR, Kretser HE, Mendez M, Grantham HS (2023) Priorities for embedding ecological integrity in climate adaptation policy and practice. One Earth 6(6):632–644
12. Villarreal-Rosas J, Rhodes JR, Sonter LJ, Possingham HP, Vogl AL (2023) Optimal allocation of nature-based solutions to achieve climate mitigation and adaptation goals. People Nat 5(3): 1034–1045
13. Rasheed N, Khan D, Gul A, Magda R (2023) Impact assessment of climate mitigation finance on climate change in South Asia. Sustain For 15(8):6429
14. Abraham Mahanga Tsoni CG, Massonini Ngoma RGT, Meng X (2022) Innovation and climate Change mitigation Technology in the Asian and African mining sector: empirical analysis using the LMDI method. Energies 15(24):9424
15. Suraci JP, Farwell LS, Littlefield CE, Freeman PT, Zachmann LJ, Landau VA, Anderson JJ, Dickson BG (2023) Achieving conservation targets by jointly addressing climate change and biodiversity loss. Ecosphere 14(4):e4490

16. Carr-Whitworth R, Barrett J, Colechin M, Pidgeon N, Styles R, Betts-Davies S, Cox E, Watson A, Wilson O (2023) Delivering net zero in the UK: twelve conditions for success. Environ Res Lett 18(7):074041
17. Diffenbaugh NS, Barnes EA, Keys PW (2023) Probability of continued local-scale warming and extreme events during and after decarbonization. Environ Res: Clim 2(2):021003
18. Khosla R, Lezaun J, McGivern AM, Omukuti J (2023) Can 'Net Zero' still be an instrument of climate justice? Environ Res Lett 18(6):061001
19. Turnell H, Marques C, Jones P, Dunham C, Revesz A, Maidment G (2023) Driving success towards zero carbon energy targets for UK's local authorities. Int J Sustain Energy Plan Manag 38:83
20. Rajbhandari S, Winyuchakrit P, Pradhan BB, Chaichaloempreecha A, Pita P, Limmeechokchai B (2024) Thailand's net-zero emissions by 2050: analysis of economy-wide impacts. Sustain Sci 19(1):189–202
21. Gupta A, Dhir A (2019) Air quality assessment of Mandi Gobindgarh city of Punjab, India. Environ Dev Sustain 21:879–893
22. Clémençon-Charles E, Baranek K (2023) Looking beyond power generation to achieve electricity sector net-zero goals. Clim Energy 39(7):20–23
23. Yang Z, Zhang J, Kintner-Meyer MC, Lu X, Choi D, Lemmon JP, Liu J (2011) Electrochemical energy storage for green grid. Chem Rev 111(5):3577–3613
24. Da Silva AF, Nathaniel J, Wong KC, Watson C, Wang H, Singh J, Chamon AA, Klein L (2022, December) Netzeroco 2, an AI framework for accelerated nature-based carbon sequestration. In: 2022 IEEE international conference on big data (big data). IEEE, pp 4881–4887
25. Alemu B (2014) The role of forest and soil carbon sequestrations on climate change mitigation. Res J Agric Environ Manag 3(10):492–505
26. Rodrigues CID, Brito LM, Nunes LJ (2023) Soil carbon sequestration in the context of climate change mitigation: a review. Soil Syst 7(3):64
27. Lin Y, Cong N, Xiao J, Kou Y, Li Y, Yu X, Qi G, Gou C, Bai Y, Ren P (2023) Projecting future aboveground carbon sequestration rate of alpine forest on the eastern Tibetan plateau in response to climate change. Front Plant Sci 14:1212406
28. Štraus H, Podvinšek S, Klopčič M (2023) Identifying optimal forest management maximizing carbon sequestration in mountain forests impacted by natural disturbances: a case study in the alps. Forests 14(5):947
29. Helferty H (2023) Harnessing nature and ancient wisdom to store carbon. Proc Natl Acad Sci 120(19):e2305214120
30. Yablonovitch E, Deckman HW (2023) Scalable, economical, and stable sequestration of agricultural fixed carbon. Proc Natl Acad Sci 120(16):e2217695120
31. Honegger M, Michaelowa A, Roy J (2021) Potential implications of carbon dioxide removal for the sustainable development goals. Clim Pol 21(5):678–698
32. Change IC (2019) Land, summary for policymakers. Intergovernmental Panel on Climate Change, Geneva, Switzerland
33. Enríquez-de-Salamanca Á (2022) Climate Change mitigation in forestry: paying for carbon stock or for sequestration? Atmos 13(10):1611
34. Smith J (2020) Carbon sequestration: a critical component of climate change mitigation. J Clim Chang 45(3):215–230
35. Kavya SR, Rani B, Banu MR, Jabin PP (2023) Carbon sequestration and stabilisation mechanisms in the agricultural soils: a review. Int J Plant Soil Sci 35:79–94
36. Kaushal S, Sharma N, Singh I, Singh H (2023) Soil carbon sequestration: a step towards sustainability. Int J Plant Soil Sci 35(11):160–171
37. Rodrigues L, Budai A, Elsgaard L, Hardy B, Keel SG, Mondini C, Plaza C, Leifeld J (2023) The importance of biochar quality and pyrolysis yield for soil carbon sequestration in practice. Eur J Soil Sci 74(4):e13396

38. Shi G, Luan L, Zhu G, Zeng Z, Zheng J, Shi Y, Sun B, Jiang Y (2023) Interaction between nematodes and bacteria enhances soil carbon sequestration under organic material amendments. Front Microbiol 14:1155088
39. Ma Z, Chen B, Pawar RJ (2023) Phase-based design of CO2 capture, transport, and storage infrastructure via SimCCS 3.0. Sci Rep 13(1):6527
40. Kang H, Staples-West A, Washington A, Turchiano C, Cooksy A, Huang J, Gu J (2023) Enhancing CO2 reduction efficiency on cobalt phthalocyanine via axial ligation. ChemCatChem 15(14):e202300576
41. van de Ven DJ, Mittal S, Gambhir A, Lamboll RD, Doukas H, Giarola S, Hawkes A, Koasidis K, Köberle AC, McJeon H, Nikas A (2023) A multimodel analysis of post-Glasgow climate targets and feasibility challenges. Nat Clim Chang 13(6):570–578
42. Ji H, Pei J, Cai J, Ding C, Guo F, Wang Y (2023) Review of recent advances in Transcritical CO2 heat pump and refrigeration cycles and their development in the vehicle field. Energies 16(10):4011
43. Han GH, Bang J, Park G, Choe S, Jang YJ, Jang HW, Kim SY, Ahn SH (2023) Recent advances in electrochemical, photochemical, and Photoelectrochemical reduction of CO2 to C2+ products (small 16/2023). Small 19(16):2370105
44. Pessarrodona A, Franco-Santos RM, Wright LS, Vanderklift MA, Howard J, Pidgeon E, Wernberg T, Filbee-Dexter K (2023) Carbon sequestration and climate change mitigation using macroalgae: a state of knowledge review. Biol Rev 98(6):1945–1971
45. Leite FFGD, Nobrega GN, Baumgärtner LC, Alecrim FB, da Silveira JG, Cordeiro RC, Rodrigues RDAR (2023) Greenhouse gas emissions and carbon sequestration associated with integrated crop–livestock–forestry (ICLF) systems. Environ Rev 31(4):589–604
46. Gao L, Li H (2023) Improving carbon sequestration capacity of forest vegetation in China: afforestation or forest management? Forests 14(6):1077
47. Qin G, Yu H (2023) Rescuing the Paris agreement: improving the global experimentalist governance by reclassifying countries. Sustain For 15(4):3207
48. Zeng X, Yang X, Zhong S, Wang Z, Ding Y, Meng D, Gao K (2023) Comprehensive evaluation of resource and environmental carrying capacity at a national scale: a case study of Southeast Asia. Sustain For 15(7):5791
49. Kaur R, Kaur N, Kumar S, Dass A, Singh T (2023) Carbon capture and sequestration for sustainable land use–a review. Indian J Agric Sci 93(1):11–18
50. Morita K, Matsumoto KI (2023) Challenges and lessons learned for REDD+ finance and its governance. Carbon Balance Manag 18(1):8
51. Marnelly TR, Dahril T, Saam Z, Nofrizal N (2023) Ecological and socio-economic factors on the rate of participation and sustainable forest management. Int J Energy Econ Policy 13(3): 92–101
52. Liu F, Hu M, Huang W, Chen CX, Li J (2023) Future carbon sequestration and timber yields from Chinese commercial forests under shared socioeconomic pathways. Forests 14(1):153
53. Pan C, Shrestha A, Innes JL, Zhou G, Li N, Li J, He Y, Sheng C, Niles JO, Wang G (2022) Key challenges and approaches to addressing barriers in forest carbon offset projects. J For Res 33(4):1109–1122
54. Leal Filho W, Salvia AL, Vasconcelos CRP, Anholon R, Rampasso IS, Eustachio JHPP, Liakh O, Dinis MAP, Olpoc RC, Bandanaa J, Sharifi A (2022) Barriers to institutional social sustainability. Sustain Sci 17(6):2615–2630
55. Carpenter C (2023) Well-integrity risk-assessment strategy applied to CO2 sequestration project. J Pet Technol 75(01):78–80
56. Katare A, Kumar S, Kundu S, Sharma S, Kundu LM, Mandal B (2023) Mixed matrix membranes for carbon capture and sequestration: challenges and scope. ACS Omega 8(20): 17511–17522
57. Alam SA, Kivinen S, Kujala H, Tanhuanpää T, Forsius M (2023) Integrating carbon sequestration and biodiversity impacts in forested ecosystems: concepts, cases, and policies. Ambio 52(11):1687–1696

58. Cropper J (2024) 'Growing a world wonder': the great Green Wall and the history of environmental decline in the Sahel, 1450–2022. Environ Hist 30(2):291–313
59. Yu Y, Lin X, Guo Y, Guan Z, Tan J, Chen D, Su Y, Li J, Qiu Q, He Q (2023) The compound forest–medicinal plant system enhances soil carbon utilization. Forests 14(6):1233
60. Udawatta RP, Gantzer CJ, Jose S (2017) Agroforestry practices and soil ecosystem services. In: Soil health and intensification of agroecosytems. Academic Press, pp 305–333
61. Melichar J, Cienciala E, Albert J, Braun Kohlová M, Máca V, Pavelčík P (2022) Dynamics and Management of Restored Forests in post-mining sites with respect to their recreation value: a matrix growth model. Forests 13(9):1519
62. White JC, Coops NC, Wulder MA, Vastaranta M, Hilker T, Tompalski P (2016) Remote sensing technologies for enhancing forest inventories: a review. Can J Remote Sens 42(5): 619–641
63. Koutika LS, Matondo R, Mabiala-Ngoma A, Tchichelle VS, Toto M, Madzoumbou JC, Akana JA, Gomat HY, Mankessi F, Mbou AT, Nzila JDD (2022) Sustaining Forest plantations for the United Nations' 2030 agenda for sustainable development. Sustain For 14(21):14624
64. Mack SK, Lane RR, Holland K, Bauer J, Cole J, Cowan R (2022) A blue carbon pilot project: Lessons learned. Carbon Manag 13(1):420–434
65. Williams M, Sutherland A (2022) Lessons learned–incorporation of action management for inherent learning and prioritisation through opportunity matrix. APPEA J 62(1):125–140
66. Singh BK, Delgado-Baquerizo M, Egidi E, Guirado E, Leach JE, Liu H, Trivedi P (2023) Climate change impacts on plant pathogens, food security and paths forward. Nat Rev Microbiol 21(10):640–656
67. Bergero C, Gosnell G, Gielen D, Kang S, Bazilian M, Davis SJ (2023) Pathways to net-zero emissions from aviation. Nat Sustain 6(4):404–414
68. Panda A, Yamano T (2023) Asia's transition to net zero: opportunities and challenges in agriculture. Asian Development Bank Economics Working Paper Series, (694)
69. Gupta A, Dhir A (2022) Assessment of air quality and chemical fingerprints for atmospheric fine aerosols in an Indian smart city. Environ Pollut Bioavailab 34(1):21–33
70. Okitasari M, Korwatanasakul U (2023) Leaving no one behind in carbon neutrality strategies: insights from developing countries in Asia and the Pacific

Part III
Water Resources and Biogeochemistry

Groundwater Potential Zones Identification in Wyra Sub-Basin in Lower Krishna Basin in South India

Raghava Rao Ch and G. K. Viswanadh

Abstract Groundwater potential zone identification is necessary for administrating each subsurface and surface water resource. Groundwater is vital to maintaining human health and the health of various aquatic and land-based organisms and their environment. The area of interest is the Wyra River basin of the lower Krishna catchment area, encompassing an area of 3443 km^2 and located between latitudes $16^0 14'$ 00“ N and 17^0 35' 00" N and 80^0 05’ 00” E and 80^0 55′ 00″ E longitude. A geographical information system (GIS) is used to generate thematic maps and identify groundwater potential zones. The multi-criteria decision analysis method and GIS were used to develop thematic maps. These thematic maps are ranked by Satty’s analytic hierarchy processes. To identify possible zones for groundwater, a total of eight thematic layers of the specified kind of digital elevation model, land use and land cover, soil map, slope, rainfall, geology, lineament density, and drainage density were planned during the analysis. The GIS platform is used to convert each of the thematic layers into a raster. Utilizing weighted overlay techniques entire Wyra basin is divided into four major classes from poor to very good. Corresponding percentages are 5.05%, 73.40%, 21.45%, and 0.08% of the study region. According to this research, groundwater planning and management can be done more effectively and economically by integrating GIS.

Keywords Potential zones · Land Use Land Cover · Wyra basin · GIS · and AHP

R. R. Ch (✉) · G. K. Viswanadh
Department of Civil Engineering, JNTUH, Hyderabad, Telangana, India

P. Pathak et al. (eds.), *Advances in Environmental Sustainability, Energy and Earth Science*, https://doi.org/10.1007/978-3-031-73820-3_17

1 Introduction

Groundwater is a vital natural resource, contributing approximately 34% of the world's freshwater supply [1]. In India, as in many other developing nations, groundwater is a crucial resource, with around 90% of the rural population and 50% of the urban population relying on it for agriculture, domestic use, gardening, fire safety, and more. Studies reveal that approximately 70% of India's agricultural sector depends on groundwater. The availability of groundwater is influenced by factors such as the geology, lithology, and geomorphology of a region [1–5]. Several factors, including rapid population growth, uneven water resource distribution, economic development, and climate change, have significantly increased the demand for freshwater, leading to shortages in many regions of the globe [6, 7]. In India, which houses 18% of the global population but only 4% of the world's renewable water resources according to National Water Policy 2012, the value of groundwater is immense. Annually, India utilizes 230 km^2 of groundwater, making it the highest groundwater user globally [8]. Over the past two decades, groundwater levels have been rapidly declining in certain areas. According to a 2018 NITI Aayog report, 0.6 million Indians face severe water stress due to poor management and insufficient freshwater resources. Consequently, nearly 75% of Indian households lack access to potable water at home. Identifying new groundwater potential zones has become crucial for meeting freshwater demands and developing effective groundwater management strategies.

Traditional methods of groundwater exploration, such as drilling and hydrogeological, geological, and geophysical techniques, are widely used but are costly and time-consuming [9, 10]. These methods often overlook the various factors that influence groundwater occurrence and flow [11]. However, advancements in technology have enabled the efficient and systematic handling of large and complex spatial data. Geographical Information Systems (GIS) and Remote Sensing (RS) have become essential tools for addressing the complexities of groundwater exploration [8, 12, 13]. Over the past two decades, numerous studies utilizing GIS and RS have been conducted globally to identify groundwater potential zones, yielding promising results [14–17]. Commonly used spatial distribution maps include topographic, geomorphological, geological, lithological, soil, lineament density, and drainage patterns maps. Some researchers have also combined RS, GIS, and geo-electrical techniques to pinpoint groundwater potential zones [18, 19]. Additionally, Multi-Criteria Decision-Making (MCDM) analysis has proven useful in water management, offering structured, accountable, and consistent decision-making [9].

This study aims to employ an integrated approach to map and demarcate groundwater potential zones, focusing on various essential parameters and methodologies. By combining advanced technologies with traditional methods, it seeks to ensure sustainable groundwater management and address the increasing demand for freshwater resources in India and other developing countries. This integrated approach not only enhances the accuracy of identifying potential zones but also facilitates the development of effective strategies for groundwater conservation and management, ultimately contributing to the long-term sustainability of water resources.

2 Study Area

The geological location of the Wyra basin watershed is situated in the landmasses of both the states Telangana and Andhra Pradesh. The Wyra catchment location, as shown in Fig. 1, is a tributary of the Krishna River. The Wyra watershed is situated between a latitude of 16^0 14′ 00“ N and 17^0 35′ 00“ N, a longitude of 80^0 05 ‘00” E, and 80^0 55′ 00″ E. Its total area is 3443 km^2. The Wyra basin has hot summer and winter temperatures, indicating a semi-arid climate. The basins land cover and land use are dominated by agricultural areas; 76% of the terrain is used for raising crops. Pulses, cotton, and paddy are the principal crops farmed in the basin. The digital elevation model of the basin downloaded from USGS website and found that the selected Wyra basin is located at a height range of 28–785 meter with respect to mean sea level (MSL).

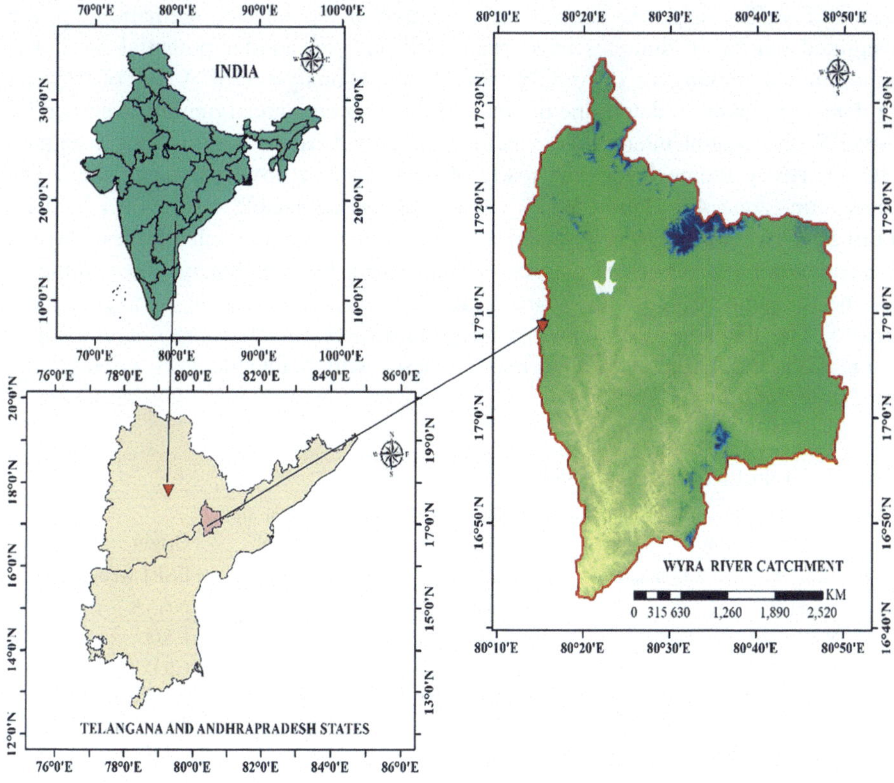

Fig. 1 Study area location map

3 Sources and Data Sets

Eight distinct thematic map types, such as geology, rainfall, soil, elevation, drainage and lineament density, and LULC of the selected basin prepared from the examination of satellite data. Digital elevation model (DEM), drainage density maps, and slope maps were developed from DEM data. LULC data was achieved from the ESRI land cover website and Information on soil from the FAO website. Geological map, Geomorphology, and lineaments downloaded from Geological Survey of India. Table 1 displays all of the input data details.

3.1 Methodology

The procedure for defining the groundwater potential zone has been represented, and the methodology's schematic Fig. 2 shows a flowchart. Eight themes, including the amount of drainage density, geology, rainfall, slope map, lineament density, soil map, study area DEM, and LULC data, have significant control over groundwater availability. To assign the respective weightage of the selected thematic map, the weighted overlay techniques have been used and the aquifer potential zone was accumulated, making use of ArcGIS 10.4.1's spatial analysis tool. A total of 10 years of data were taken to determine the map of the average annual rainfall. Adopted by ArcGIS, the spatial interpolation method is known as inverse distance weighted (IDW). Every theme layer was ready, rasterized, and projected using the UTM projection, zone 44, Datum WGS 1984, with an adjustable resolution of 30 m. Further, a weighted overlay analysis is used to combine each of these layers. Researchers have suggested classes used to assign the weightages and ranking to all the thematic maps of the Wyra basin. The class of the scale is ranging from 1 to 5. The thematic map showing the groundwater potential zones is developed using GIS and grouped to four major classes which include very good, good, moderate, and poor.

Table 1 Distinct input parameters of the region

S. N.	Representation	Resolution of data and scale	Year and range	Origin
1	Sentinel-2 (LULC map)	10 m	2021	ESRI website
2	ASTER DEM	High data resolution	2018	NASA
3	Soil map	1:5000000	–	FAO
4	Rainfall data	High-resolution gridded datasets	2011–2020	CRU TS v.407 data variables
5	Geology, geomorphology, and lineament data	1:250,000	–	The India Geological Survey

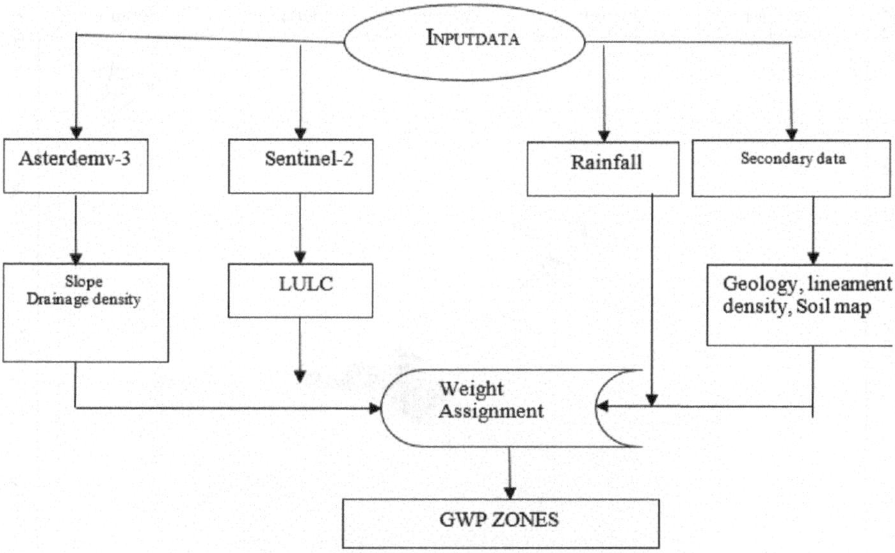

Fig. 2 Delineating Groundwater Potential Zone (GWPZ) using a methodological flowchart

4 Results and Discussions

4.1 Thematic Layers Related to Hydrology

The occurrence and distribution of groundwater are influenced by various factors, including topography, soil type, slope, precipitation, land use and land cover (LULC), drainage patterns, and secondary porosity. These factors are interrelated and contribute to the presence of groundwater [1, 3, 11, 20]. This research identifies eight key elements—drainage density, slope, geology, rainfall distribution, lineament density, DEM, soil characteristics, and LULC to determine areas with potential for groundwater in the study region.

4.1.1 Elevation Model

The most basic type of DEM provides surface elevation data, devoid of vegetation, structures, or other non-round objects, to a datum above MSL. DEM data is utilized in the production of a slope map, drainage density map, fill map, stream network, flow accumulation, and river basin analysis. DEM is classified into five categories, such as 28–86, 87–130, 131–206, 207–362, and 363–786 m, respectively. Ranks are assigned 1 to 5 Scales. Figure 3 shows the DEM map of the Wyra basin.

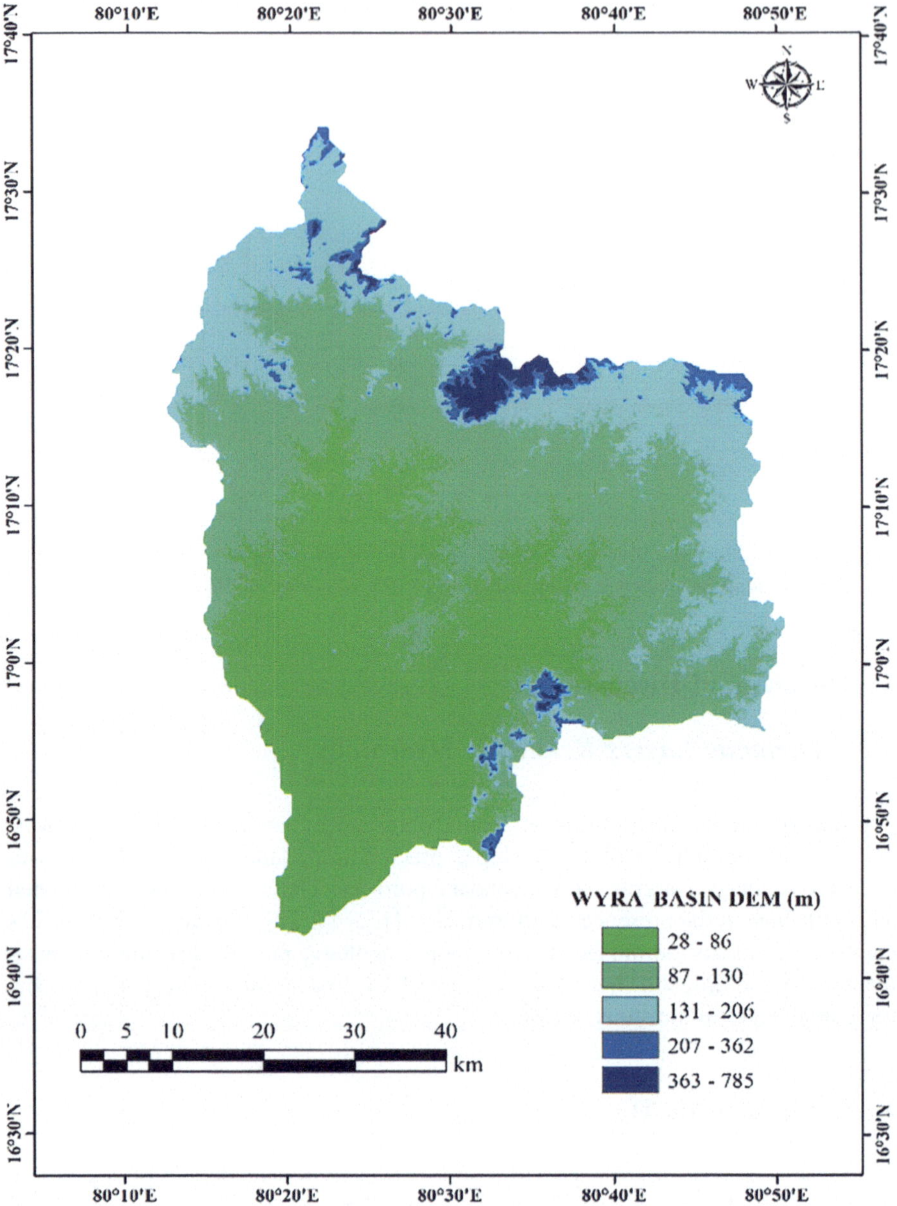

Fig. 3 Map showing the variation in elevation over the Wyra basin

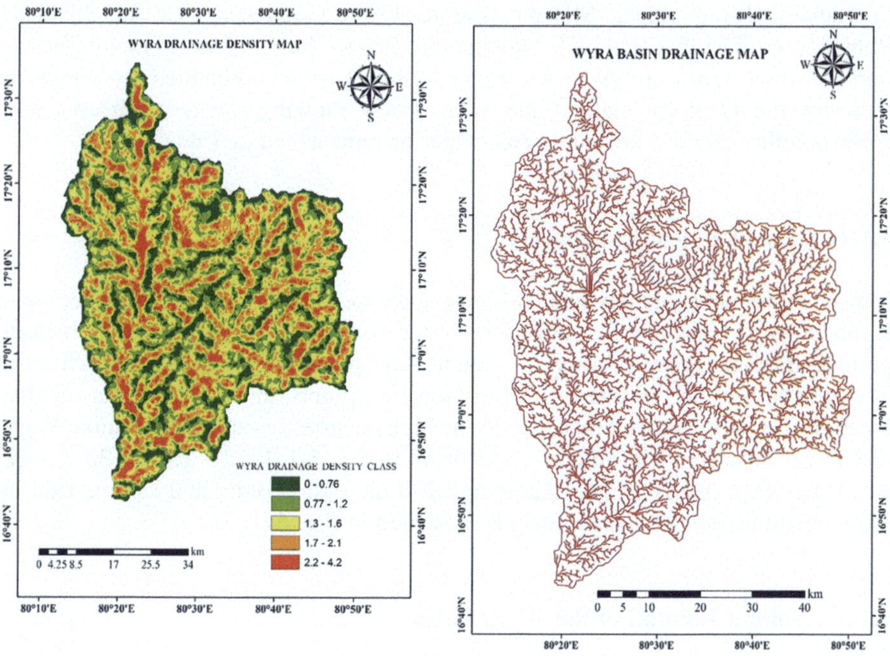

Fig. 4 Spatial variation of drainage density and drainage pattern over the Wyra basin

4.1.2 Drainage Density of the Wyra Basin

The drainage map displays a system of tributaries, streamlets, and drainage lines. When multiple tributaries of varying orders are combined, a drainage network is produced. The drainage map is produced from DEM, with the GIS software. The watershed forms a dendrites-drainage pattern, as demonstrated in Fig. 4. The permeability of the soil and rock below the surface is depicted on the drainage density map and categorized into five main groups including 0–0.76, 0.77–1.2, 1.3–1.6, 1.72.1, and 2.2–4.2 Km/Km^2 as demonstrated in Fig. 4. Groundwater potential zones are inversely correlated with drainage density. An enhanced density of drainage values is suitable for runoff and limited potential for groundwater. Hence, there are five Ranks, ranging from 1 to 5. The highest Seepage was found in the main River portion. On the other hand, the study region's border is the result of the lowest drainage density. Figure 4 shows the variation of drainage density of the Wyra basin.

4.1.3 Geology Map

The groundwater development and its fluctuations are influenced by the studied region's geology. The Wyra catchment contains various geology features such as the Charnockite Gneissic complex, Khondalite Gneissic complex, upper Gondwana

(Yerrapalli, Terani, Kota, Maleri) Triassic, lower Gondwana (Talchir, Barakar, Raniganj, etc.) Mulug (Pakhal), Mallampalli, Peninsular Gneissic Complex formation, in which ranks are given according to their hydraulic conductivity. Figure 5 indicates the Geology map of the Wyra basin showing the research area and corresponding features area and percentages as represented in Table 2.

4.1.4 Map of Lineament Density

Lineament features, namely Faults, Joints, and Cracks, are crucial to hydrogeology. It functions as a passage for subsurface-water-motions, which influences enhanced permeability and hence promotes a groundwater potential inside the research area. The area across the lineament and their junction points are recognized as suitable locations for subsurface storage, due to the high occurrence rate of infiltration. Wyra catchment density of lineaments varies from 0–5.3, 5.4–15, 16–25, 26–38, 39–68 Km/Km^2, respectively. Ranks are assigned in increasing order and summarized in Table 3 and associated density map is presented in Fig. 6.

4.1.5 Annual Rainfall of the Wyra Basin

The rainfall is a crucial parameter to be considered when assessing the feasible the groundwater information over the selected area. The influencing factors such as Rainfall, Percolation, and Infiltration, which directly manage the aquifer groundwater table [21, 22], are a good correlation between variations in the groundwater table and Rainfall. The Wyra basin reflects the average rainfall range between (1012–1255) mm according to Fig. 7. There are five categories for the rainfall map, and the scores range from 1 to 5, which are 1012–1094 mm, 1095–1131 mm, 1132–1166 mm, 1167–1200 mm, and 1201–1255 mm, and the scores have been allotted as 1 to 5, respectively, as shown in Table 3.

4.1.6 Slope Map

The gradient of a slope indicates the surface elevation of the ground, as shown in Fig. 8. Slope inclination directly impacts the percolation and infiltration of rainfall. Consequently, slope information can be utilized to determine zones with groundwater potential [23]. A raised slope indicates quick runoff and is inversely correlated with infiltration and percolation rates, especially in moderately sloping or level areas. The slope of the Wyra basin ranges from 0° to 53°, as shown in Fig. 8. Areas with a level slope reflect greater potential for infiltration and percolation compared to steep slopes. This information was considered while assigning weightage or scores to different slope classes. Wyra basin slopes have been divided into five categories based on their slope cover. Almost level slopes, ranging from 0° to 3°, are assigned the highest score of 5, indicating the greatest potential for infiltration and percolation

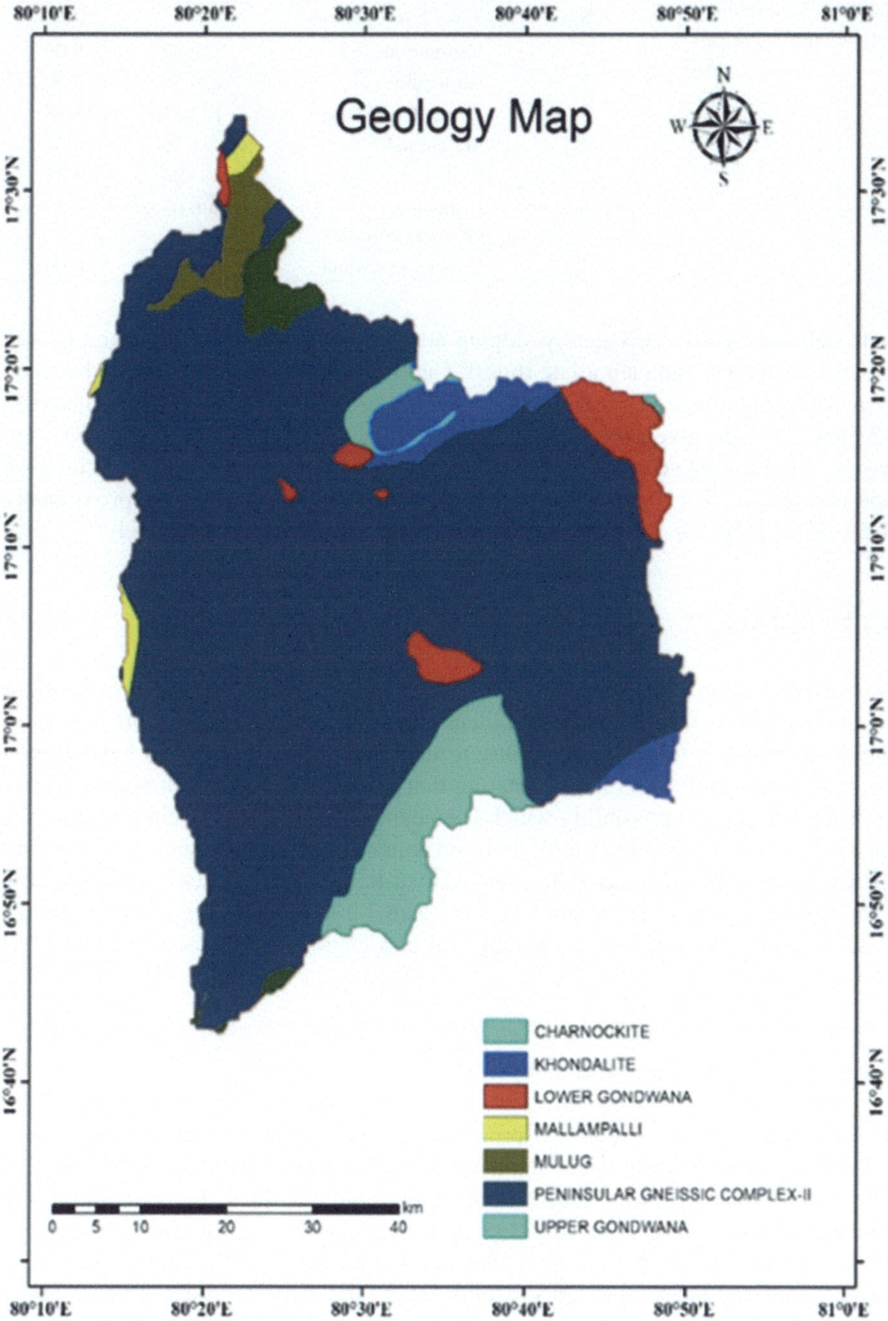

Fig. 5 Geology map representing the research field

Table 2 Distribution and formation of geology

S.N.	Wyra basin geology	Area	Area (%)
1.	Charnockite	312.22	9.06
2.	Khondalite	120.50	3.50
3.	Lower Gondwana	123.13	3.58
4.	Mallampalli	21.36	0.62
5.	Mulug	63.86	1.85
6.	Peninsular Gneissic complex	2800.02	81.32
7.	Upper Gondwana	2.04	0.06

with delayed runoff. Very gently sloping areas, with slopes between 4° and 6°, are given a score of 4, indicating late runoff. Gently sloping areas, with slopes between 7° and 12°, are assigned a score of 3. Moderately sloping areas, with slopes between 13° and 22°, are given a score of 2. Steep slopes, ranging from 23° to 53°, are assigned the lowest score of 1 due to fast runoff, resulting in almost 0% infiltration and percolation. The slope cover percentages for the Wyra basin are approximately 44%, 36%, 15%, 4%, and 2% of the entire watershed region, respectively.

4.1.7 Soil Map

Soil map represents the type or texture of the soil, and soil texture performs a major function in subsurface recharge and identification of GWPZ. The runoff pattern of any basin depends on the soil's ability to hold water. Runoff features have a good correction alongside GWPZ [24], stating that groundwater replenishment in loamy soils has the greatest possibility which is evidence of a more appropriate area for the possibility of groundwater. The Wyra basin contains various soils such as Clay loam, Loam, Sandy-Clay-loam, and Clay, and its distribution is as follows Clay-loam 28%, Clay soil 46%, sandy-clay-loam 23%, and loam 3.2%, respectively, in Fig. 8. In this field of study, most of the soil is clay soil and clay-loam type according to FAO soil data.

4.1.8 LULC Mapping of the Wyra Basin

The permeability and percolation processes, or infiltration, regulate LULC. It also influences numerous geological and hydrological processes in the hydrologic circle, like surface runoff, infiltration, evapotranspiration, etc. [25]. In agricultural and forested terrain, runoff is usually less and there is greater infiltration. In areas of settlement, the rate of infiltration is generally lower. As a result, vegetation cover in agricultural land areas is assigned a high score, and settlements are given a lower score. Wyra catchment: Large part of the land use is cropland, encompassing 2607.03 km^2 (76%), area high rank 4, trees covering 37.86 km^2 (11%), built-up area covering 165.94 km^2 (5%), water bodies covering 87.72 km^2 (3%) area height

Table 3 Each feature's scale values and impacts

Thematic layers	Weightage	Significance	Scale (1 to 5)
Drainage density	20	0–5	1
		6–15	2
		16–25	3
		26–38	4
		39–68	5
Geology data	20	Charnockite	1
		Khondalite	1
		Lower Gondwana	3
		Mallampalli	2
		Mulug	2
		Peninsular Gneissic complex-ii	1
		Upper Gondwana	2
Lineament density	10	0–5.1	1
		5.2–15	2
		16–24	3
		25–37	4
		38–68	5
Rainfall	10	1012–1094	1
		1095–1131	2
		1132–1166	3
		1167–1200	4
		1201–1255	5
Slope	10	0–3	5
		4–6	4
		7–12	3
		13–22	2
		23–53	1
DEM	15	28–86	1
		87–130	2
		131–206	3
		207–362	4
		363–785	5
Soil map	5	Clay soils	1
		Clay-loam	3
		Sandy-clay-loam	2
		Loam	3
LULC	10	Water	5
		Trees	3
		Flood vegetation	4
		Crops	4
		Built up	1
		Bare land	2
		Rangeland	2

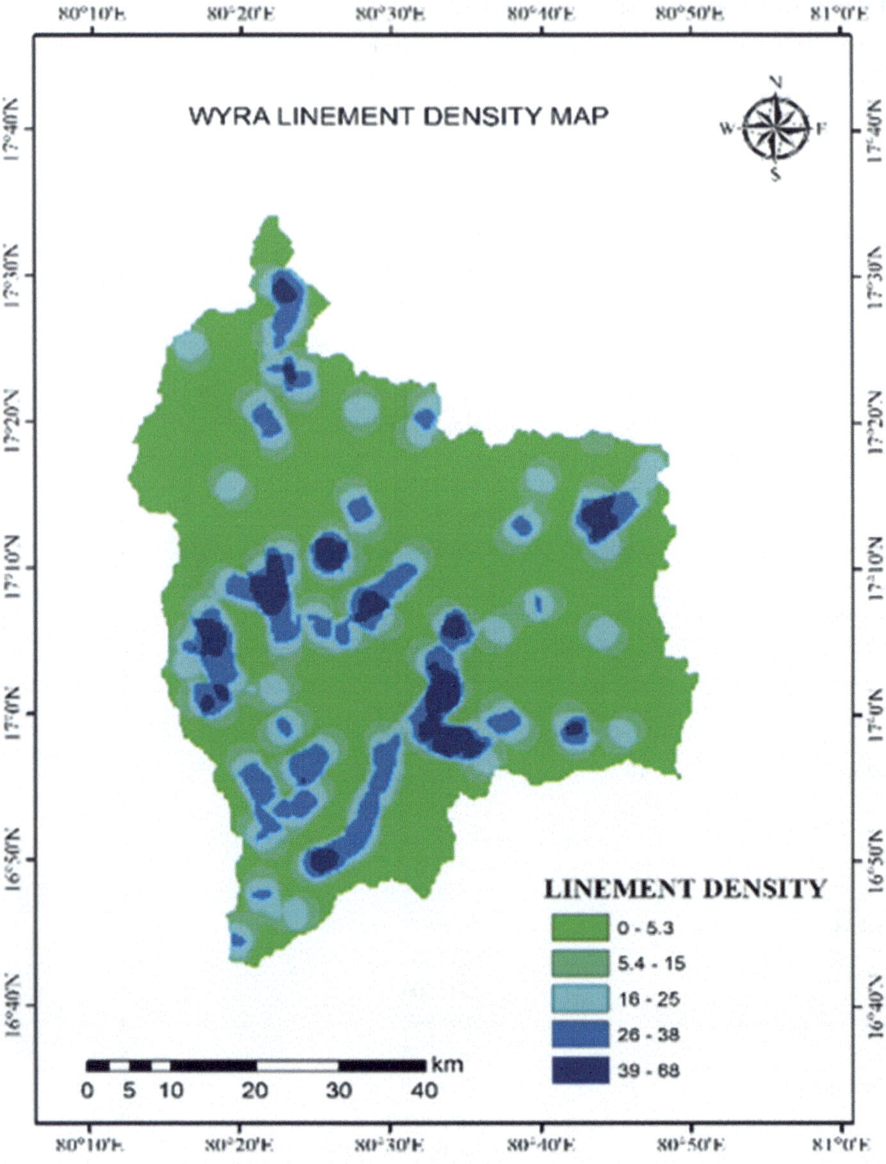

Fig. 6 Spatial variation of lineament density map of the Wyra basin

rank, bare ground/range land covering 203.87 km^2 (6%), area rank 3 assigned in Table 3 represents Lulc weightages ,significance and scale(1 to 5). And Table 4 represents Lulc class, Area (Km2) and Area (%). Figure 9 represents the LULC of the research region.

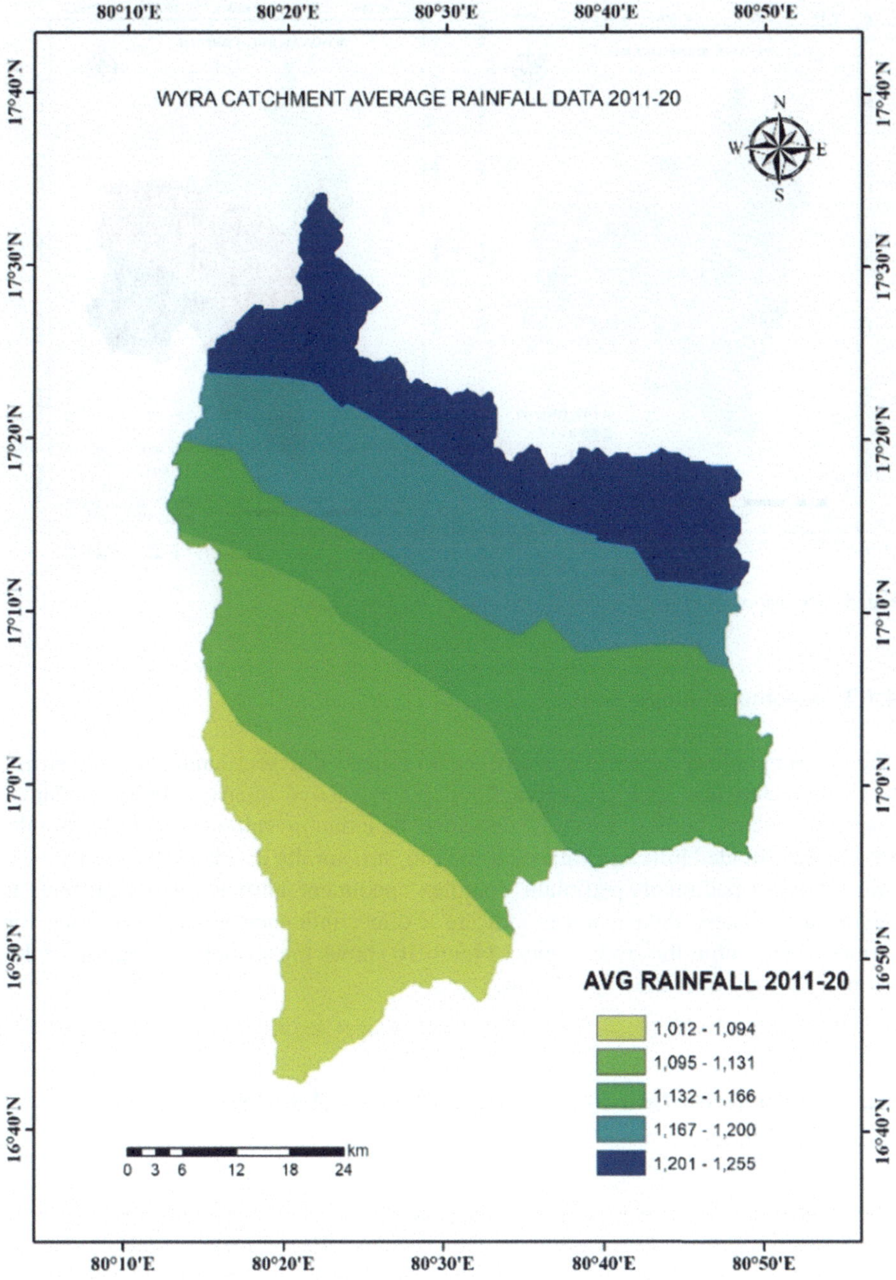

Fig. 7 Average rainfall map of the investigation region

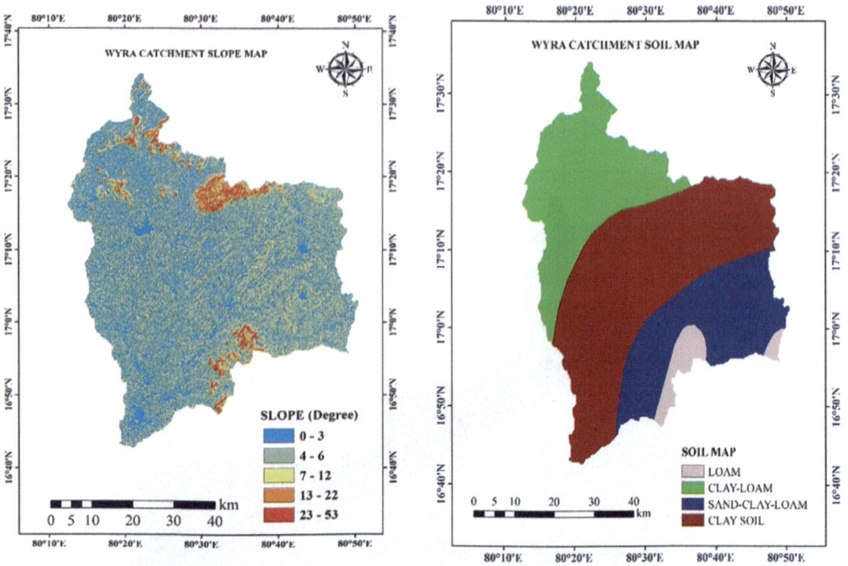

Fig. 8 The variation of slope and soil elements of the Wyra basin

4.1.9 Geomorphology Map

The geomorphology around the study region reflects different landforms and structural characteristics such as active flood plains, active quarries, bajadas, dams, reservoirs, water bodies, severely dissected denudation valleys and hills, highly structurally divided hills and valleys, low-lying structurally divided hills and valleys, mine dumps, pediments-pediplain complex, pediment alluvial plains present in Wyra basin. These data features indicate a direct influence prospective zone for groundwater within the study region. Figure 10 shows the spatial distribution of the geomorphology over the Wyra basin.

4.2 Implementation of Weighted Overlay Analysis (WOA) Technique

The pixel-based weighted overlay analysis is a type of analysis. Different thematic categories have distinct measurement units such as DEM in meters, rainfall layer in mm, the degree of the slope layer, and drainage density per km. LULC Map 10 m. For WOA, all the layers are on the same scale; therefore, scores 1 to 5 were allotted to every layer.

All layers were given the same index after scoring, and they were then combined into one. To be specific, a raster has a score of 5, representing a gradient of 0^0 to 3^0; a

Table 4 The Wyra Basin's land usage and land cover

S.N.	LULC class	Area (km^2)	Area (%)
1	Water	87.72	2.55
2	Trees	378.66	11.00
3	Flooded vegetation	0.48	0.01
4	Crops	2607.03	75.71
5	Built-up	165.94	4.82
6	Bare ground	0.7736	0.02
7	Range-land	202.87	5.89

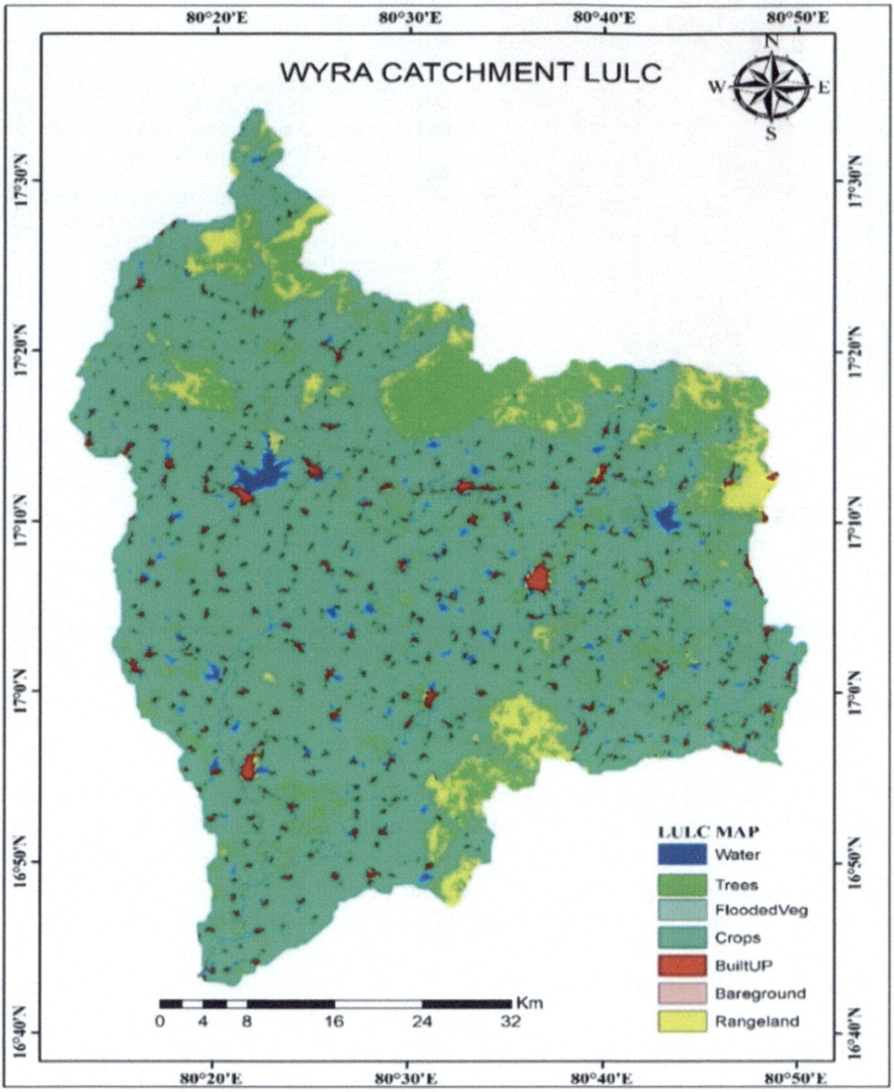

Fig. 9 Map showing LULC

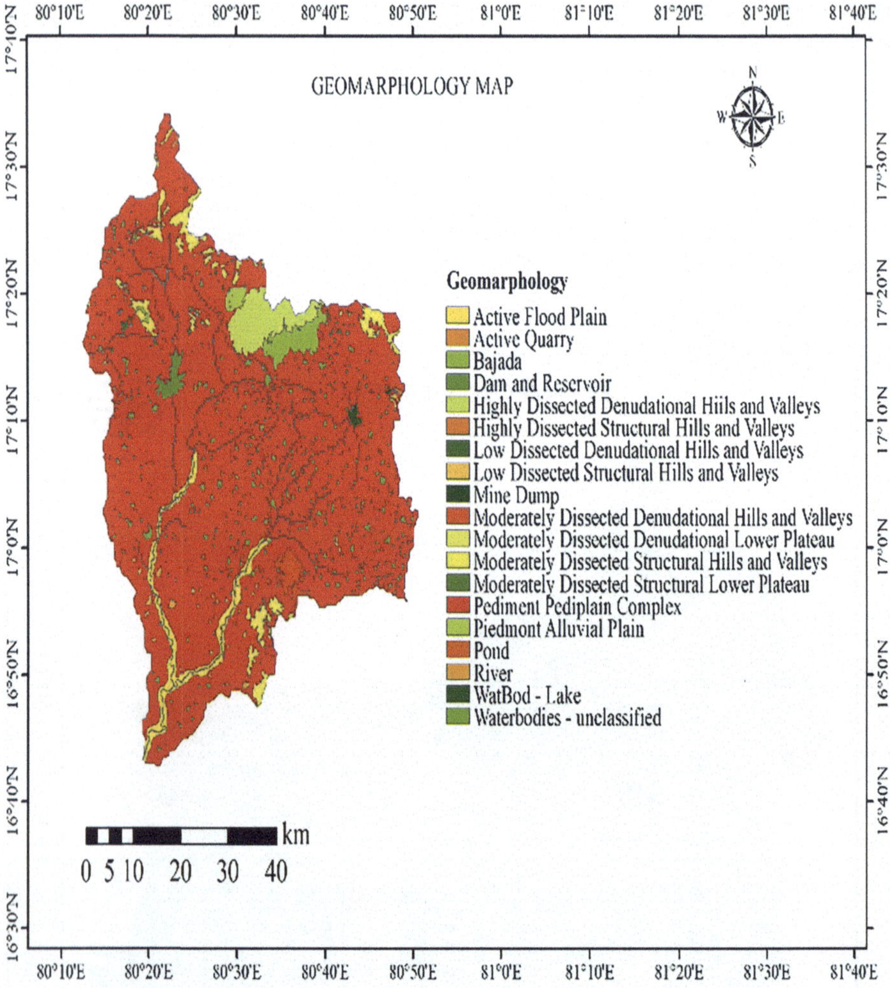

Fig. 10 Geomorphology map of the Wyra basin

score of 5 describes a drainage density of 3.9–2.13 Km/Km^2. A total score of 5 describes water bodies in LULC, and an overall rating of 5 describes the rainfall map of 1201–1255 mm Fig 11. Groundwater Potential zones of Wyra basin and Table 5 Groundwater potential zone area and percentages of areas of Wyra basin. The more weighted values were provided for groundwater potential zones [26].

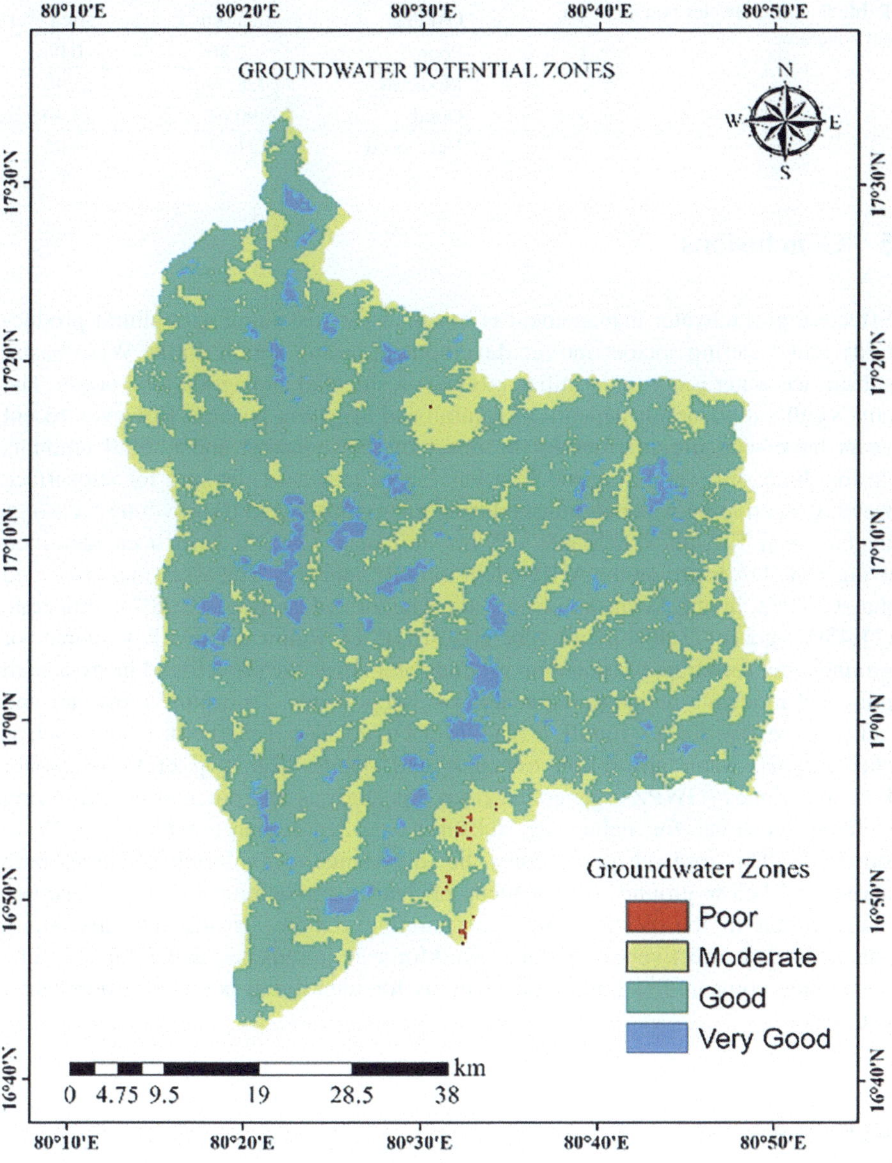

Fig. 11 Groundwater potential zones of Wyra basin

Table 5 Groundwater potential zone area

S.N.	GWPZ	Area (km^2)	Area (%)
1	Poor	2.80	0.08
2	Moderate	728.09	21.46
3	Good	2490.63	73.40
4	Very good	171.63	5.06

5 Conclusions

Effective groundwater management is essential for sustaining agricultural productivity and fostering socioeconomic development in any region. In the Wyra basin, subsurface water is currently utilized to meet residential and agricultural needs. The area's cultivation largely depends on rainfall and irrigation systems. However, recent years have seen irregularities in the onset of the monsoon and rainfall amounts during June, July, and August. This has led to increased demand for subsurface water, exacerbating water shortages during the monsoon and Rabi (winter) seasons. In the lower Krishna Catchment's Wyra sub-basin, GWPZs have been identified using a MCDA based on the AHP, integrating RS and GIS. The assessment revealed that GWPZs in the Wyra sub-basin are categorized as poor (0.08%), moderate (21.45%), good (73.40%), and very good (5.05%). Rainwater is a key source for groundwater recharge, yet potential groundwater zones are often found in areas with poor and moderate recharge, especially on steep slopes. To capture rainwater and reduce excessive surface runoff, it is advisable to develop recharge structures such as check dams, water absorption trenches, and ponds. The map of Groundwater Potential Zones (GWPZs) generated from this study is instrumental in identifying suitable locations for enhancing subsurface water recharge within the Wyra sub-basin. This research underscores the importance of exploring resources both above and below ground, with a focus on land use planning and water resource management in the Wyra basin. Incorporating remote sensing into this study enhances cost effectiveness, reduces workforce requirements, and mitigates time constraints compared to traditional methods for identifying potential groundwater zones.

References

1. Acharya T, Kumbhakar S, Prasad R, Mondal S, Biswas A (2019) Delineation of potential groundwater recharge zones in the coastal area of North-Eastern India using geoinformatics. Sustain Water Resour Manag 5:533–540
2. Avtar R, Singh CK, Shashtri S, Singh A, Mukherjee S (2010) Identification and analysis of groundwater potential zones in Ken–Betwa river linking area using remote sensing and geographic information system. Geocarto Int 25(5):379–396
3. Pradhan B (2009) Groundwater potential zonation for basaltic watersheds using satellite remote sensing data and GIS techniques. Cent Eur J Geosci 1:120–129

4. Singh S, Singh C, Kumar K, Gupta R, Mukherjee S (2009) Spatial-temporal monitoring of groundwater using multivariate statistical techniques in Bareilly district of Uttar Pradesh, India. J Hydrol Hydromech 57(1):45–54
5. Singh SK, Srivastava PK, Pandey AC (2013) Fluoride contamination mapping of groundwater in northern India integrated with geochemical indicators and GIS. Water Sci Technol Water Supply 13(6):1513–1523
6. Jasrotia AS, Kumar A, Singh R (2016) Integrated remote sensing and GIS approach for delineation of groundwater potential zones using aquifer parameters in Devak and Rui watershed of Jammu and Kashmir, India. Arab J Geosci 9:1–15
7. Selvam S, Magesh NS, Chidambaram S, Rajamanickam M, Sashikkumar MC (2015) A GIS based identification of groundwater recharge potential zones using RS and IF technique: a case study in Ottapidaram taluk, Tuticorin district, Tamil Nadu. Environ Earth Sci 73:3785–3799
8. Arkoprovo B, Adarsa J, Animesh M (2013) Application of remote sensing, GIS and MIF technique for elucidation of groundwater potential zones from a part of Orissa coastal tract, Eastern India. Res J Recent Sci 2277:2502
9. Jha MK, Chowdary VM, Chowdhury A (2010) Groundwater assessment in Salboni block, West Bengal (India) using remote sensing, geographical information system and multi-criteria decision analysis techniques. Hydrogeol J 18(7):1713–1728
10. Razandi Y, Pourghasemi HR, Neisani NS, Rahmati O (2015) Application of analytical hierarchy process, frequency ratio, and certainty factor models for groundwater potential mapping using GIS. Earth Sci Inf 8:867–883
11. Oh HJ, Kim YS, Choi JK, Park E, Lee S (2011) GIS mapping of regional probabilistic groundwater potential in the area of Pohang City, Korea. J Hydrol 399(3–4):158–172
12. Arkoprovo B, Adarsa J, Prakash SS (2012) Delineation of groundwater potential zones using satellite remote sensing and geographic information system techniques: a case study from Ganjam district, Orissa, India
13. Jha MK, Chowdhury A, Chowdary VM, Peiffer S (2007) Groundwater management and development by integrated remote sensing and geographic information systems: prospects and constraints. Water Resour Manag 21:427–467
14. Ghosh PK, Bandyopadhyay S, Jana NC (2016) Mapping of groundwater potential zones in hard rock terrain using geoinformatics: a case of Kumari watershed in western part of West Bengal. Model Earth Syst Environ 2:1–12
15. Gnanachandrasamy G, Zhou Y, Bagyaraj M, Venkatramanan S, Ramkumar T, Wang S (2018) Remote sensing and GIS based groundwater potential zone mapping in Ariyalur District, Tamil Nadu. J Geol Soc India 92:484–490
16. Gogu R, Carabin G, Hallet V, Peters V, Dassargues A (2001) GIS-based hydrogeological databases and groundwater modelling. Hydrogeol J 9:555–569
17. Golla V, Etikala B, Veeranjaneyulu A, Subbarao M, Surekha A, Narasimhlu K (2018) Data sets on delineation of groundwater potential zones identified by geospatial tool in Gudur area, Nellore district, Andhra Pradesh, India. Data Brief 20:1984–1991
18. Choudhari PP, Nigam GK, Singh SK, Thakur S (2018) Morphometric based prioritization of watershed for groundwater potential of Mula river basin, Maharashtra, India. Geol Ecol Landsc 2(4):256–267
19. Kumar A, Krishna AP (2018) Assessment of groundwater potential zones in coal mining impacted hard- rock terrain of India by integrating geospatial and analytic hierarchy process (AHP) approach. Geocarto Int 33(2):105–129
20. Shahid S, Nath S, Roy J (2000) Groundwater potential modelling in a soft rock area using a GIS. Int J Remote Sens 21(9):1919–1924
21. Singh PK, Kumar S, Singh UC (2011) Groundwater resource evaluation in the Gwalior area, India, using satellite data: an integrated geomorphological and geophysical approach. Hydrogeol J 19(7):1421

22. Das S (2019) Comparison among influencing factor, frequency ratio, and analytical hierarchy process techniques for groundwater potential zonation in Vaitarna basin, Maharashtra, India. Groundw Sustain Dev 8:617–629
23. Maity DK, Mandal S (2019) Identification of groundwater potential zones of the Kumari river basin, India: an RS & GIS based semi-quantitative approach. Environ Dev Sustain 21:1013–1034
24. Singh LK, Jha MK, Chowdary VM (2018) Assessing the accuracy of GIS-based multi-criteria decision analysis approaches for mapping groundwater potential. Ecol Indic 91:24–37
25. Zomlot Z, Verbeiren B, Huysmans M, Batelaan O (2015) Spatial distribution of groundwater recharge and base flow: assessment of controlling factors. J Hydrol Reg Stud 4:349–368
26. Kaliraj S, Chandrasekar N, Magesh NS (2014) Identification of potential groundwater recharge zones in Vaigai upper basin, Tamil Nadu, using GIS-based analytical hierarchical process (AHP) technique. Arab J Geosci 7:1385–1401

Enhancing Water Quality of Panchaganga River in Kolhapur, Maharashtra: A Study on the Efficacy of Vertical Flow Constructed Wetland System

S. S. Joshi, M. Kothuri, and M. M. Yadav

Abstract Wastewater is a major threat to the natural water streams, damaging their ecological balance and aquatic environments. Untreated wastewater streams from both domestic and industrial effluents pollute natural streams, which further causes several health and environmental concerns for human society. This study explores the efficacy of the Vertical Flow Constructed Wetland (VFCW) system for the benefit of recovering the Panchaganga river passing through the city of Kolhapur, Maharashtra, India. Jayanthi Nallah is the major contributor to stream pollution. Hence, the pollution load has been evaluated for the grab integrated samples from six different sampling locations by analyzing physicochemical parameters such as biochemical oxygen demand (BOD), chemical oxygen demand (COD), total suspended solids (TSS), total phosphorus (TP), and total Kjeldahl nitrogen (TKN) in this study. A lab-scale model of VFCW with the dimensions of 0.62 m*0.70 m*0.45 m was developed for phytoremediation using three different combinations of locally available plant species (*Canna indica*, *Canna indica* & *Typha,* and *Canna indica* & *Cyperus rotundus*) and their treatment efficiencies were evaluated for three different hydraulic loading rates of 0.12 m^3/m^2/hr., 0.18 m^3/m^2/hr., and 0.24 m^3/m^2/hr. The treated water from all three plant combinations had all the parameters well below the regulating standards when the hydraulic loading rate was fixed at 0.18 m^3/m^2/hr. The VFCW balances treatment efficiency and cost, making it a sustainable wastewater treatment technology.

Keywords Constructed wetland · Hydraulic loading rates · Sewage treatment · Sustainable wastewater treatment · Phytoremediation

S. S. Joshi · M. Kothuri (✉) · M. M. Yadav
Department of Civil and Environmental Engineering, Kolhapur Institute of Technology's College of Engineering (Autonomous), Kolhapur, Maharashtra, India

P. Pathak et al. (eds.), *Advances in Environmental Sustainability, Energy and Earth Science*, https://doi.org/10.1007/978-3-031-73820-3_18

1 Introduction

Rivers are crucial in sustaining ecosystems and providing essential resources for human communities. However, rapid urbanization, population growth, and inadequate sewage treatment infrastructure have led to the deterioration of water quality in many rivers worldwide, including those in India. The Panchaganga River in Kolhapur, India, exemplifies this trend, facing severe pollution primarily due to sewage contamination. Like other polluted rivers in India, such as the Ganga and Yamuna, the Panchaganga River suffers from elevated levels of pollutants from domestic sources, posing risks to the environment and human health. The influence of domestic sewage on surface water bodies is a critical environmental concern due to its significant impact on water quality. Studies conducted on the Ganga and Yamuna rivers have highlighted the ongoing challenges of sewage pollution, emphasizing the urgent need for effective sewage treatment and management practices. Approximately 85% of the pollution in the Yamuna River stems from domestic sources, mainly urban areas such as Delhi, with heightened ammonium concentrations in the water because of sewage contamination, highlighting the significant impact of sewage on water quality [1, 2].

Several studies have highlighted the detrimental effects of untreated or poorly treated domestic sewage on surface water bodies. The discharge of domestic sewage directly into rivers and lakes significantly affects the living environment. Furthermore, domestic sewage has been identified as a primary influence on water quality parameters in surface water. The chemical composition of river water is significantly influenced by domestic sewage [3]. Moreover, the impact of domestic sewage on water quality is evident in various water bodies, including rivers, lakes, and reservoirs. Studies have reported that the distribution and sources of sedimentary sterols in water bodies indicate significant contamination from human domestic sewage. The study involved collecting sediment samples from various locations and extraction and analysis using gas chromatography-mass spectrometry (GC-MS). Sedimentary sterols, especially those derived from human sewage, can harm the environment by promoting eutrophication, which leads to oxygen depletion and harm aquatic life. For human health, these sterols can indicate the presence of pathogens and other harmful substances associated with untreated sewage [4].

Furthermore, the discharge of domestic sewage into aquatic environments has been linked to reduced dissolved oxygen, increased turbidity, and changes in pH, which collectively affect water quality and the survival of organisms [5]. Additionally, domestic sewage has been identified as a significant contributor to groundwater pollution due to nutrients, detergents, and human and animal excreta [6].

The pollution in the Panchaganga River is likely exacerbated by increasing urbanization, population growth, and inadequate sewage treatment infrastructure, leading to untreated or partially treated sewage being discharged into the river. High levels of pollutants from domestic sources significantly degrade water quality, impacting aquatic ecosystems and posing risks to human health. Hence, the efforts to address sewage pollution in the Panchaganga River would require comprehensive strategies that focus on improving sewage treatment facilities, implementing proper waste management practices, and enforcing regulations to prevent further contamination of the river. Jayanthi nallah, a stream channel, is reported as the major contributor to Panchaganga river pollution [7–9]. Hence, treating the wastewater from the Jayanthi nallah is more beneficial before the water reaches the river stream.

Constructed wetlands emerged as a promising approach for sewage treatment, offering both efficiency and sustainability. They are artificial ecosystems designed to mimic the functions of natural wetlands in treating wastewater and providing various ecosystem services. Constructed wetlands can be classified into different types based on their hydrologic modes, including free water surface, horizontal subsurface flow, and vertical flow wetlands [10]. In tropical countries like India, constructed wetland systems offer a solution for sewage treatment and for deploying treated effluent in the agricultural sector [11]. Horizontal flow constructed wetlands offer potential for the decentralized treatment of community domestic wastewater despite requiring large surface areas compared to conventional treatment units [12]. However, vertical flow constructed wetlands have several advantages over horizontal flow constructed wetlands due to their lower demand for land, indicating a potential advantage in terms of space utilization and land requirements [13]. Furthermore, the removal capability of total nitrogen in vertical flow constructed wetlands has been reported to exceed that of horizontal flow systems, suggesting a potential superiority in nitrogen removal [14].

Design parameters of constructed wetlands, such as depth, hydraulic loading rates, and substrate loading rates, are critical to achieving customized pollutant removal levels, thus facilitating the attainment of sewage discharge or reuse standards in a more scientifically grounded manner [15]. The hydraulic loading rate is a critical factor in the design and performance of constructed wetlands for wastewater treatment. It significantly impacts the treatment efficiency of constructed wetlands. For instance, it has been observed that constructed wetlands processing high hydraulic loads are economically feasible [16]. The optimal hydraulic loading rates for removing agricultural pollutants from wetlands are 10–30 cm/d with a hydraulic retention time of 6–8 days [17]. Furthermore, the hydraulic characteristics of the layered filling structure are significantly better than those of single-layer structured subsurface flow constructed wetlands, indicating the importance of hydraulic design in achieving efficient pollutant removal [18].

Moreover, the removal efficiency for total nitrogen in constructed wetlands significantly correlates with the hydrologic loading rate and hydraulic retention time [19]. Studies have also indicated that the treatment efficiency of pollutants in a constructed wetland can be improved by decreasing the hydraulic loading or increasing the hydraulic retention time [20]. Furthermore, the mean removal rates of ammonia, BOD, and Volatile Suspended Solids (VSS) in a constructed reed wetland were found to be influenced by the hydraulic loading, with optimal removal rates obtained within specific ranges of hydraulic loading and aeration time [21, 22]. Proper management of hydraulic loading rates, in conjunction with hydraulic retention time, can significantly enhance the removal of pollutants and improve the overall treatment efficiency of constructed wetlands.

Substrates are another major influencer governing constructed wetland systems' overall performance [18, 23]. The choice of substrate significantly influences the lifetime and efficiency of constructed wetlands, particularly in removing phosphorus and nitrogen [24, 25]. Studies have shown that the effective porosity of the substrate during the initial operation of constructed wetlands is essential for strong interception and adsorption effects, which are vital for pollutant removal [26]. Selecting suitable substrates reduces the required area, improves the purification performance, and prolongs the service life of constructed wetlands. Moreover, the substrate's role in transporting oxygen to the wetland is essential for nitrification [27].

Selecting suitable vegetation is also a crucial parameter in deciding the performance of constructed wetlands. *Canna indica* belongs to the tropical, monocotyledonous *Canna* family and is well recognized for its rapid growth rate, large biomass, and capability for nutrient removal, making it a commonly used plant for wetland establishment in various countries [28]. It has been extensively studied for its potential for treating wastewater with high nitrogen and phosphorus, showcasing its potential for efficient nutrient removal [29]. Moreover, the constructed wetland systems using *Canna indica* have demonstrated high removal efficiency for chemical oxygen demand, fats, oils, grease, and total suspended solids [30]. Furthermore, field-scale studies have shown that *Canna indica* exhibited higher efficiency in removing BOD than other plant species in vertical flow constructed wetlands [31].

Typha is another plant species extensively studied and utilized in vertical flow constructed wetlands for wastewater treatment. Research has shown that *Typha latifolia* is widely used as an emergent plant in constructed wetlands, demonstrating its potential for pollutant removal and transformation processes [32, 33]. Moreover, *Typha* management strategies, including biomass harvesting and aquatic connectivity treatments, have enhanced the diversity and functionality of wetlands, indicating the potential benefits of incorporating *Typha* in wetland management practices [34]. Additionally, *Typha latifolia* is known for its high efficiency in removing inorganic nutrients and organic matter from wastewater, further emphasizing its suitability for wastewater treatment in constructed wetlands [35]. Furthermore, the feasibility of using *Typha angustifolia* in vertical flow constructed wetlands for the treatment of heavy metals from effluents of the pulp and paper industry has been demonstrated, highlighting the potential of Typha species in addressing challenges caused by heavy metals such as iron, copper, manganese, nickel, and cadmium in wastewater [36].

Vertical flow constructed wetlands have gained attention for their effectiveness in wastewater treatment, particularly in removing nitrogen and phosphorus. The combination of vertical and horizontal flow constructed wetlands has been explored to achieve higher treatment efficiency, especially in removing ammonia and total nitrogen [37]. Vertical subsurface flow wetlands utilizing *Cyperus alternifolius* have efficiently removed COD, TP, and NH^{4+}-N from polluted river water. The substrate consisted of a sandy layer with varying diameters, larger sands near the reactor's tap, and medium to small sands toward the end, and some clay was also added. Wastewater was retained for 4 days in the wetlands before sampling for parameter analysis. The reactor with *Cyperus alternifolius* resulted in higher removal efficiencies than the control [38]. Another study demonstrated the efficiency of *Cyperus alternifolius* in removing total phosphorus from domestic wastewater in vertical flow constructed wetlands [39]. *Cyperus alternifolius* and *Canna indica* were used simultaneously to reduce organic matter and phosphate in domestic wastewater using subsurface flow constructed wetlands [40].

Through the systematic review of existing literature, this study seeks to propose VFCW to mitigate the pollution caused by sewage contamination in the Jayanthi Nallah to improve the water quality in the Panchaganga River.

2 Materials and Methods

Composite samples from six different locations were collected along the stretch of Jayanthi nallah. BOD, COD, TSS, TP, and TKN were analyzed for the raw wastewater screened through mesh to remove coarse material like leaves and toilet paper. A lab-scale model of VFCW with the dimensions of 0.62 m*0.70 m*0.45 m was developed. The substrate media contained four different layers of river sand, crushed sand, fine gravel, and coarse gravel, placed one below the other in the same order (Fig. 1). The substrate was compacted uniformly, ensuring consistent porosity and hydraulic conductivity throughout the wetland. Each layer contained particles of increasing sizes, and the thickness of the layers was 10 cm each for sand and 12 cm each for gravel. The vegetation was provided at 10 cm spacing. Five different conditions were selected by altering the vegetation and hydraulic loading rates, as shown in Table 1.

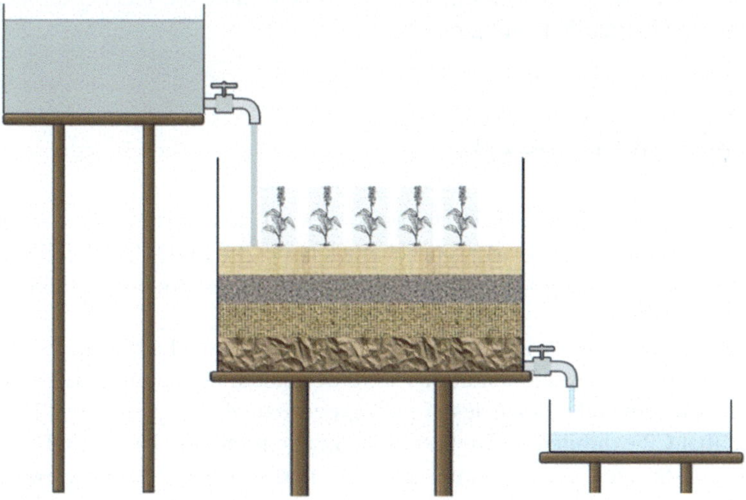

Fig. 1 Experimental setup

Table 1 Experimental conditions

Code	Vegetation	Hydraulic loading rate
C1	*Canna indica*	0.12
C2	*Canna indica*	0.18
C3	*Canna indica*	0.24
C4	*Canna indica* & *Typha*	0.18
C5	*Canna indica* & *Cyperus rotundus*	0.18

Table 2 Wastewater characteristics

Characteristics	Range	Desirable limits
pH	6.4–7.2	5.5–9
BOD	113–238 mg/L	< 30 mg/L
COD	339–541 mg/L	< 250 mg/L
TSS	219–331 mg/L	< 100 mg/L
TP	7.10–19.53 mg/L	< 5 mg/L
TKN	60–75 mg/L	< 100 mg/L

Table 3 Removal efficiencies

Code	BOD	COD	TSS	TP	TKN
C1	75.27	80.74	71.33	50.93	67.50
C2	81.23	86.53	75.86	58.34	75.72
C3	70.90	78.55	67.79	45.52	61.27
C4	78.05	85.50	80.55	57.36	75.05
C5	77.41	84.06	81.06	76.76	78.61

The BOD, COD, TSS, TP, and TKN were monitored at six-hour intervals for an acclimatization period of 8 days. The treatment efficiencies were evaluated for each experiment and compared with one another to determine the ideal combination of vegetation and hydraulic loading rate.

3 Results and Discussion

Raw water characteristics: The characteristics of the wastewater samples are presented in Table 2. The wastewater characteristics of BOD, COD, TSS, and TP were higher than the disposal standards prescribed by the American Public Health Association (APHA).

The laboratory tests were performed, and the removal efficiencies for each parameter in all five testing conditions were evaluated, as shown in Table 3. Upon comparing the removal efficiencies of each experimental test, as shown in Fig. 2, it is observed that C2 exhibited the highest efficiency in removing BOD, COD, and TSS from the wastewater. At the same time, C5 performed exceedingly well in the removal of total phosphorus. Lower efficiencies are acceptable since the TKN levels in the raw wastewater are within the desirable disposal standards.

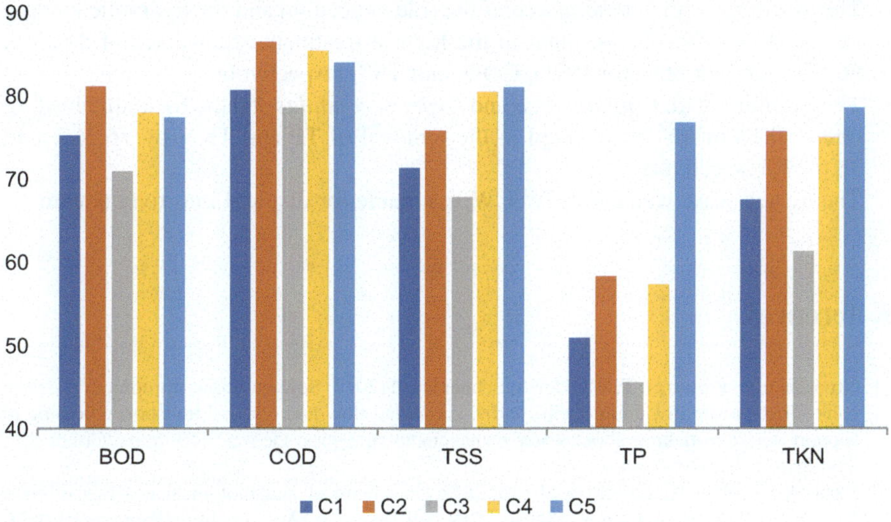

Fig. 2 Comparison of test conditions

Higher BOD and COD removal efficiencies in C2 are observed because of the large rhizomatous root network of *Canna indica*, which are thick, fleshy underground stems that grow horizontally. The root system of *Canna indica* is relatively shallow, providing extensive atmospheric oxygen transfer for bacterial growth. The overall TSS, TP, and TKN removal efficiencies were relatively higher in the multispecies VFCW due to the diverse root network provided by *Typha/Cyperus rotundus* with a fibrous root system that consists of many thin, branching roots. They offer a large surface area for bacterial attachment and create aerobic and anoxic microzones in wetland medium. However, unlike *Typha*, *Cyperus rotundus* also produces tubers or rhizomes underground, further enhancing the overall TP and TKN removal efficiency. The root exudates and oxygen transfer from the *Cyperus rotundus* roots promote the growth of phosphorus-accumulating bacteria. These bacteria play a crucial role in phosphorus uptake and removal. They enhance the breakdown and assimilation of phosphorus compounds.

4 Conclusions

The following conclusions and recommendations were made based on the experimental results:

- Removal efficiencies of all the combinations considered in this study were attaining the disposal standards according to the APHA.

- The assembly with *Canna indica* as the sole vegetation and the hydraulic loading rate of 0.18 $m^3/m^2/hr$. resulted in the highest treatment efficiencies of 81.23%, 86.53%, and 75.86% for BOD, COD, and TSS, respectively.
- The assembly with *Canna indica* and *Cyperus rotundus* and the hydraulic loading rate of 0.18 $m^3/m^2/hr$. resulted in the removal of TP and TKN by 76.76% and 78.61%, respectively.
- The treated wastewater from VFCW is suitable for disposal into river stream.

References

1. Groeschke M, Frommen T, Winkler A, Schneider M (2017) Sewage-borne ammonium at a river bank filtration site in Central Delhi, India: simplified flow and reactive transport modeling to support decision-making about water management strategies. Geosci 7:48. https://doi.org/10.3390/geosciences7030048
2. Joshi P, Chauhan A, Dua P et al (2022) Physicochemical and biological analysis of river Yamuna at Palla station from 2009 to 2019. Sci Rep 121(12):1–19. https://doi.org/10.1038/s41598-022-06900-6
3. Su C, Li Z, Wang W et al (2022) Key factors dominating the groundwater chemical composition in a grain Production base: a case study of Muling–Xingkai Plain, Northeast China. Water 14:2222. https://doi.org/10.3390/w14142222
4. Wen X, Bai Y, Zhang S et al (2020) Distributions and sources of sedimentary sterols as well as their indications of sewage contamination in the Guanting reservoir, Beijing. J Chem 1:3050687. https://doi.org/10.1155/2020/3050687
5. Froehner S, Martins RF, Errera MR (2009) Assessment of fecal sterols in Barigui River sediments in Curitiba, Brazil. Environ Monit Assess 157:591–600. https://doi.org/10.1007/S10661-008-0559-0
6. Vijay R, Samal D, Mohapatra PK et al (2011) GIS based identification and assessment of groundwater quality potential zones in Puri City, India. J Water Resour Protect 3:440–447. https://doi.org/10.4236/jwarp.2011.36054
7. Thorvat AR, Sonaje NP, Mujumdar MM (2015) Regression modeling and impact assessment of Panchaganga River water quality in Kolhapur City. Int J Adv Found Res Sci Eng 2:456. https://api.semanticscholar.org/corpusid:14884244
8. Joshi SS, Shivapur AV, Karjinni VV et al (2020) A review paper on characterization of domestic wastewater from nallah. Int Res J Eng Technol 7(5) https://www.irjet.net/archives/v7/i5/irjet-v7i5242.pdf
9. Mohite NA (2020) A study of physico-chemical characteristics of Jayanti Nalla water along with its tributaries with special emphasis on quality of Panchganga River. Int J Res Appl Sci Eng Technol 8:616–620. https://doi.org/10.22214/ijraset.2020.1112
10. Nakano K, Nakamura K, Yano T, Nishimura O (2014) First 3 years of performance in a planted and unplanted five-stage vertical flow constructed wetland treating milking parlor wastewater. J Water Environ Technol 12:177–189. https://doi.org/10.2965/jwet.2014.177
11. Usharani B, Vasudevan N (2021) Pollutant removal from sewage in tropical climate by constructed wetland system: an asset for irrigation. Int Lett Nat Sci 82:20–43. https://doi.org/10.56431/p-vicv20
12. Kadaverugu R (2016) Modeling of subsurface horizontal flow constructed wetlands using OpenFOAM®. Model Earth Syst Environ 2:1–10. https://doi.org/10.1007/s40808-016-0111-0
13. Kraiem K, Kallali H, Ammeri RW et al (2021) Effects of partial saturation on nitrogen removal and bacterial community in vertical-flow constructed wetlands. Int J Environ Chem 5:38–44. https://doi.org/10.11648/j.ijec.20210502.14

14. Chen AX, Ma Y, Liu XR, Chen JW (2011) Effect research of nitrogen removal out of sewage in constructed wetland. In: ISWREP 2011 – Proc 2011 Int Symp Water Resour Environ Prot, vol 2, pp 1279–1282. https://doi.org/10.1109/iswrep.2011.5893252
15. Soti A, Singh S, Verma V et al (2022) Designing the vertical flow constructed wetland based on targeted limiting pollutant. Bioresour Technol 351:127068. https://doi.org/10.1016/j.biortech.2022.127068
16. Turcios AE, Papenbrock J (2014) Sustainable treatment of aquaculture effluents—what can we learn from the past for the future? Sustain 6:836–856. https://doi.org/10.3390/su6020836
17. Tang Z, Wood J, Smith D et al (2021) A review on constructed treatment wetlands for removal of pollutants in the agricultural runoff. Sustain 13:13578. https://doi.org/10.3390/su132413578
18. Wang R, Xu L, Xu X et al (2021) Simulation and optimization of hydraulic performance of small baffled subsurface flow constructed wetland. Water Sci Technol 84:632–643. https://doi.org/10.2166/wst.2021.249
19. Land M, Granéli W, Grimvall A et al (2016) How effective are created or restored freshwater wetlands for nitrogen and phosphorus removal? A systematic review. Environ Evid 5:1–26. https://doi.org/10.1186/s13750-016-0060-0
20. Juang DF, Chen PC (2007) Treatment of polluted river water by a new constructed wetland. Int J Environ Sci Technol 4:481–488. https://doi.org/10.1007/bf03325984
21. Xie B, Cui Y, Yuan Q, Li J (2009) Pollutants removal and distribution of microorganisms in a reed wetland of Shanghai Mengqing Park. Environ Prog Sustain Energy 28:240–248. https://doi.org/10.1002/ep.10333
22. Wang H, An X, Yang Y et al (2016) Analyzing the removal effect of Nitrogen before and after enhanced Aeration in constructed wetlands. Pol J Environ Stud 25:2161–2166. https://doi.org/10.15244/pjoes/62711
23. Rajan RJ, Sudarsan JS, Nithiyanantham S (2019) Microbial population dynamics in constructed wetlands: review of recent advancements for wastewater treatment. Environ Eng Res 24:181–190. https://doi.org/10.4491/eer.2018.127
24. Babatunde AO, Zhao YQ (2009) Phosphorus removal in laboratory-scale unvegetated vertical subsurface flow constructed wetland systems using alum sludge as main substrate. Water Sci Technol 60:483–489. https://doi.org/10.2166/wst.2009.384
25. Li T, Jin Y, Huang Y (2022) Water quality improvement performance of two urban constructed water quality treatment wetland engineering landscaping in Hangzhou, China. Water Sci Technol 85:1454–1469. https://doi.org/10.2166/wst.2022.063
26. Xu Q, Cui L (2019) Removal of COD from synthetic wastewater in vertical flow constructed wetland. Water Environ Res 91:1661–1668. https://doi.org/10.1002/wer.1168
27. Qian C, Li Z, Ding Y, et al (2016) Substrates for pollution control in constructed wetland systems, pp 66–69. https://doi.org/10.2991/icseee-16.2016.11
28. Zhang Y, Hu C, Luo WS (2014) Effects of pH manipulation, biological reduction, and plant growth on Cu2+ and Zn2+ removal from mine drainage using iron-oxide-enriched red earth soils. CLEAN—Soil Air Water 42:1272–1279. https://doi.org/10.1002/clen.201300440
29. Haritash AK, Sharma A, Bahel K (2015) The potential of canna lily for wastewater treatment under Indian conditions. Int J Phytoremediation 17:999–1004. https://doi.org/10.1080/15226514.2014.1003790
30. Rahmadyanti E, Audina O (2020) The performance of hybrid constructed wetland system for treating batik wastewater. J Ecol Eng 21:94–103. https://doi.org/10.12911/22998993/118292
31. Barya MP, Gupta D, Thakur TK et al (2020) Phytoremediation performance of *Acorus calamus* and *Canna indica* for the treatment of primary treated domestic sewage through vertical subsurface flow constructed wetlands: a field-scale study. Water Pract Technol 15:528–539. https://doi.org/10.2166/wpt.2020.042
32. Maddison M, Mauring T, Remm K et al (2009) Dynamics of Typha latifolia L. populations in treatment wetlands in Estonia. Ecol Eng 35:258–264. https://doi.org/10.1016/j.ecoleng.2008.06.003

33. Malyan SK, Yadav S, Sonkar V et al (2021) Mechanistic understanding of the pollutant removal and transformation processes in the constructed wetland system. Water Environ Res 93:1882–1909. https://doi.org/10.1002/wer.1599
34. Lishawa SC, Schrank AJ, Lawrence BA et al (2023) Aquatic connectivity treatments increase fish and macroinvertebrate use of Typha-invaded Great Lakes coastal wetlands. Freshw Biol 68: 1462–1477. https://doi.org/10.1111/fwb.14141
35. Camacho A, Picazo A, Rochera C et al (2018) Serial use of helosciadum nodiflorum and typha latifolia in mediterranean constructed wetlands to naturalize effluents of wastewater treatment plants. Water 10:717. https://doi.org/10.3390/w10060717
36. Arivoli A, Mohanraj R, Seenivasan R (2015) Application of vertical flow constructed wetland in treatment of heavy metals from pulp and paper industry wastewater. Environ Sci Pollut Res 22:13336–13343. https://doi.org/10.1007/s11356-015-4594-4
37. Vymazal J (2010) Constructed wetlands for wastewater treatment. Water 2:530–549. https://doi.org/10.3390/w2030530
38. Ebrahimi A, Taheri E, Ehrampoush MH et al (2013) Efficiency of constructed wetland vegetated with *Cyperus alternifolius* applied for municipal wastewater treatment. J Environ Public Health 2013:815962. https://doi.org/10.1155/2013/815962
39. Cui L-H, Zhu X-Z, Ouyang Y et al (2011) Total phosphorus removal from domestic wastewater with Cyperus alternifolius in vertical-flow constructed wetlands at the microcosm level. Int J Phytoremediation 13:692–701. https://doi.org/10.1080/15226514.2010.525552
40. Trifando RY, Sutanto HB, Prihatmo G (2022) The reducing of organic loading and phosphate (Po4) in domestic wastewater treatment by constructed wetland system using canna Indica and Cyperus Alternifolius. BIOLINK (Jurnal Biol Lingkung Ind Kesehatan) 9:95–105. https://doi.org/10.31289/biolink.v9i1.6837

Groundwater Quality Assessment for Drinking and Irrigational Purposes Using Water Quality Index (WQI) of Dubbak (M), Siddipet District, Telangana, India

G. Subba Rao, G. Sri Harsha, Syed Hamim Jeelani, Syed Shams Rizvi, S. K. Roy, and M. S. Manjaly

Abstract Hydrogeochemical methods have been carried out in Dubbak (M) of Siddipet district, Telangana, to understand the limitations of major cations viz., sodium (Na^{+}), potassium (K^{+}), calcium (Ca^{2+}), magnesium (Mg^{2+}), and major anions carbonate (CO_3^{-}), bicarbonates (HCO_3^{-}), chloride (Cl^{-}) fluoride (F^{-}), nitrate (NO_3-), and sulfate (SO_4^{2-}) present in the groundwater. Groundwater sampling has been done from the 14 borewell locations in the study area and examined at TSGWD, Hyderabad. The analysis has been done based on BIS 2012. The highest concentrations of fluoride have been recorded in Thimmapur (1.9 ppm), Dubbak (1.81 ppm), Lachhapet (1.65 ppm), and Boppapur (1.54 ppm) whereas the highest concentrations of nitrate were marked at Akaram (125 ppm) and Thimmapur (55 ppm). Higher concentrations of F and N are due to weathering of fluoride-bearing minerals and anthropogenic sources, respectively. The US Salinity diagram depicts the C3-S1 zone. Gibbs's plot indicates that rock-weathering processes control the geochemistry of the study region. Wilcox's plot showed that 57% of the total samples belong to the good to permissible, whereas 43% belong to the excellent to good category. The Water Quality Index (WQI) represents 35% of the total samples belonging to a very poor category, 57% belonging to an unsuitable category, and 0.07% belonging to the poor category.

G. S. Rao
Department of Civil engineering, Nalla Narasimha Reddy Group of Institutions, Hyderabad, Telangana, India

G. S. Harsha (✉)
Department of Civil Engineering, Maturi Venkata Subba Rao Engineering College, Nadergul, Telangana, India

S. H. Jeelani
Department of Civil Engineering, KL Deemed to be University, Guntur, India

S. S. Rizvi · S. K. Roy · M. S. Manjaly
Department of Geology, Central University of Karnataka, Kalaburagi, Karnataka, India

P. Pathak et al. (eds.), *Advances in Environmental Sustainability, Energy and Earth Science*, https://doi.org/10.1007/978-3-031-73820-3_19

Keywords Contamination · Fluoride · Nitrate · Water quality · Water Quality Index (WQI) · Quality assessment

1 Introduction

Water is a natural resource and a prerequisite for life as water makes up 65% of the human body and offers home for a wide variety of aquatic species in rivers, lakes, and the ocean. Rainfall or yearly precipitation replenishes water supplies. It is estimated that there are 1.37 × 108 million hectare-meters (ha-m) of water resources in the world. Remarkably, 2.8% of the world's water resources are always accessible as fresh water, while the remaining 97% of water resources are salty, mostly found in seas. Just 0.6% of the 2.8% total amount of water stated is subsurface water, while around 2.2% is surface water. In addition to significantly increasing water usage, the growth of agriculture and industrial development has also had an impact on the quality of the water. Water has a high dissolution capacity, which makes it easily contaminated. Raindrops are accumulating contaminants even before they make contact with the ground. Once on the ground, objects quickly settle in the water, which then becomes contaminated due to the release of fertilizers, pesticides, and other substances into the subsurface due to host rock fissures or joints. Water quality may be determined by measuring its physical and chemical characteristics, which is crucial in determining whether or not it is suitable for agricultural use. The characteristics of the water, how it interacts with the host rock it comes into touch with, and the processes that take place in both.

The industrial sector uses a lot of water and pollutes it in the process. Water is contaminated by residential sources and agriculture in addition to industry. Moreover, chemicals and biological organisms that contaminate water are released into the environment. Since groundwater is being contaminated, which affects aquatic life and human health, water pollution has become a significant problem in India throughout time. If the water quality of a pond declined significantly, fishing would have to be discontinued completely and the types of fish cultivated there would need to be changed. Humans have health issues as a result of harmful germs found in groundwater.

Although the groundwater is usually drinkable, issues arise from the increased ion concentrations. Anthropogenic activities (domestic and agricultural) and geographic elements (rock–water interactions) intensified due to the fast urban population expansion and agricultural development, particularly in hard rock terrain. Recent studies have demonstrated the significance of examining the effects of different variables on hydrogeochemistry. In Ghatkesar Mandal, the groundwater contribution for drinking and agricultural purposes is around 95% in rural regions without water treatment or transportation. When the soil is excellent, crops, especially paddy, yield well. The agricultural sector's appropriateness for irrigation water is determined by several factors, including climate, drainage, soil type, salt tolerance, and water quality. Considering the Ghatkesar Mandal groundwater problems.

Due to urbanization and the growth of agricultural operations, there is a demand for groundwater in the Dubbak Mandal region of Siddipet District, especially in places with hard rock terrain. Several factors include the way that rocks and water interact, as well as residential, agricultural, and other activity processes involving hydrogeochemistry. A recent experiment made clear how important it is to look at how various factors affect hydrogeochemistry [1]. In Dubbak Mandal, groundwater provides around 95% of the water utilized for agriculture and consumption in rural areas where water treatment and transportation are lacking. When the soil in the research area is healthy, crops like paddy yield well. A number of variables, including drainage, climate, and water quality, affect how suitable a body of water.

1.1 Research Area

The Dubbak Mandal study area is a part of the Siddipet revenue division, and the Dubbak town is around 24 kilometers from the Siddipet district headquarters. Dubbak Mandal was reorganized from Medak (Dist.) to Siddipet (Dist.) as part of their organization of Telangana's districts, and it now exists as a municipality. A total of 26.40 sq. km falls under the purview of Dubbak Mandal. The research area covers the survey of India Toposheet E44Gl2 and is located between latitude 18° 6′ 17.91" to 18°15' 6.87" N and longitudes 78° 32′ 56.82" to 78° 44' 48.81"E (Fig. 1).

The Siddipet-Ramayampet-Medak Road, which runs across the southern half of the area, provides access to the area. The area has hillocks including Nalla Gutta (524′) and other isolated hillocks (573′), which rise from the general level farmed

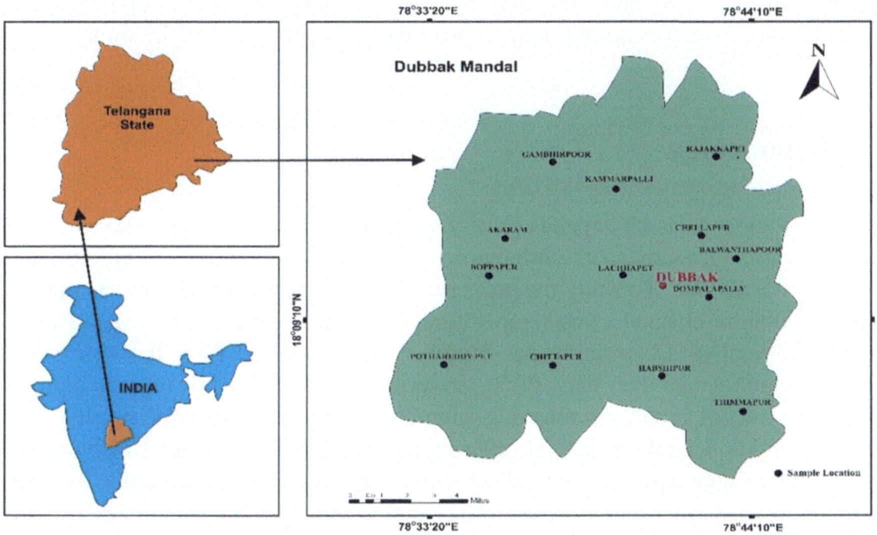

Fig. 1 Groundwater samples from the study area

Table 1 Rainfall data of Dubbak Mandal

S.No.	Year	Rainfall in mm		
		Actual	Normal	Deviation
1	2011–2012	623.4	859.6	−27.5
2	2012–2013	692.6	859.6	−19.4
3	2013–2014	9.0	859.6	−99.0
4	2014–2015	437.4	859.6	−49.1
5	2015–2016	516.8	859.6	−40.0
6	2016–2017	784.9	859.6	−9.0
7	2017–2018	679.5	859.6	−21.0
8	2018–2019	608.3	859.6	−29.0
9	2019–2020 (upto Feb 2020)	1065.6	811.4	3.10

soil at about 450′ feet. These mounds are made up of granite outcrops. The prominent ridges in the study area created these hillocks, which served as a watershed for a few streams and rivulets that flowed south to northwest and formed tributaries to the river Kodli. The exposures are from the Archaean era, which occupies the basement rock, is seen and predominantly composed of crystalline rocks namely Granite, which also serves as a significant limit. The region is tropical and hot, sub-humid temperatures, and is primarily covered by granitic topography that dates to the Archaean. The temperatures range from 18 °C to 42 °C and the area experiences rainfall mostly during the months of July, August, September, and October every year. Because of depressions brought on by the Bay of Bengal, there are only a few days per year with high rainfall.

As shown in Table 1, the average precipitation from 2011–2012 to 2018–2019 ranged from 437 to 784 mm. Annual average precipitation up to Feb 2020 is reported as 1065.6 mm, which is higher than the average precipitation of 859.6 mm.

2 Methodology

Hydrogeological research depends greatly on the quality of the water. The chemical composition of groundwater is influenced by geology, topography, structure, and agricultural activities. To study the properties like groundwater recharge, its mixing areas, circulation channels, and temperature at depth, water chemistry data can be utilized. A total of 14 groundwater samples were collected from the bore wells between January and February 2020, covering the entire study region. Normal sampling protocol involves collecting samples in pre-cleaned plastic bottles. The samples were collected for analysis after meticulous labeling. Figure 1 displays the places where water samples were collected in the research region and Table 2 shows the analytical data of water samples from Dubbak Mandal. The Chemistry Laboratory of the Telangana State Groundwater Department in Hyderabad examined the collected samples. The parameters EC and pH were measured on the sport itself and other parameters were chemically evaluated according to [2] the procedures to

Table 2 Analytical data of water samples of Dubbak Mandal

S. No.	Village	pH at 25 degrees C	**EC 25°Cmic. sie**	TDS calculated mgL	Na (mg/L)	K (mg/L)	Ca (mg/L)	Mg (mg/L)	CO3 (mg/L)	HCO3 (mg/L)	Cl (mg/L)	SO4 (mg/L)	Nitrate (mg/L)	F (mg/L)	Total Hardness as CaCO3Cl (mg/L)
BIS permissible limits		6.50 to 8.50	<750 = good 750–1500 = safe 1500–3000 permissible	500–2000	No guidelines	No guidelines	75–200	30–100	200–600	200–600	200–1000	200–400	45	1.0–1.5	200–600
1	Pothareddy pet	6.71	1030	659	97	3	72	39	0	390	90	16	15	1.05	340
2	Chittapur	7.4	1321	845	119	6	80	44	0	310	210	19	9	1.07	380
3	Boppapur	7.32	1733	1109	207	2	56	63	0	450	250	27	4	1.54	400
4	Akaram	7.26	1893	1212	106	3	200	39	0	410	270	21	125	0.97	660
5	Gambhirpoor	7.75	866	554	37	2	96	29	0	270	100	12	9	1.36	360
6	Kammarpalli	7.51	785	502	34	2	72	39	0	320	50	11	6	0.64	340
7	Lachhapet	7.77	1163	744	108	3	88	29	0	350	150	8	0.4	1.65	340
8	Dubbak	7.21	1415	906	98	4	120	49	0	380	220	11	12	1.81	500
9	Chellapur	7.07	881	564	38	2	120	15	0	340	40	16	30	1.35	360
10	Rajakkapet	7.34	713	456	20	2	120	15	0	350	20	8	22	0.89	360
11	Balwanthapoor	6.8	611	391	27	2	64	34	0	240	50	11	40	1.36	300
12	Dompalapally	7.56	1211	775	84	3	136	24	0	360	160	12	25	1.10	440
13	Habshipur	7.8	845	541	45	3	112	24	0	340	70	11	27	0.94	380
14	Thimmapur	7.24	1744	1116	82	1	112	58	0	440	130	10	55	1.90	520

identify the chemical composition of each parameter [3]. Major cations, anions, pH, EC, and TDS were tested using standards set by BIS [4].

3 Results and Discussions

3.1 pH

The pH of water measures its acidity or basicity and ranges between 6.5 and 8.5 are suitable for drinking (BIS, 2012). A pH value close to 7.0 at 25 °C is neutral like pure water. pH ranges from 6.71 in Pothareddy Pet village to 7.80 in Habshipur village, indicating that the pH is within acceptable limits. Figure 2 depicts the geospatial distribution of pH concentration within the study region.

3.2 Specific Conductance

The ability of water to carry electric current determines its suitability for various purposes. Conductance depends on the concentration of ions dissolved in the

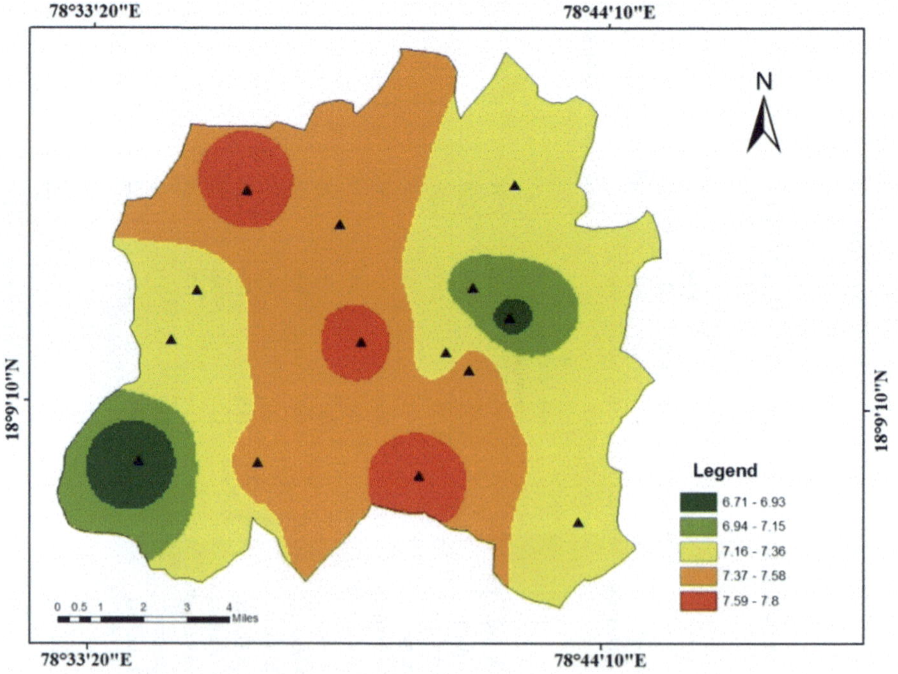

Fig. 2 Geospatial distribution map for pH variation

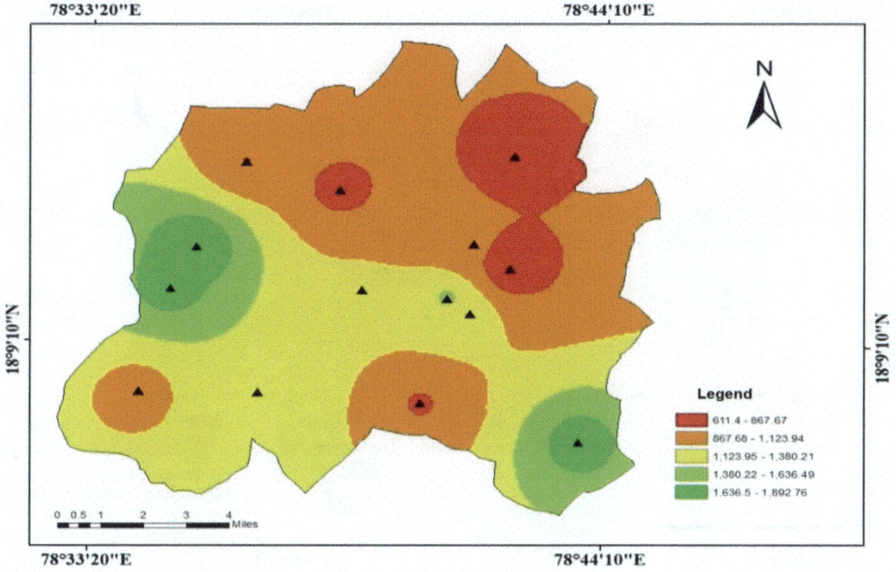

Fig. 3 Geospatial distribution map of specific conductance

solutions [5]. The EC will be higher the more ionizable solids there are. The EC within 300 μSiem is desirable for drinking purposes. A minimum of 611 μSiem (Balwanthapoor village) to a maximum of 1893 μSiem (Akaram village) is measured from the study area. All 14 villages are observed to fall out of the desirable limits. Figure 3 depicts the geospatial distribution of specific conductance concentration within the study region.

3.3 *Total Dissolved Solids*

The total mass of dissolved cations or anions is known as the total dissolved solids (TDS). The charges on cations and anions should be equal. A higher TDS indicates the presence of more cations and anions in the water and an increase in salinity. A TDS above 500 mg/L is not recommended for drinking water and water with a TDS between 1500 and 2600 mg/L is considered troublesome for irrigation use, specifically for crops with low salt tolerance [6]. High TDS concentrations cannot be suitable for bathing and washing [7]. The TDS from the study area ranges from 391 mg/L (Balwanthapoor village) to 1212 mg/L (Akaram village) which points out the water is suitable for agriculture, but except for Balwanthapoor (391 mg/L) and Rajakkapet (456 mg/L) villages, the other 13 samples are not recommended for drinking purposes. Figure 4 depicts the geospatial distribution of TDS concentration within the study region.

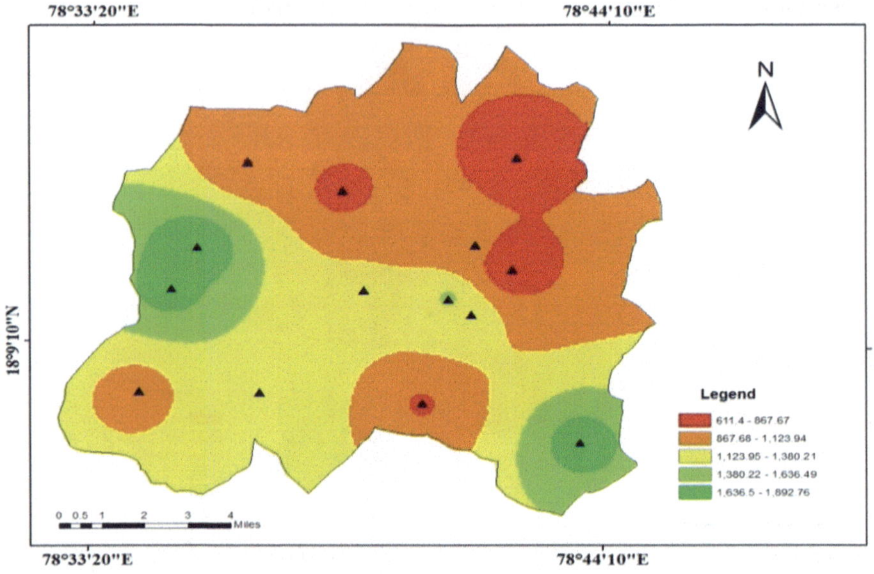

Fig. 4 Geospatial distribution map of TDS

3.4 *Calcium and Magnesium*

Ca and Mg concentrations are related to water hardness, and rocks such as limestone and dolomite is possibly the secondary sources in surface and subsurface waters for the ions in addition to cation exchange. Magnesium has been found to come from all three types of rocks: Igneous rocks (basalt, dunite, and pyroxene), metamorphic rocks (amphibolites, talc, tremolite, and shale), and sedimentary rocks (dolomite, gypsum) [8]. Figures 5 and 6 depict the geospatial distribution of Ca and Mg concentration within the study region. The Akaram village (200 mg/L) has a maximum concentration of Ca^{2+} and the minimum is found in Boppapur Village (56 mg/L). Whereas the magnesium content ranges between 15 mg/L (Chellapur and Rajakkapet) and 63 mg/L (Boppapur). These cations are important and involved in many enzymatic systems that are essential for several biological processes, such as blood formation, proper functioning of the heart, and prevention of neoplastic disease. However, an increase or decrease in their levels can be detrimental to life. All magnesium samples are within the safety limits of WHO whereas calcium 50% of the samples lie outside the WHO limits [9].

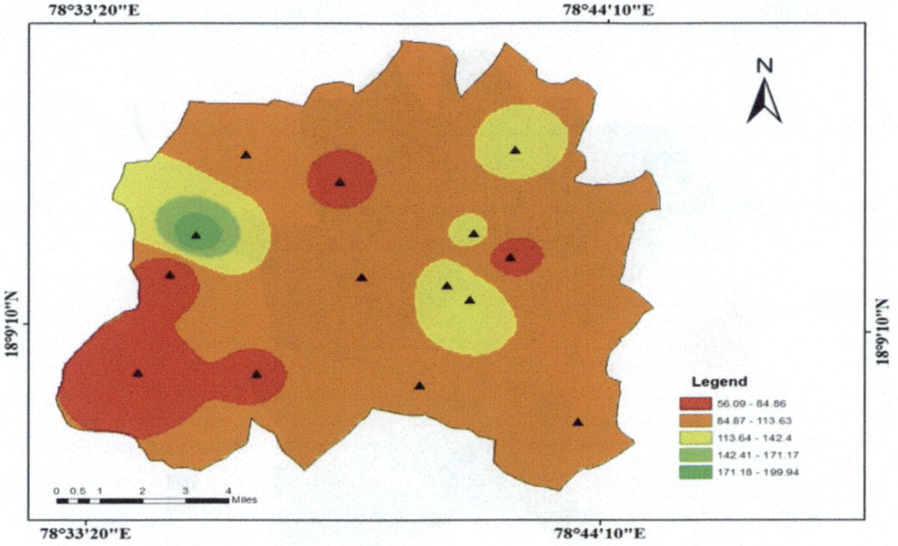

Fig. 5 Geospatial distribution map of Ca

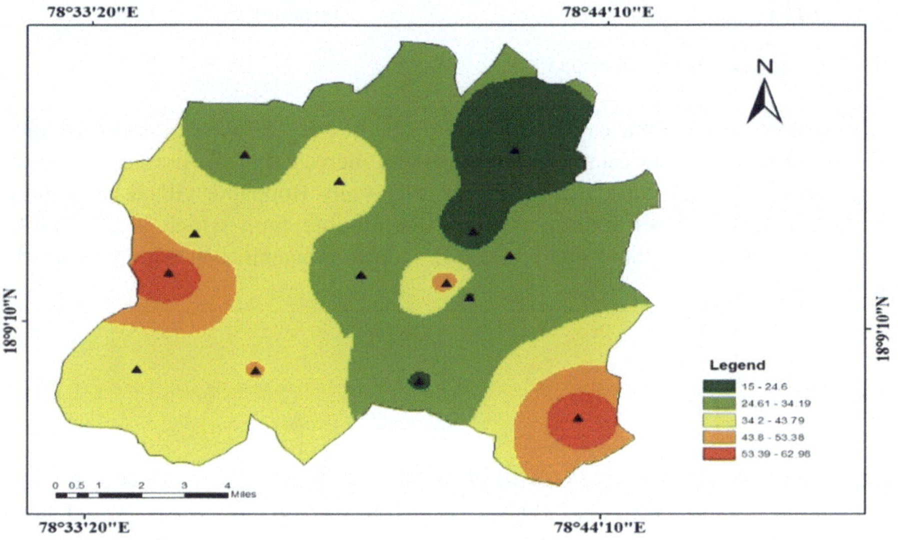

Fig. 6 Geospatial distribution map of Mg

3.5 *Sodium*

Long-term intake of heavy metals in water can lead to serious illnesses such as Bladder cancer, lung cancer, and skin cancer. Sodium (Na) is naturally present in

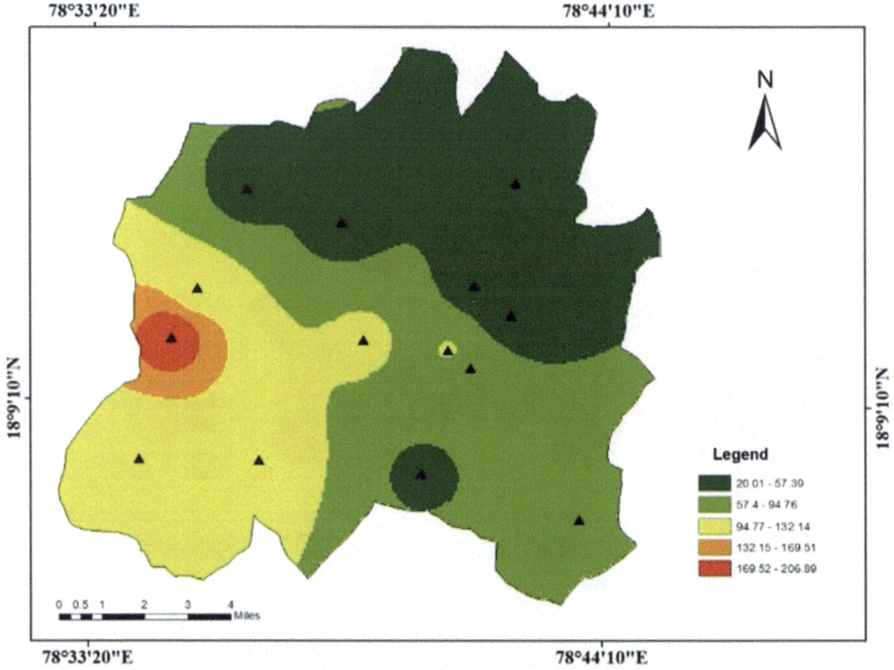

Fig. 7 Geospatial distribution map of Na

water that we use and it is essential for us to stay healthy, however, excess amounts of sodium in water can cause hypertension and increased blood pressure. Sodium ranges from 20 mg/L to 207 mg/L in this area. Only Boppapur village has a little higher concentration of sodium than the permissible limit which is 207 mg/L. Figure 7 depicts geospatial distribution of sodium concentration within the study region.

3.6 Potassium

Potassium is required for the growth of plants and animals. The excess amount of potassium can cause extensive kidney damage if it is already there, usually high level of potassium does not affect everybody. It varies between 1 mg/L and 6 mg/L in the study area within the permissible limit of 12 mg/L. Figure 8 depicts the geospatial distribution of potassium concentration within the study region.

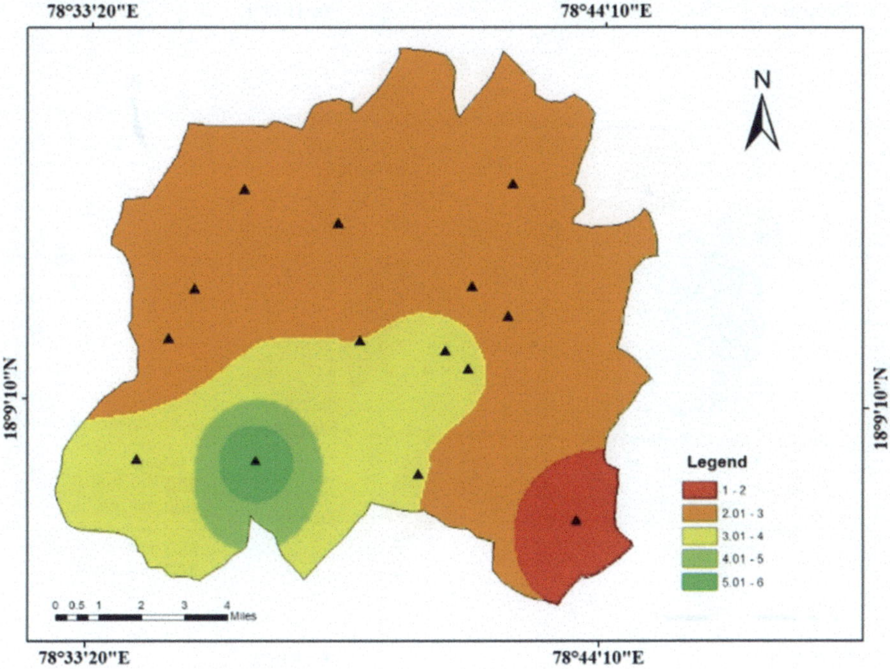

Fig. 8 Geospatial distribution map of K

3.7 Chloride

High chloride concentrations make water more corrosive, affecting human health and causing eye and nose irritation and stomach problems. Irrigation water with chloride concentrations above 350 mg/L is cropped. In the present area, concentrations ranged from 20 mg/L to 270 mg/L. Samples from the villages of Chittapur (210 mg/L), Dubbak (220 mg/L), Boppapur (250 mg/L), and Akaram (270 mg/L) belonged to the undesirable category classified, require proper treatment before use. Figure 9 depicts the geospatial distribution of chloride concentration within the study region.

3.8 Fluoride

Fluoride (F) concentration in groundwater depends on factors such as TDS, Alkalinity, Hardness, and pH of water. According to BIS, Fluoride concentrations in water should be less than 1.5 mg/L [4] and have set safe exposure limits for fluoride in drinking water between 0.5 and 1.5 mg/L [9]. If water containing excess fluoride is consumed for a long period, it may cause dental fluorosis, skeletal fluorosis, and

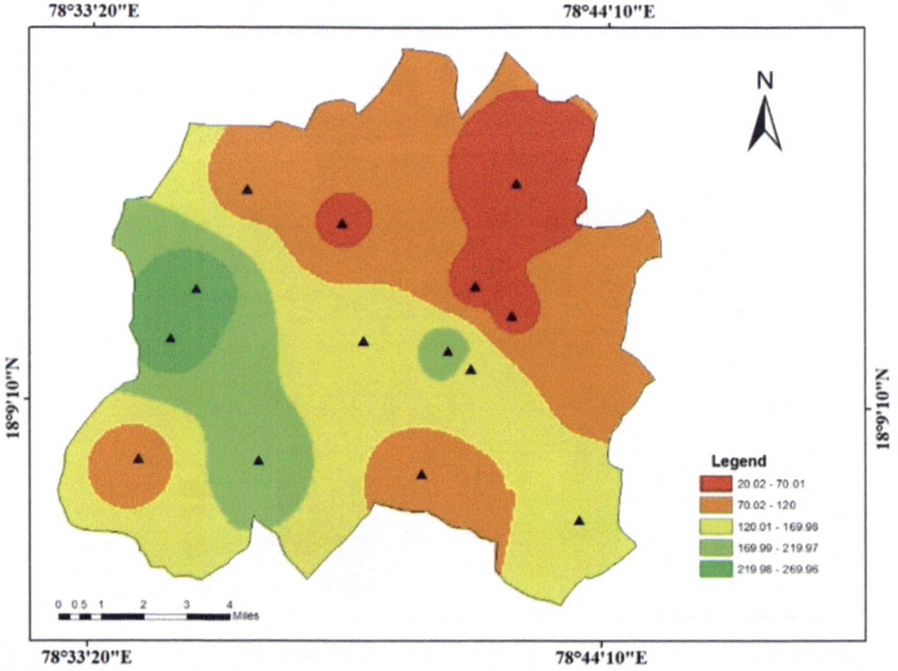

Fig. 9 Geospatial distribution map of chloride

non-skeletal fluorosis. The concentration of Fluoride ranges between 0.64 mg/L and 2.74 mg/L in the study area. Four out of 15 villages namely, Lachhapet (1.65 mg/L), Dubbak (1.81 mg/L), and Thimmapur (1.9 mg/L), show excess fluoride content. Figure 10 depicts the geospatial distribution of fluoride concentration within the study region.

3.9 *Sulfate*

Sulfate is a fully ionized form of sulfur combined with metals or hydrogen ions and is present in water that we use for drinking. The acceptable limit for this ion by BIS is 200 mg/L [4]. People with chronic heart or lung disease can experience decreased lung function, worsening asthma symptoms, and death if $_{SO4}$ concentrations in their drinking water are too high. The concentration ranges from 8 mg/L to 27 mg/L which is under the desirable limit in the study area. Figure 11 depicts the geospatial distribution of sulfate concentration in the area.

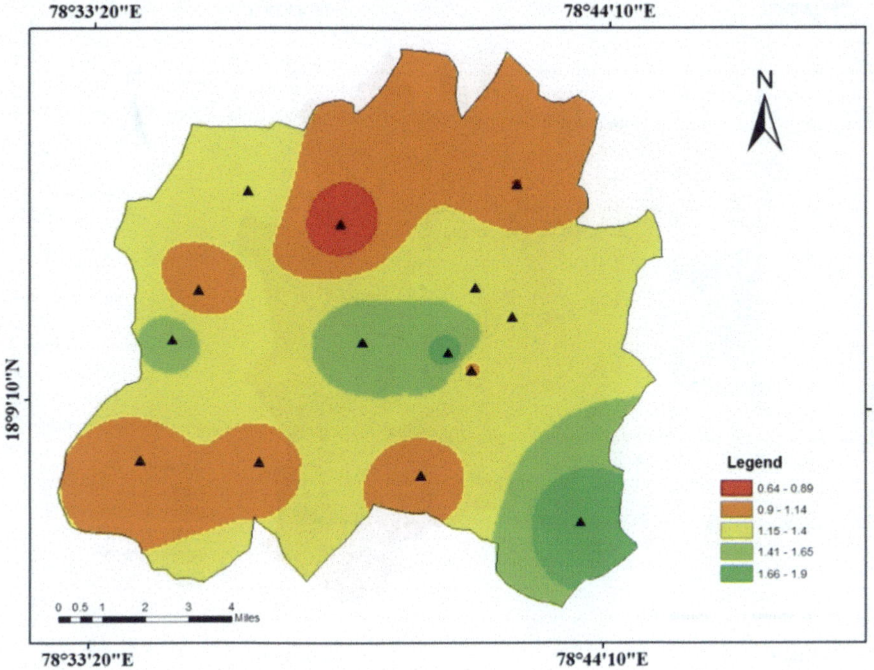

Fig. 10 Geospatial distribution map of F

3.10 Nitrate

Nitrate (NO_3) is a stable combined form of nitrogen for oxygenated systems, and it occurs naturally in water. According to BIS standards the level of nitrate should not exceed 45 mg/L [4]. If water containing excess nitrate is consumed for a long period, then it may cause diseases such as methemoglobinemia, cancer, increased infant mortality, hypertension, diabetes, and respiratory tract infection [10, 11]. Nitrate concentration ranges from 0.4 mg/L to 238 mg/L. A small increase in Nitrate levels from Thimmapur village (55 mg/L) to immense levels in Akaram (125 mg/L) villages can cause serious health issues for the population of the area. Figure 12 depicts the geospatial distribution of nitrate concentration within the study region.

3.11 Carbonate and Bicarbonate

Figure 13 depicts the geospatial distribution of HCO_3 concentration in the study area.

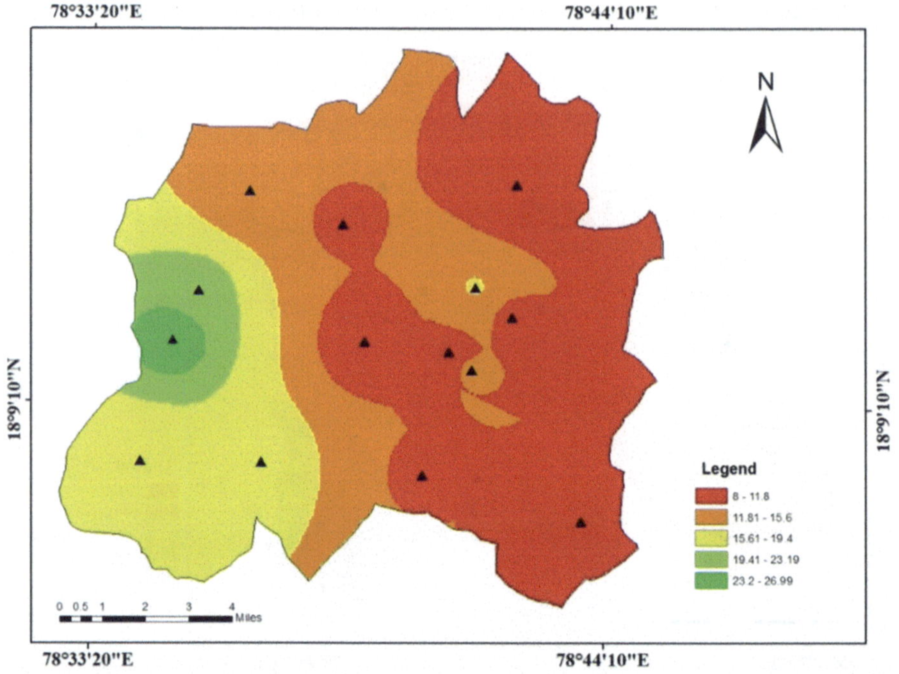

Fig. 11 Geospatial distribution map of SO_4

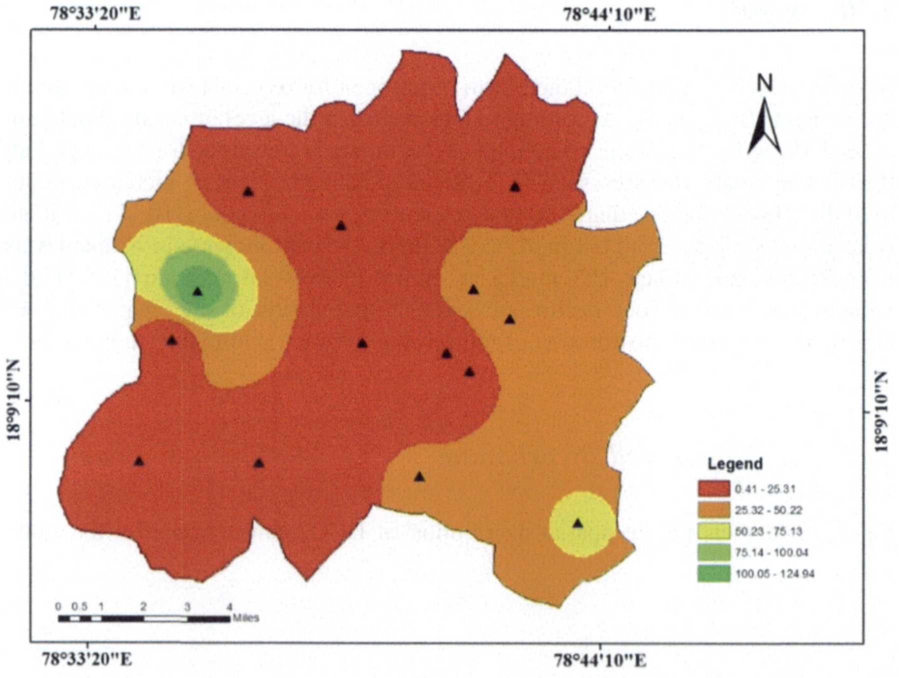

Fig. 12 Geospatial distribution map of NO_3

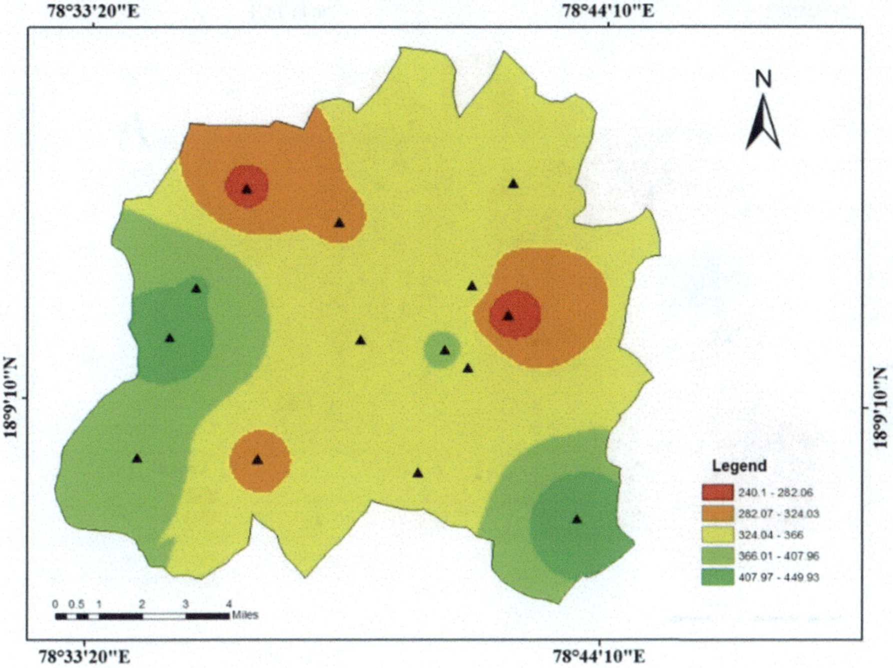

Fig. 13 Geospatial distribution map of HCO_3

3.12 *Total Hardness*

Water becomes hard when CA and Mg are present in excess quantity. Hard waters usually originate in places covered by thick topsoil or limestone formations. TH ranges from 300 mg/L (Balwanthapoor village) to 660 mg/L (Akaram village) in the study area. Figure 14 depicts the geospatial distribution of HCO_3 concentration. Table 3 represents the drinking water compatibility of Dubbak Mandal.

3.13 *Water Quality Index*

WQI is used to assess the suitability of water for agriculture, and it was developed by Horton [12] and the mathematical formula developed allows us to represent water quality by considering multiple parameters by a single value. For the calculation of WQI, 10 parameters were used (pH, Specific conductance, TDS, Ca, Mg, HCO_3, so_4, NO_3, F, Total Hardness, and Mg). For the calculation of WQI, the given formula was used. Table 4 represents the status categories of WQI whereas Table 5 represents the WQI of the collected samples.

For the calculation of WQI, the formula was used as shown in Eq. (1)

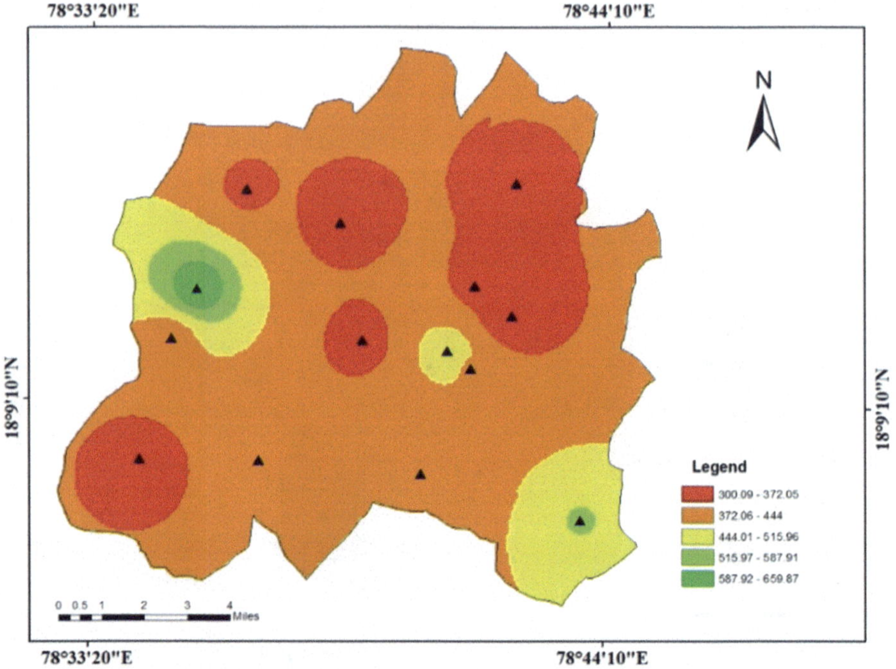

Fig. 14 Geospatial distribution map of TH

$$WQI = \frac{W_n Q_n}{\sum W_n} \tag{1}$$

where, $Wn = \frac{K}{S_n}$; $K = \frac{1}{\sum \frac{1}{S_n}}$; $Q_n = \frac{V_n}{S_n}\ x\ 100$;

K is constant for each parameter, S_n is the recommended standard by BIS, V_n is the mean value. WQI varies from 62.26 (the lowest) to 168.08 (the highest). Of all the samples Kammarpalli with 62.26 WQI is suitable for irrigation purposes and the remaining samples are unsuitable. Figure 15 depicts the geospatial distribution of WQI within the study region.

3.14 Agricultural Suitability

The suitability of agricultural water is evaluated by indicators of salinity, and chlorine content. Higher concentrations of sodium consistently affect soil irrigation capacity. Sodium in water replaces calcium in the soil which causes a decrease in soil

Table 3 Drinking water compatibility of Dubbak Mandal as per BIS-2012 [4]

S.No.	Parameters	Range	No. of samples	% of sample	BIS Standard (IS-10500)
1	pH	<6.5 6.5–8.5 >8.5	14	100.00	6.5–8.5
2	EC	<500 500–1000 >1000	6 8	42.85 57.14	–
3	TDS (mg/L)	<500 >500	2 12	14.28 85.71	500
4	Ca (mg/L)	<75 75–200 >200	4 10	28.57 71.43	75
5	Mg (mg/L)	<30 30–100 >100	6 8	42.85 57.14	30
6	Sodium (mg/L)	<200 >200	13 1	92.86 7.14	–
7	Potassium (mg/L)	<40 >40	14	100.00	–
9	Chloride (mg/L)	<250 >250	13 1	92.86 7.14	250
10	Fluoride (mg/L)	<1 1–1.5 >1.5	4 6 4	28.57 42.85 28.57	1.0
11	Nitrate (mg/L)	<45 >45	12 2	85.71 14.29	45
12	Sulfate (mg/L)	<200 >200	10 4	71.43 28.57	200
13	Bicarbonate (mg/L)	<200 >200	14	100.00	200

Table 4 Status categories of WQI

WQI	Water quality
0–25	Excellent
26–30	Good
31–75	Poor
76–100	Very poor
Above 100	Unsuitable

permeability. Excess Na^+ can affect plants either by toxic effects on leaf/root uptake; or K^+ and Ca^{2+} deficiencies can occur when soil or irrigation water has high Na^+ concentrations [13]. Sodicity is estimated using SAR, SSP, %Na, and RSC figures and Table 6 represents the classification of groundwater based on irrigation water quality parameters.

Table 5 Water Quality Index of the collected samples

S. No.	Village name	WQI	Inference
1	Pothareddy pet	94.65	Very poor
2	Chittapur	97.58	Very poor
3	Boppapur	136.94	Unsuitable
4	Akaram	92.84	Very poor
5	Gambhirpoor	122.14	Unsuitable
6	Kammarpalli	62.26	Poor
7	Lachhapet	141.62	Unsuitable
8	Dubbak	157.96	Unsuitable
9	Chellapur	116.41	Unsuitable
10	Rajakkapet	80.55	Very poor
11	Balwanthapoor	119.06	Unsuitable
12	Dompalapally	100.93	Unsuitable
13	Habshipur	88.98	Very poor
14	Thimmapur	168.08	Unsuitable

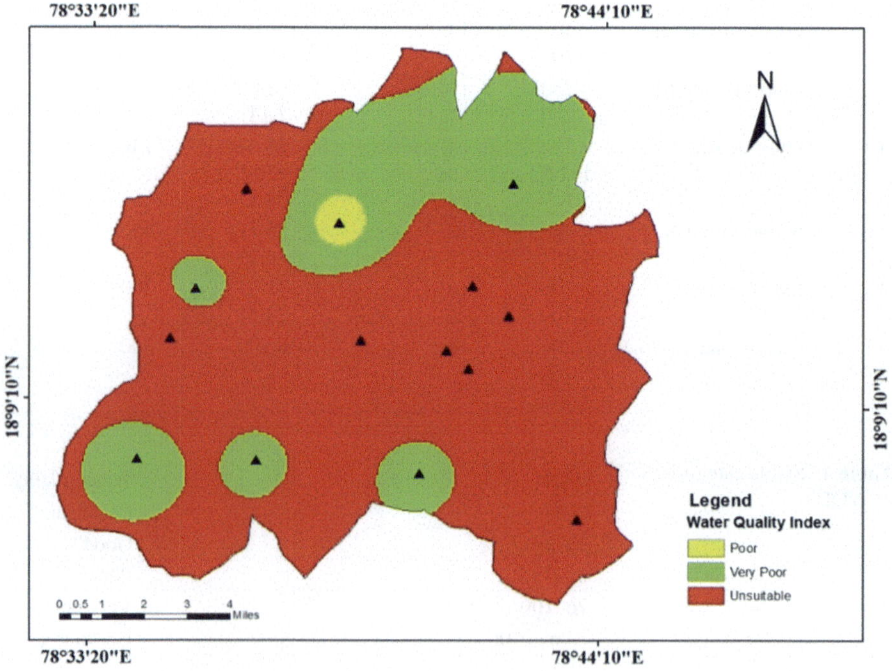

Fig. 15 Geospatial Distribution map of WQI

Table 6 Groundwater classification

Parameters	Scale	Inference	Total samples	% of samples	Reference
Electric conductivity (EC)	0–250 251–750 750–2250 2251–6000	Low Medium High Very high	0 1 13 0	7.14 92.86	Handa(1969)
Total dissolved solids (TDS) in mg/L	<1000 1000–1500	Fresh water Brackish water	11 3	78.57 21.43	Carroll(1962)
SSP (soluble sodium percentage) in meq/L	<=50 >50	Good quality Unsuitable	13 1	92.86 7.14	Carroll(1962)
Sodium absorption ratio (SAR) in meq/L	<10 10–18 18–26 >26	Excellent Good Permissible Doubtful	14	100.00	Richards (1954)
Residual sodium carbonate (RSC) in meq/L	<1.25 1.25–2.5 >2.5	Safe Marginal Unusable	14	100.00	Eaton (1950)
Percentage sodium (%Na)	<20 20–40 40–60 60–80 >80	Perfect Good Permissible Doubtful Unusable	5 6 3	35.71 42.86 21.43	Wilcox(1955)
Kell's Index (KI) in meq/L	<1 >1	Suitable Unsuitable	13 1	92.86 7.14	Kelly (1940)
PI (permeability index) in meq/L	Class III (<25%) Class II (25–75%) Class I (>75%)	Unsuitable Suitable Suitable	0 14 0	100.00	Doneen(1964)
Magnesium hazards(MH)	<50 >50	Suitable Unsuitable	13 1	92.86 7.14	Szabolcs and Darab(1964)

3.14.1 Sodium Adsorption Ratio (SAR)

During the calculation of SAR, only alkali and alkaline earth ions are used. The excess amount of Sodium in the soil could reduce the infiltration and permeability of the soil. SAR values were calculated using the formula [14] as shown in Eq. (2) and concentrations of Ca^{2+}, Mg^{2+}, and Na^{+} ions were expressed in mEq/L.

$$SAR = \frac{Na}{\sqrt{\frac{Ca+Mg}{2}}} \tag{2}$$

The range of 0.4578 mEq/L to 4.5088 mEq/L (< 10 mEq/L) indicates that 100% of the samples are within the good agricultural suitability SAR category.

3.14.2 Soluble Sodium Percentage

Soluble Sodium Percentage (SSP) below 50 is considered a good quality for irrigation purposes. SSP values above 50 are not acceptable for irrigation due to low permeability. Eq. (3) is used to calculate SSP:

$$SSP = \frac{Na^{+} + K^{+}}{Ca^{2+} + Mg^{2+} + Na^{+} + K^{+}} \times 100 \tag{3}$$

where the ion concentration is in meq/L.

Boppapur village with SSP of 53.16 is an exceptional case and 93.3% of samples are of good quality in terms of SSP classification.

3.14.3 Residual Sodium Carbonate

This was evaluated [15] as per the formula: Residual Sodium Carbonate (RSC) index = [HCO_3 + CO_3] − [Ca + Mg]. The concentration of each ion is mentioned in mEq/L. Negative RSC values are the indicator of higher Ca and Mg concentrations in groundwater [16], and high concentrations of carbonates in water lead to the precipitation of calcium and magnesium as bicarbonates [17]. 100% of the collected samples show a significant presence of calcium and magnesium with RSCs ranging from −8.6008 mEq/L to −0.4088 mEq/L.

3.15 Sodium Percentage

Sodium Percentage (% Na) is calculated using Eq. (4) (Wilcox 1955):

$$\%Na = \frac{(Na + K)}{(Ca + Mg + Na + K)}\ x\ 100 \tag{4}$$

where Na, K, Ca, and Mg are calculated in mEq/L.

3.16 Kelly's Index

The effect of sodium on water quality for agricultural use is determined using the Kelly ratio [18]. Kelly's Index (KI) was calculated using Eq. (5):

$$KI = \frac{Na^{+}}{Ca^{2+} + Mg^{2+}} \tag{5}$$

where Na^{+}, Ca^{2+}, and Mg^{2+} are expressed in mEq/L.

If KI ≤ 1 water can be used for agricultural purposes, where it exceeds KI ≥ 1 is not recommended at all due to the risk of alkalinity [8, 19]. Boppapur village shows anomalies in both the SSP (53.16) and KR (1,12) areas. Another 93.3% of the samples are suitable for irrigation purposes based on the Kelly ratio.

3.17 Potential Salinity

A Potential Salinity (PS) value <3 mEq/L is suitable for irrigational use and calculated using Eq. (6) [20].

$$PS = Cl + 0.5\,(SO_4) \tag{6}$$

Ion concentrations are in mEq/L. Almost 47% of the samples are in the suitable categorization based on PS.

3.18 Magnesium Hazard

Excess magnesium concentration in water makes the soil alkaline and causes a reduction in crop yield [13, 21]. This is evaluated by Eq. (7) [22].

$$MH = \frac{Mg^{2+}}{Ca^{2+} + Mg^{2+}} \times 100 \tag{7}$$

The range of 10.9011 mEq/1 to 64.9683 mEq/L is calculated from data collected from the Dubbak Mandal area. For irrigational purposes, Magnesium Hazard (MH) >50 cannot be recommended.

3.19 Permeability Index

Permeability Index (PI) was used to classify irrigation water quality and was calculated using Eq. (8) of [20]. All ion concentrations were given in mEq/L.

$$PI = \frac{Na^{+} + \sqrt{HCO_3^{-}}}{Ca^{2-} + Mg^{2+} + Na^{-}} x\ 100 \tag{8}$$

The PI showed that groundwater from the study area is suitable for agricultural purposes.

3.20 Domestic Suitability

Water hardness can determine the suitability of domestic water, and Eq. (9) was used for calculating hardness [23].

$$DS = (\mathrm{Ca} + \mathrm{Mg} - \mathrm{HCO_3}), \quad (9)$$

Ca^{2+}, Mg^{2+}, and HCO_3 are specified in mg/L.

3.21 Industrial Suitability

LSI, RSI, and Corrosivity Ratios are calculated for suitability for industrial sector. Langelier Saturation Index (LSI) determines the tendency of water to cause calcium carbonate scaling and is calculated using Eq. (10).

$$LSI = pH(measured) - pHs \quad (10)$$

where,

$$pHs = (9.3 + A + B) - (C + D)$$

$$A = (Log_{10}TDS - 1)/10$$

$$B = -13.12xLog(°C + 273) + 34.55$$

$$C = Log[Ca_2 + as\ CaCO_3] - 0.4$$

$$D = Log[alkalinity\ as\ CaCO_3]$$

Table 7 represents the LSI range and inference.

For a more accurate prediction of calcium carbonate scaling Ryzner Stability Index was introduced as shown in Eq. (11). Table 8 represents the RSI range and inference.

$$RSI = 2pHs - pH(measured) \quad (11)$$

Table 7 LSI range and inference

RANGE (concerning calcium carbonate ($CaCO_3$))	Inference
<0	Water is under saturated
=0	Water in a neutral state
>0	Water is supersaturated

Table 8 RSI range and inference

Range	Inference
RSI < 5.5	The heavy scale will form
5.5 < RSI < 6.2	A moderate scale will form slight
6.2 < RSI < 6.8	Non-scaling or corrosive water
6.8 < RSI < 8.5	Corrosive water
RSI > 8.5	Very corrosive water

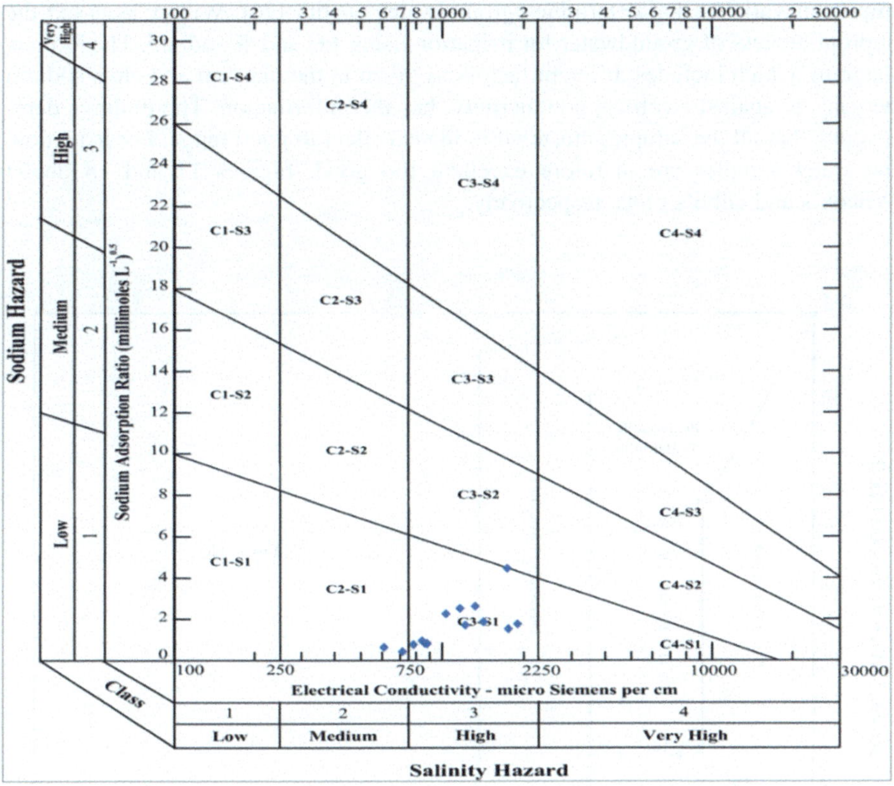

Fig. 16 USSL diagram of the study area

3.22 *Classification Based on the USSL Diagram*

The chemistry of water presented using the USSL diagram signifies the collected samples are of high to very high salinity that is not recommended for irrigational purposes. When compared to salinity levels, sodium content is low with one sample (C4S2) possessing a serious threat in terms of both SAR and EC. Figure 16 depicts the USSL diagram [24].

3.23 Gibbs and Wilcox's Plot

Na + K/(Na + K + Ca) and $Cl^{-}/(Cl^{-} + HCO_3^{-})$ ratio act as a function of TDS are shown in the graphs of Gibbs [25], showing dissolved chemical elements such as precipitation dominance and rock mass dominance. Water–rock contact has a significant impact on groundwater samples. Aquifers and water–rock interactions are both important factors in determining water chemistry. Geological location is an important variable that determines groundwater quality [26]. Wilcox assessed the appropriateness of groundwater for irrigation using EC and % sodium. The Wilcox diagram, which includes different factors as given in the diagram and plots sodium percentage against electrical conductivity, has this information. The findings demonstrate that all the samples clustered in the excellent to good range. For irrigation, the water samples are therefore excellent too good. Figures 17 and 18 depict Wilcox's and Gibb's plots, respectively.

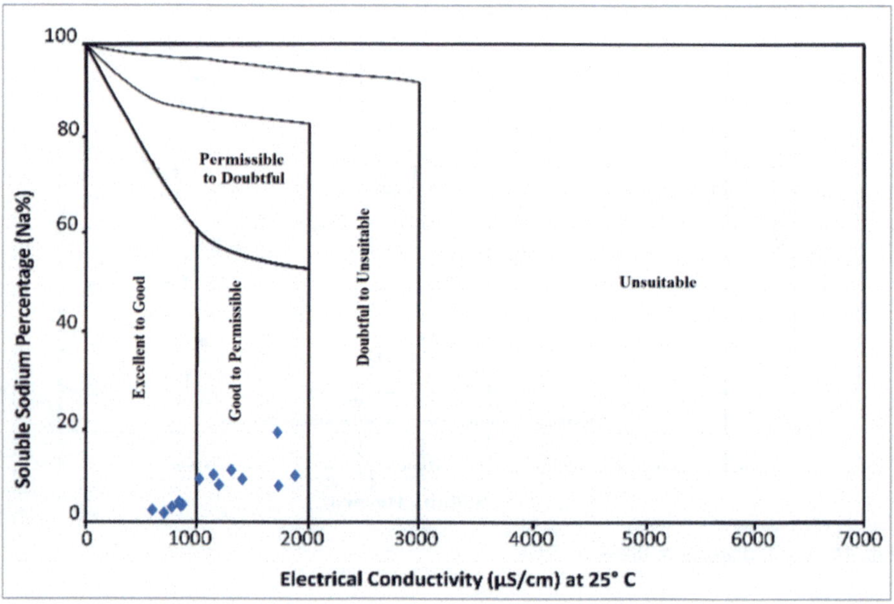

Fig. 17 Willcox diagram of the study area

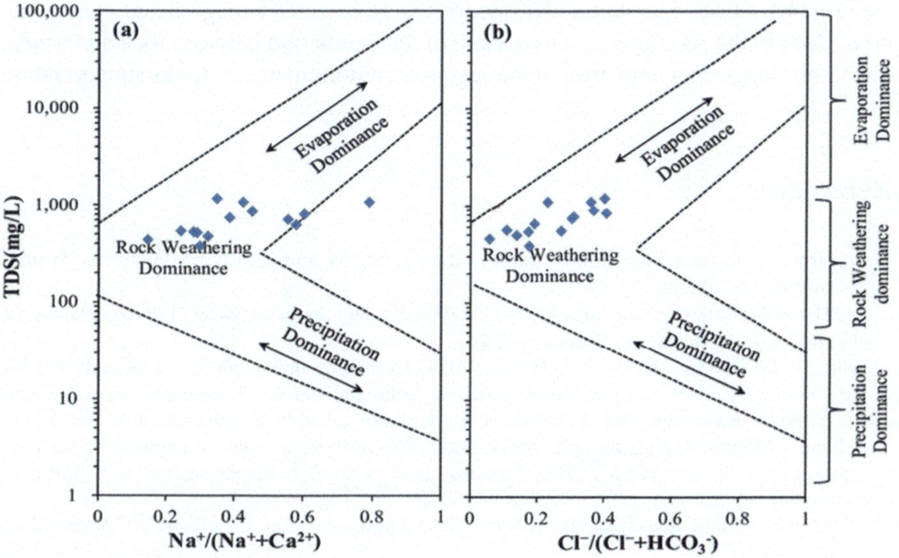

Fig. 18 Gibb's diagram

4 Conclusions

By permitting access to fresh water during the rainy season and managing soil and surface water contamination, it would be possible to improve the quality of groundwater. According to BIS Permissible limits, the pH range that can be utilized in a residential context is 6.5–8.5. The pH of the groundwater in the research area ranges between 6.71 and 7.80, making it safe to drink. The pH values from the groundwater tests are within allowable limits, but the other components are highly concentrated. Since the Specific Conductance (EC) of water makes the determination of total dissolved solids (TDS) more quickly, it is used to gauge the quality of the water. The ability of a medium to carry electrical current is described by its specific electrical conductance. The connection between conductance and dissolved solids becomes nonsensical in fluids with extremely high concentrations of dissolved solids. The research region's documented specific conductivity, which is good to safe for drinking and agricultural use, varies from 611 to 1321, with the exception of the villages of Akaram, Thimmapur, and Boppapur. The Total Dissolved Solids (TDS) measurements in the research region fall between 500 and 2000 ppm, the WHO permitted range. The high TDS measurements, especially from Akaram (1212 ppm), Thimmapur (1116 ppm), and Boppapur (1109 ppm), might be due to the presence of minerals in granitic terrain that are resistant to dissolving. Too high a TDS level may make the water unfit for swimming and washing. In water samples, the concentration of K ions ranges from 1 to 6. Higher fluoride (F) concentration in the area is caused by weathering of rocks, which occurs as a result of phosphate use in agriculture or the presence of sewage sludge. Nitrate ions were detected in significant concentrations (238) in the Akaram areas as a result of the interaction between rock and water. The Gibbs diagram's depiction of the region's predominance of rocks supports this.

References

1. Central groundwater board report, Southern Region (2013). http://cgwb.gov.in/District_Profile/Telangana/Medak.pdf
2. APHA (1995) Standard methods for examination of water and wastewater, 19th edn. American Public Health Association, Washington, DC
3. Harsha GS, Rao GS, Jeelani SH, Naresh (2024) Evaluating the suitability of groundwater for irrigational purposes using electrical resistivity technique and the implementation of artificial recharge at Kalugotla (V), Veldurthy (M), Kurnool District, Andhra Pradesh. In: Vinod Chandra Menon N, Kolathayar S, Sreekeshava KS (eds) Environmental engineering for ecosystem restoration. IACESD 2023. Lecture notes in Civil Engineering, vol 464. Springer, Singapore. https://doi.org/10.1007/978-981-97-0910-6_19
4. Bureau of Indian Standards, IS 10500 (2012) Drinking water (cgwb.gov.in) Accessed in September 2022
5. Bashir I, Lone FA, Bhat RA, Mir SA, Dar ZA, Dar SA (2020) Concerns and threats of contamination on aquatic ecosystems. In: Harter T (ed) Bioremediation and biotechnology. Springer, Cham, pp 1–26
6. Harter T (2003) Groundwater quality and groundwater pollution. https://doi.org/10.3733/ucanr.8084
7. Kumar M, Kumar R (2013) Assessment of Physico-chemical properties of ground water in granite mining areas in Goramachia, Jhansi, UP, India. Int Res J Environ Sci 2(1):19–24
8. Karanth K (1987) Ground water assessment: development and management. Tata McGraw-Hill Education
9. WHO (1997) World Health Organization (WHO) (2011) valuing water, valuing livelihoods: guidance on social cost-benefit analysis of drinking water interventions, with special reference to small community water supplies. In: John C, Paul H, Paul J, Katherine P (eds) . IWA, London, England
10. Gupta SK, Gupta RC, Gupta AB, Seth AK, Bassin JK, Gupta A (2000) Recurrent acute respiratory infections in areas with high nitrate concentrations in drinking water. Environ Health Perspect l08:363–36610753096
11. Rizvi SS, Mohammed-Aslam MA (2019) Microbiological quality of drinking water in Amarja reservoir catchment, Karnataka, Aland Taluk, Karnataka, India. Curr Sci 117(1):114–121. https://doi.org/10.18520/cs/v117/i1/114-121
12. Horton RK (1965) An index number system for rating water quality. J Water Pollut Control Fed 37(3):300–306. https://doi.org/10.1007/sl3201-020-01320-l. https://doi.org/10.1016/j.chemosphere.2006.05.065
13. Rawat KS, Singh SK, Gautam SK (2018) Assessment of groundwater quality for irrigation use: a peninsular case study. Appl Water Sci 8:233. https://doi.org/10.1007/s13201-018-0866-8
14. Rizvi SS et al (2022) Geochemical evaluation of fluoride and nitrate contamination in groundwater using graphical method, Karnataka State, India. In: AIP conference proceedings, vol 2454. 050042, pp 1551–7616. https://doi.org/10.1063/5.0078404
15. Eaton FM (1950) Significance of carbonates in irrigation waters. Soil Sci 69(2):123–134
16. Adimalla N, Wu J (2019) Groundwater quality and associated health risks in a semi-arid region of South India: implication to sustainable groundwater management. Hum Ecol Risk Assess Int J 25(1–2):191–216

17. Ravi Kumar P, Venkatesh Raju K, Prakash KL et al (2011) Geochemistry of groundwater and groundwater prospects evaluation, Anekal Taluk, Bangalore Urban District, Karnataka, India. Environ Monit Assess 179:93–112. https://doi.org/10.1007/s10661-010-1721-z
18. Kelly WP (1940) Permissible composition and concentration of irrigated waters. Proc Am Soc Civ Eng 66:607–613
19. Ramesh K, Elango L (2012) Groundwater quality and its suitability for domestic and agricultural use in Tondiar river basin, Tamil Nadu, India. Environ Monit Assess 184:3887–3899. https://doi.org/10.1007/s10661-011-2231-3
20. Doneen LD (1964) Notes on water quality in agriculture, Published as a water sciences and engineering. Department of Water Sciences and Engineering, University of California, paper, p 4001
21. Gowd SS (2005) Assessment of groundwater quality for drinking and irrigation purposes: a case study of Peddavanka watershed, Ananthapur District, Andhra Pradesh, India. Environ Geol 48(6):702–712
22. Szabolcs I (1964) The influence of irrigation water of high sodium carbonate content on soils. Agrokémiaéstalajtan 13(sup):237–246
23. Durfor CN, Becker E (1964) Public water supplies of the 100 largest cities in the United States, 1962 (No. 1812). US Government Printing Office
24. U.S. Salinity Laboratory Staff (1954) Diagnosis and improvement of saline and alkali soils: U.S. Dept Agric Handbook No.60, 160
25. Gibbs RJ (1970) Mechanisms controlling world water chemistry. Science 170(3962): 1088–1090
26. Wilcox LV (1955) Classification and use of irrigation water. USDA circular 969, p 19. Washington, DC, USA

Solar Powered Water Purification Using a Natural Coagulant

P. N. Jyothi and H. R. Rashmi

Abstract Accessing clean, safe drinking water is still a problem for many people worldwide. Numerous technologies for filtering and purifying water demand more electricity, energy, or resources. As the environment continues to change, these difficulties will only get worse. Due to the depletion of water supplies and the fear mongering surrounding environmental degradation, water treatment is currently one of the main research areas. This has forced the upgrading of conventional water treatment technologies toward recycling and reuse. They function by neutralizing the charge by polymeric bridging, an adsorption process. The work offers particular information about the coagulation properties of natural coagulants, fabricating a solar powered water purification system at lower maintenance cost, where sand, pebbles, and charcoal are used for physical impurity filtration, water is preheated using heating coils, where moringa seeds act as a natural coagulant and solar energy is used to power the heating coils and UV light source. Total Coliform bacteria causing illness were completely absent and parameters like TDS, pH, and Turbidity have reduced from 309.6 mg/L, 7.4, 10.8 NTU to 274.4 mg/L, 7.2, 1.2 NTU, respectively. This is the cheapest water purification system that can be used in rural areas with household containers, Moringa (drumstick) seeds, and solar panels. The use of natural coagulants, which are widely accessible, affordable, simple to use, biodegradable, non-toxic, environmentally friendly, efficient, and produce less volumes of sludge, is a sustainable substitute.

Keywords Filtration · Coagulants · Moringa seeds · Solar panel · Sustainable

P. N. Jyothi
Department of Mechanical Engineering, KSSEM, Bangalore, India

H. R. Rashmi (✉)
Department of Civil Engineering, KSSEM, Bangalore, India
e-mail: rashmihr@kssem.edu.in

P. Pathak et al. (eds.), *Advances in Environmental Sustainability, Energy and Earth Science*, https://doi.org/10.1007/978-3-031-73820-3_20

1 Introduction

In developing countries like India, one of the major issues is to get clean drinking water. Presently in rural areas still people use untreated pond/well water for their domestic purposes. In India, due to increase in population, degradation of the environment with faster infrastructural development is also in getting increased, this urbanization forces conversion of water wells/ponds into residential facilities and shopping complexes leading to depletion of water resources. Freshwater quality is coming down day by day due to pollution by domestic, agricultural, and industrial wastes. These wastes are directly/indirectly discharged into freshwater, causing a threat to water supplied to urban and rural areas. Drinking contaminated water and insufficient supply of water for personal hygiene results in poor sanitation causing many deadly diseases and increasing death rates every year [1]. To overcome this problem, the Government is making heavy investments in supplying safe and clean drinking water.

In order to get away from many germs and contaminants that are present, drinking water needs to be treated. Guidelines on water quality provided by the World Health Organization (WHO) serve as the foundation for regulations and standards across the globe. By supporting the creation of water safety plans, these Guidelines for Drinking Water Quality (GDWQ) work to protect public health [2]. Reverse osmosis (RO), ultraviolet (UV), ultrafiltration (UF), activated carbon, and sediment filter are some of the methods used to clean water. The market offers water purifiers that use the aforementioned methods, with prices ranging from 8000 to 30,000 rupees. The combination of technologies employed to purify the water determines the cost. Numerous researchers are developing various techniques for purifying water.

Solar water purification is the process of using solar energy in a variety of methods to clean water for drinking and domestic use. Since solar energy is typically a low-tech method of capturing solar heat and energy to make water cleaner and healthier, its use in water treatment has increased. For consumption and usage by humans rural areas benefit most from solar water treatment because they lack the infrastructure for other types of water purification and, more significantly, the electricity needed to power such facilities. The fact that fuel is not needed for solar water filtration is its best feature [3]. Applications of solar energy are better than traditional energy sources since they don't create pollution that causes ozone depletion, acid rain, or global warming. They also don't offer any health problems. This is specifically due to the lack of fuel needed for solar energy. This is precisely because solar energy does not require fuel. The four main techniques for treating solar water are sun water pasteurization, sun distillation, solar disinfection, and solar water treatment systems. While some of these technologies are relatively new, most are based on solar energy principles and have been around for a while [4]. These technologies often have a proven track record, are affordable, and are easy to use.

2 Materials and Methodology

Using Renewable sources and natural coagulants, i.e., Moringa seeds, which can be easily replaced even by an unskilled person. The present project uses the following elements:

2.1 *Solar Panel*

One important renewable energy source that has the ability to help humanity overcome a number of its problems is solar energy. Due to its many advantages for both people and the environment, as well as its versatility, this power source is growing in favor.

Clean, noiseless, and without any hazardous byproducts, electricity is produced. Solar energy is transformed into a practical amount of direct current (DC) electricity using an array of solar cells. The output voltage is increased when solar cells are joined in series. Series connected cells form what is called solar PV modules or PV panels as shown in Fig. 1.

2.2 *Steel Containers*

There are four chambers in the apparatus, each with a specific function as shown in Fig. 2. It is supported by a firm leg frame made from Mild Steel. The apparatus uses a solar panel to generate sufficient electricity, convert the same and store it in a set of batteries placed underneath. This electricity is used to power the water pump, heating coils, the UV light source, and the Peltier cooling system.

Fig. 1 Watts's solar panel

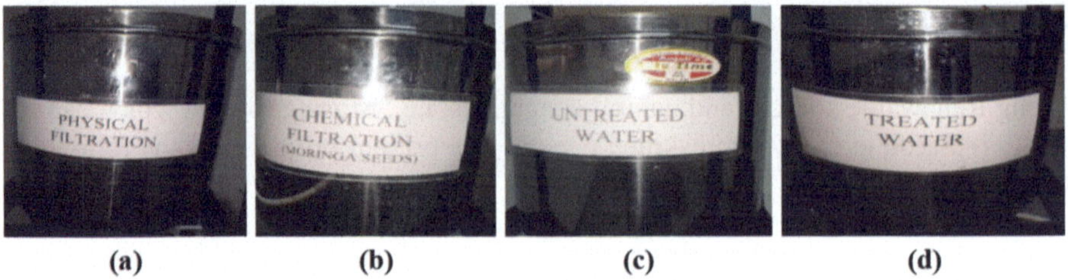

Fig. 2 Stainless steel containers (**a**) Physical filtration, (**b**) Chemical filtration, (**c**) Untreated water, and (**d**) Treated water

Fig. 3 (**a**) Sand grain size 4 mm, (**b**) Activated charcoal, and (**c**) Sand grain size 6 mm

2.3 *Sand and Charcoal*

Sand and Charcoal mixture is used to filter out the physical impurities present in the water sample. We have used sands of different grain sizes in this project, as shown in Fig. 3a, b, and c which show 4 mm grain size of sand, activated charcoal, and 6 mm grain size of sand, respectively.

2.4 *Moringa Seeds*

In the process of production of drinkable water, a variety of coagulants are employed in water treatment procedures. The most widely used coagulant is alum, but because of concerns that using aluminum salts could cause Alzheimer's disease, natural coagulant made from moringa seeds has taken its place. According to earlier research, using natural coagulants for water treatment is favored for the wellbeing and health of the human population.

The plant *Moringa oleifera* is also known as the horseradish tree, the drumstick tree, the miracle tree, and the ben oil tree [5]. Because of the health advantages and medicinal qualities of moringa, it has been utilized for millennia. Its other qualities include antifungal, antiviral, depressive, and anti-inflammatory. Figure 4 depicts the sun-dried moringa seeds in two different states: (a) before they are peeled and (b) after they are peeled but before they are crushed and submerged in water.

(a) (b)

Fig. 4 Moringa seeds (**a**) with shells, (**b**) peeled

UV Bulb Casing UV Bulb UV Bulb SMPS

Fig. 5 A schematic of the UV bulb set

2.5 UV Bulb

UV radiation from a UV bulb will inactivate microorganisms. A UV bulb can purify large volumes of water without adding chemicals to the water tanks. UV rays at a calculated distance from the bulb with a determined contact time can disrupt microorganisms that are unseen by the human eye and make the water drinkable. A pre-filter screen placed at the top of the settling tank will separate larger particles before the water is pumped into the UV Bulb casing.

The casing is a tubular plastic structure outside and a glass tube inside, within which the UV bulb and the electrical connections are given to the bulb as shown in Fig. 5.

2.6 Pipes, Tubes, and Valves

Three types of valves used here are PVC flow control valves shown in Fig. 6a, b. The pipes and other tubes used here in our project are ½ inch PVC pipes shown in

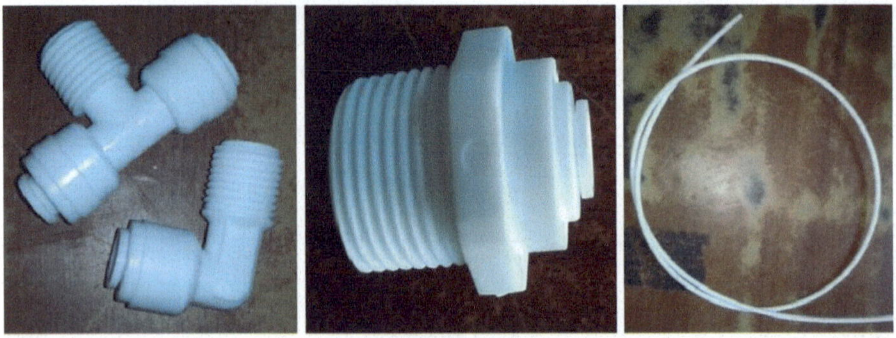

(a) (b) (c)

Fig. 6 Connectors and pipes (**a**) T and L joints, (**b**) Straight joint, (**c**) PVC pipe

Fig. 6c. We have used PVC material here because of its flexibility to bend and attain any desired shape. Mechanical floats are used to stop the flow of water into the container once it gets filled completely.

3 Result and Discussion

Drinking water also known as potable water, is the water that is safe to drink and also can be used for preparing food. For healthy and safe living, water quality is most fundamental and by providing access to safe water, good health can be promoted. Water safety and quality are most important to human development and wellbeing. Ensuring that people have access to clean water is one of the best ways to promote health and combat poverty [6].

WHO leads international efforts to stop the spread of waterborne infections since WHO is the world's foremost authority on public health and water quality. To do this, governments should be encouraged to enact health-based rules. Collaborating with partners, families, communities, and water suppliers should be encouraged to establish effective risk management practices. The Bureau of Indian Standards (BIS) in India has established guidelines for the quality of drinkable water (BIS 10500-1991) [7]. The following objectives have been established by the Bureau of Indian Standards:

- To assess the quality of water resources.
- To check the effectiveness of water treatment which is supplied by the concerned authorities.

They are applicable everywhere in the nation to drinking water provided by various Authorities, Agencies, and Departments of State and Federal Governments. Additionally, they apply to water provided for human consumption nationwide by non-governmental or private agencies.

In addition to elemental chemicals including iron, manganese, sulfate, nitrate, chloride, fluoride, arsenic, chromium, copper, cyanide, lead, mercury, zinc, and coliform bacteria, other characteristics included include color, odor, pH value, total dissolved solids, hardness, and alkalinity. Different attributes are considered necessary or desirable for the standard categories. It outlines the desired limit and provides background information so that the implementing authorities can use their discretion while taking the public's health, the quality of their care, and other factors into consideration. Routine examinations of all crucial traits are necessary [8]. In the present work, the following parameters which influence water quality are considered and studied:

1. *Total Coliforms*: Coliforms belong to a broad class of bacteria. These bacteria are found freely in our environment. The human feces as well as other warm-blooded animals are a source of these bacteria. Coliforms are primarily used as a method for indicating water quality. When coliform bacteria are detected in drinking water; it may be indicative of the potential presence of other harmful, disease-causing organisms. Testing for bacteria is the only way to know if your water is safe. Just by look, taste, or smell of the water, we cannot detect if disease-causing organisms are in it or not.
2. *pH Value*: A helpful technique to determine if the water is soft or hard is to look at the pH number. Pure water has a pH of 7. Water is generally classified as basic if its pH is greater than 7 and acidic if it is less than 7. Surface water systems typically have a pH between 6.5 and 8.5, while groundwater systems typically have a pH between 6 and 8.5 [9]. The ability of the water to withstand a pH shift that would normally cause the water to become more acidic is known as alkalinity. To ascertain how corrosive the water is, alkalinity and pH measurements are required.
3. *Turbidity*: The degree to which suspended particulates cause water to lose its transparency is measured as turbidity. The water appears murkier and has a greater turbidity level when there are more total suspended solids in it. Turbidity is regarded as a reliable indicator of water quality.
4. *TDS*: Total dissolved solids, or TDS for short, is the total amount of dissolved material in water. Inorganic salts and a trace amount of organic stuff make up TDS. Common inorganic salts found in water are carbonates, nitrates, bicarbonates, chlorides, and sulfates, which are all anions, and calcium, magnesium, potassium, and sodium, which are all cations. Generally, a high concentration of dissolved materials poses a health risk [10].
5. *Total Hardness*: The water hardness is the amount of dissolved calcium and magnesium in the water. Hard water is high in dissolved minerals, both calcium and magnesium. Hard water also contains bicarbonates and sulfates but calcium and magnesium are the most abundant minerals in hard water. The hardness of the water mainly depends upon the concentration of calcium and magnesium in the water. If the concentration of the calcium is less than or equal to 160 ppm (Parts Per Million) then water is referred to as soft water but if the concentration of the calcium in water is above 160 ppm, then it is regarded as hard water [11].

Table 1 Results of chemical analysis of the untreated and treated water

Parameters	Sample 1 (untreated water)	Sample 2 (treated water)	BIS 10500:2012 Standards	
			Desirable limits	Acceptable limits
Total coliforms	6 MPN/100 mL	Not detected	Not detected	
pH value	7.4	7.2	6.5	8.5
TDS	309.8	274.4 mg/L	$<$ 500 mg/L	2000 mg/L
Turbidity	10.8 NTU	1.2 NTU	1 NTU	5 NTU
Total hardness	148 mg/L	156 mg/L	$<$ 200 mg/L	600 mg/L

For all the parameters discussed above, BIS has set quality standards as shown in Table 1 and values measured for the water should be within the limit, if not water is not recommended for drinking.

In the present work, water is purified at different stages as shown in the design.

1. *Physical Treatment*: A mixture of Sand and Charcoal is used as a filter, to remove the physical impurities present in the water sample. Sand grains of different sizes are used, i.e., 4 mm sand and 6 mm sand grain size. This is used along with the activated charcoal. The quantity of the sand and charcoal mixture depends upon the value of the TDS in the sample collected before purification. A layer of activated charcoal is placed first in the container over which 4 mm and 6 mm grain size sand layers are placed one above the other, respectively. Water first passes down through 6 mm sand, then through 4 mm sand, and later through the charcoal. Because of three-layer filtration, turbidity of the water is reduced from 10.8 NTU to 1.2 NTU, which is well within the acceptable limit ranging between 1 NTU and 5 NTU [12].
2. *Chemical Treatment*: The natural coagulant used in the present work is the most economical and easily available, i.e., Moringa seeds powder. Moringa powder's water-clarifying ability is due to a positively charged protein called the Moringa Oleifera Cationic Protein. After the seeds are crushed and added to water, the water must be shaken or stirred vigorously to activate the moringa seeds. This process is called flocculation. The protein will kill some of the microbial organisms and cause them to clump together and settle to the bottom of the container. This process of breaking down and clumping is called coagulation.

As a result, these seeds are a superior substitute for purifying water. In addition, these seeds offer additional health advantages such as lowering the risk of ulcers occurring in people or accelerating their occurrence, avoiding cancer, promoting weight reduction, etc. Prior to treatment, the total coliform bacteria were measured to be 6 MPN/100 mL. As demonstrated in Fig. 7a, b, c, d, the addition of these crushed seeds resulted in zero total coliform bacteria, a pH value reduction from 7.4 to 7.2, and a TDS value reduction from 309.6 mg/L to 274.4 mg/L.

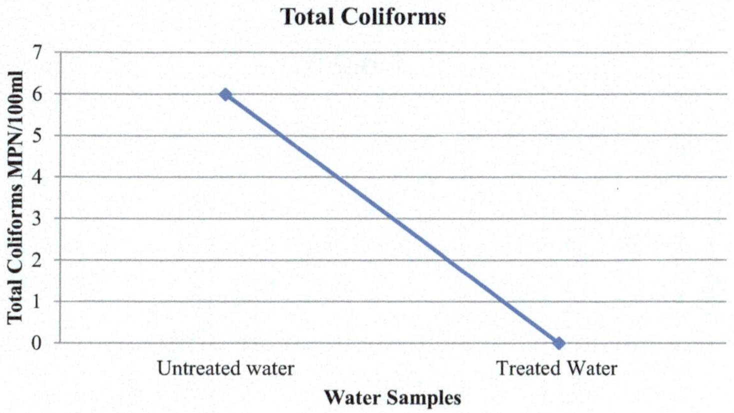

a) Total Coliform

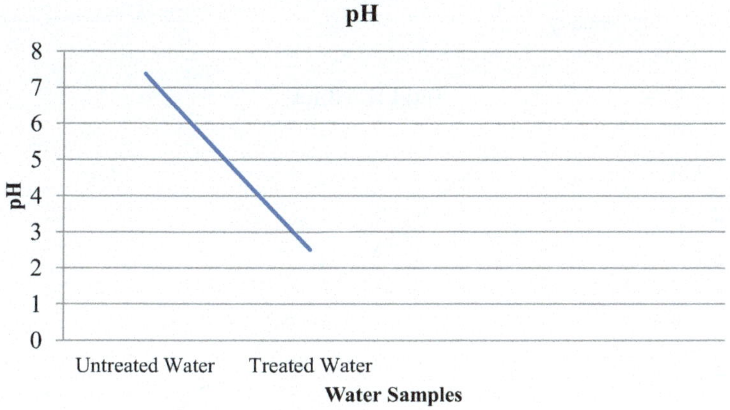

b) pH

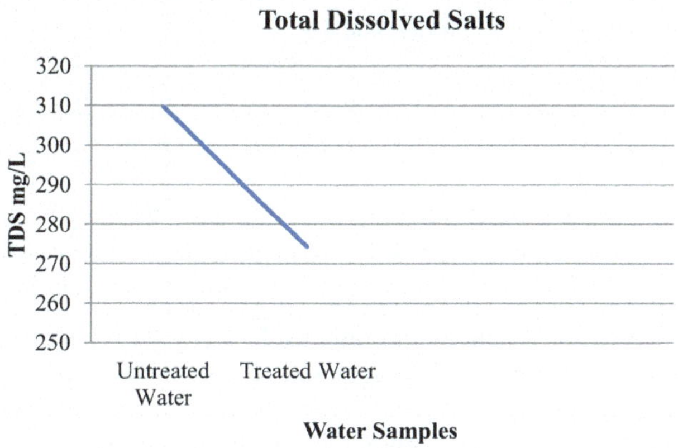

c) TDS

Fig. 7 Effect of water treatment on various parameters (**a**) Total Coliforms, (**b**) pH, (**c**) TDS, (**d**) turbidity, and (**e**) Total hardness

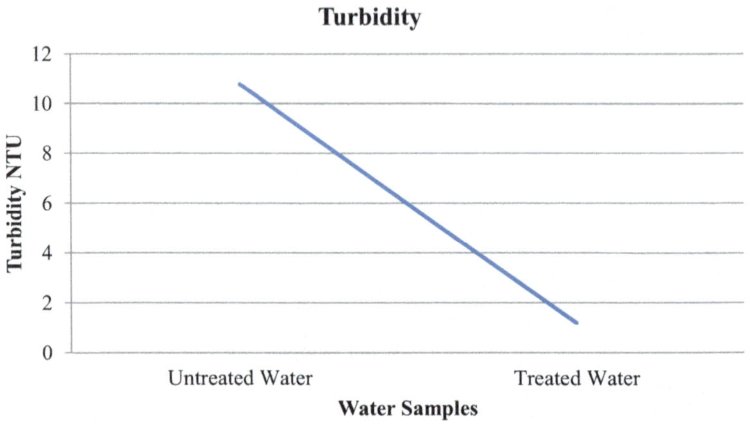

d) Turbidity

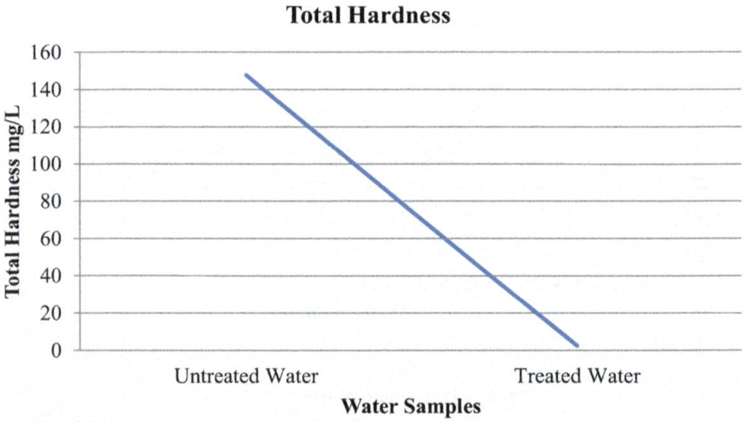

e) Total hardness

Fig. 7 (continued)

However, the primary issue brought about by the addition of these moringa seeds is the rise in elemental values, such as fluoride and chloride, which makes the water harder, as shown in Fig. 7e. Entire design of water treatment is shown in Fig. 8.

3. *UV Treatment*: After the treatment by moringa seeds, the water is passed into the UV casing for treatment through a UV light source due to which bacteria gets killed reducing the turbidity as seen in Fig. 1c in Chap. 5. The casing is a tubular plastic structure on the outside and a glass tube on the inside, within which the UV bulb is positioned. There will be gaps between the bulb and the inlet and outlet through which water enters the casing, gets purified, and exits the casing.

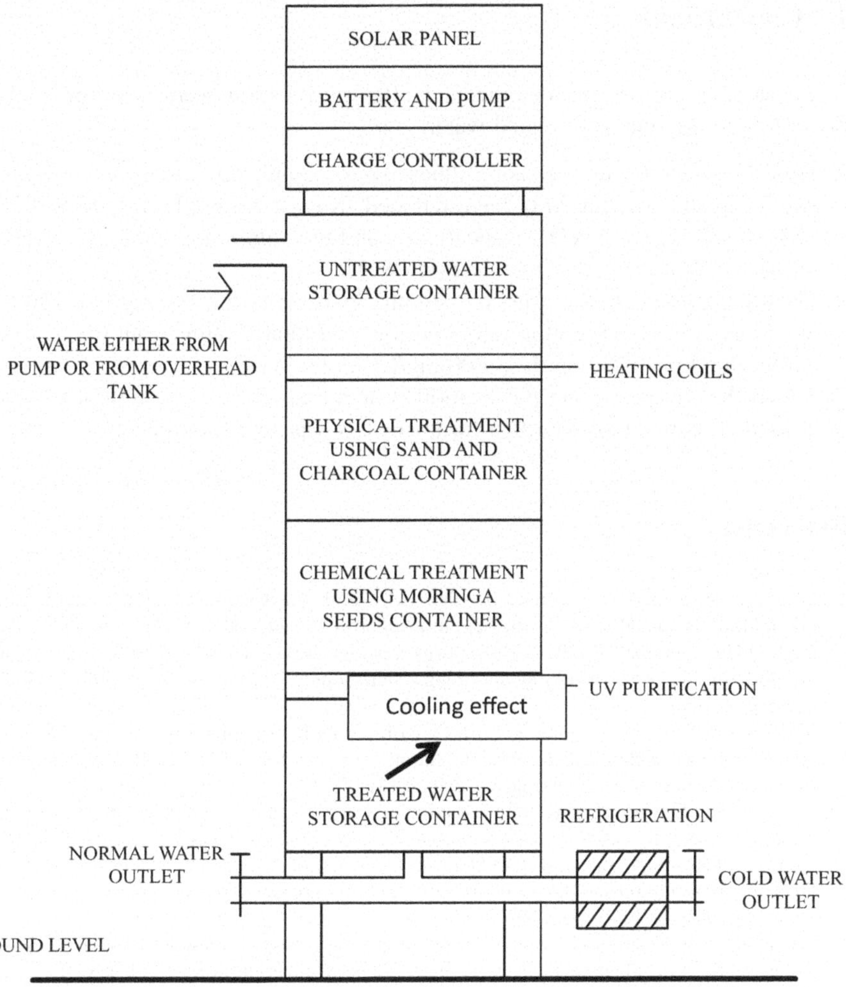

Fig. 8 A schematic of waste treatment design

4. *Refrigeration*: This is an additional element to the design only to provide cold water. A thermal Peltier is used here to cool the water temperature. A heat sink is attached to the inner wall of the container. The cold side of the Peltier is faced toward the container and is stuck with the aid of a thermal paste. Next to it, a heat sink with a fan is placed to cool the hot side of the Peltier. This again is stuck by the thermal paste. All the electrical connections are made to run the entire setup and obtain cold water.

4 Conclusions

Water samples before purification and after purification were sent for analysis. According to the findings, treated water:

1. Total Coliform bacteria causing illness were completely absent and parameters like TDS, pH, and turbidity have reduced from 309.6 mg/L, 7.4, 10.8 NTU to 274.4 mg/L, 7.2, 1.2 NTU, respectively, and the values are within the standards set by the World Health Organisation (WHO).
2. The total hardness of the water has slightly increased from 148 mg/L to 156 mg/L due to a slight increase in nitrate, fluoride, and total alkalinity but the values are within the limits (200 mg/L – 600 mg/L) set by WHO.
3. This is the cheapest water purification system that can be used in rural areas with household containers, Moringa (drumstick) seeds, and solar panels.

References

1. Welch J, Simmons V, Meléndez E, Sees M, Gold Y, Heider EC (2020) Assessment of non-anthropogenic addition of uric acid to a water treatment wetlands. Environments 7:60
2. Rashmi HR, Devatha CP (2019) Phosphorus recovery as induced struvite from deep dewatered liquors using magnesium chloride as a magnesium source. https://doi.org/10.1088/1757-899X/561/1/012010
3. Mohammed A-H, Hussein AH, Yeboah D, Al Khaddar R, Abdulhadi B, Shubbar AA, Hashim KS (2020) Electrochemical removal of nitrate from wastewater. In: IOP conference series: materials science and engineering, vol 888, p 12037
4. Rashmi HR, Devatha CP (2020) Experimental investigation on sludge dewatering using granulated blast furnace slag as skeleton material. Environ Sci Pollut Res 27:11870–11881
5. Guo Y, Zelekew OA, Sun H, Kuo D-H, Lin J, Chen X (2020) Catalytic reduction of organic and hexavalent chromium pollutants with highly active bimetal cubios oxysulfide catalyst under dark. Sep Purif Technol 242:116769
6. Muhammad Burhanuddin B (2021) Recent advances on coagulation-based treatment of wastewater: transition from chemical to natural coagulant. Curr Pollut Rep 7:379–391
7. Rashmi HR, Devatha CP (2021) Dewatering performance of sludge using coconut shell biochar modified with ferric chloride (sludge dewatering using bio-waste). Int J Environ Sci Technol 19(7):6033–6044
8. Panhwar AA, Almani KF, Kandhro AA (2020) Environmental degradation by textile industry; performance of chemical coagulants and activated carbon for removal of cod, bod. Tech J 25: 16–20
9. Wageningen AJ, Coa YC, Metcalf RH (2000) The effects of direct sunshine and heat on rotavirus viability in water. In: Abstracts of the general meeting of the American Society for Microbiology, vol 100, p 604
10. Antov MJ (2012) Evaluation of the efficiency of natural coagulant obtained by ultrafiltration of common bean seed extract in water turbidity removal. Ecol Eng 49:48–52
11. Kunduru KR, Nazarkovsky M, Farah S, Pawar RP, Basu A, Domb AJ (2017) Nanotechnology for water purification: applications of nanotechnology methods in wastewater treatment. In: Water Purification, pp 33–74
12. Bolisetty S, Mezzenga R (2016) Amyloid–carbon hybrid membranes for universal water purification. Nat Nanotechnol 11(4):365

Assessment of Water Surcharge Conditions in an Urban Area Using SEWERGEMS: A Case Study

K. Veerendra Gopi, N. Anish, A. Harshith Reddy, P. Sanjay, K. Pavan Datta, and B. Vaishnavi

Abstract The most frequent issues with combined sewer distribution systems in urban areas are the ones related to water surcharges. Rapid urbanization leading to an increase in impervious areas, high-intensity rainfall, and clogging of drains due to improper maintenance of sewer systems are the most common causes of surcharge. The main objective of the study is to measure and assess the overflow conditions and locate the potential flood-effected areas in a part of Pragathi Nagar of Hyderabad city, considering the existing sewer drainage system and with the inclusion of Low Impact Development (LID) techniques. To model the scenario for the study area, a computational SEWERGEMS was developed. SEWERGEMS model was developed with and without consideration of LID approaches for various design rainfall intensities for 2-, 5-, 10-, and 30-year return periods. The outcomes implied the effectiveness of the computational model in explaining multiple scenarios and can support the municipal authorities in effectively managing the combined sewer distribution network.

Keywords Combined sewer distribution system · Water surcharge · SEWERGEMS · Low impact developments

1 Introduction

Urban floods are one of the most frequently occurring and have impact on the day-to-day events of the people living in urban areas. Stormwater and sewer drains are the most important structures in draining stormwater and wastewater. In most of the urban areas, sewer drains are used as a combined drain. A combined sewer distribution system is considered essential infrastructure in urban areas because it prevents flooding and helps maintain hygiene conditions by disposing of sewage waters [1]. With frequent changes in urbanization and rapid growth, impervious areas are

K. V. Gopi (✉) · N. Anish · A. H. Reddy · P. Sanjay · K. P. Datta · B. Vaishnavi
Department of Civil Engineering, VNR Vignana Jyothi Institute of Engineering and Technology, Hyderabad, India

P. Pathak et al. (eds.), *Advances in Environmental Sustainability, Energy and Earth Science*, https://doi.org/10.1007/978-3-031-73820-3_21

increasing faster when associated with the development of distribution systems [2–4]. Urban floods are mostly considered to be due to human intervention rather than natural hazards. The intensity of flash floods or urban floods has increased in recent times mainly in municipal areas, which is attributed to blockage of drainage channels by solid waste, unplanned development in low—lying areas, and varying rainfall intensities due to climate change [5]. An increase in rural people's mitigation to urban areas has increased the problem of urban floods to a greater extent [6]. Urban flood modelling has gained interest among researchers to assess flood behavior and how to mitigate it because of increasing flood frequencies [7–9]. For heterogenous urban areas, spatially distributed modelling is a difficult task due to the least availability of digitally distributed data [10]. Researchers have raised concerns about the use of conventional methods (like sewer enlargement and detention ponds) in urban flood management because of their limited capacity and flexibility in coping with instant urbanization and changing climate [11–14].

It is identified that there is a change in rainfall patterns because of climate change. Researchers and metropolitan planners are evaluating various strategies that can reduce the flood risk brought about by climate change. Attention was drawn to a need for sustainable stormwater management due to increased flood frequencies and intensity of storm flood occurrence [1, 15, 16]. Increasing focus has been given to the use of Low impact development (LID) structures to increase the infiltration of water into sub-surface [12, 13, 17, 18]. While most research focuses on evaluating the hydraulic and hydrological performance of LID in controling flooding, limited knowledge is available about the effectiveness of LID in reducing hydrological impact at the catchment scale [19, 20]. The need to include climate change conditions, effect of extreme events along with LIDs to understand the changes in flood peaks is gaining interest [21–23]. Implementation of mitigation strategies to control the impact of urbanization and climate change extremes by incorporating more than one LID can play a significant role [23–25].

The present study aims at developing the combined sewer distribution system in SewerGEMS to identify possible surcharge locations, estimation of peak flood discharges and evaluate the effectiveness in minimizing the flood magnitude with inclusion of LID methods in the model. The existing sewer distribution system was simulated for design storm intensities with and without considering LID practices to check the efficiency in reducing the peak floods. The model is tested using green roofs (GF) and porous pavements (PP) as LID practices.

2 Study Area and Methodology

2.1 Study Area

A part of the Pragathi Nagar area is considered for the study, and it is a rapidly expanding locality of Nizampet municipality of Hyderabad city in India. The study area is located at 17.5208° N latitude and 78.3922° E longitude. Greater Hyderabad

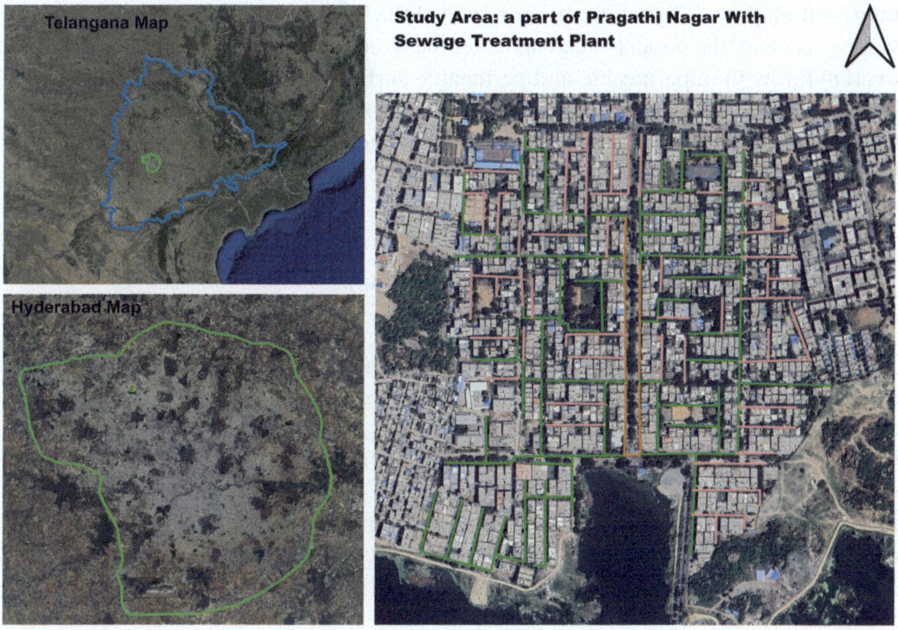

Fig. 1 Aerial view of the study area with an existing drainage system

municipality corporation has varying elevations and is divided into 16 sub-catchment zones based on topography and drainage flow patterns. Pragathi Nagar is a part of Zone 13, and for the study, a part of Pragathi Nagar which has a sewage treatment plant of capacity 2.5 MLD, and its surface flow accumulation into Pragathi Nagar Lake. The study area covers an area of 0.515 km^2 or 51.5 hectares, and slopes vary between 578 m and 603 m. The study area consists of a combined sewer distribution system of varying diameters between 8 inches, 1-, and 2-feet size, and circular section pipes, as presented in Fig. 1. In the study area map, orange color indicates 2-feet diameter sewer drain and green color indicate 1-feet diameter drain.

2.2 Sewer GEMS Model

For analyzing flood conditions and modelling runoff volume with the use of LID methods, SewerGEMS software was used. SewerGEMS is a fully dynamic model developed by Bentley to model, design, and analyze both sewage water and stormwater distribution systems (And to simulate the LID controls in urban areas). Implicit SewerGEMS dynamic wave model which included the full St. Venan equation, was used for hydraulic calculation. Using the GIS interface, sub-catchment information such as percentage slope of the study area, percentage impervious area, and curve number values are extracted from Land use land cover map, soil map, and digital elevation model (DEM) map. While the other parameter values used in the model, such as roughness coefficient for sewer systems and overland flow in impermeable and permeable surfaces, surface retention parameters for permeable and impervious areas are taken from literature Rossman LA [26]. The Curve Number method is used to determine the rate of infiltration into the catchment, Unit hydrograph method is used to determine the outflow peak/ hydrograph.

2.3 *Intensity Duration Frequency Curve*

The study area is a completely developed urban area with the least amount of open space for parks and vegetation, which indicates high impervious conditions and high surface flow even for a short-duration rainfall event. Because high impervious conditions indicate high surface flow even for a short period of rainfall. Urban catchments require sub-hourly rainfall data to assess the flood magnitude or intensity that can occur in the catchment. Sub-hourly rainfall depth required for the study is downscaled from 24-hour rainfall data using Eq. (1) which was developed by the Indian Meteorological Department (IMD) [27].

$$P_t = P_{24}\left[\frac{t}{24}\right]^{\frac{1}{3}} \tag{1}$$

where, t indicates the specific duration during which rainfall depth is required, P_{24} is the 24-hour rainfall depth in mm, and P_t is the rainfall depth in mm for a specific time period t hours. Daily rainfall data is downscaled for every 5-min interval duration up to 2 h duration and rainfall intensities are calculated for the same. From the goodness of fit test, Gumbel's distribution is identified as the best-fitted distribution for the sub-hourly rainfall data based on the Chi-square test. Design storm intensities are determined based on Gumbel's frequency analysis for return periods 2-, 5-, 10-, and 30-year.

2.4 *Low-Impact Development Techniques*

Existing drainage system is simulated from various design storm events of 2-, 5-, 10, 30-year return periods. Model is simulated for the above storm events with and without considering the LID methods. The preliminary objective of the LID or standard methods is to control the flood volume flow toward the drainage system [17].

Green roof (GF) LID retains certain parts of storm and increases the length of flow which further reduces the runoff from impermeable surfaces. While porous pavement (PP) can be used to replace paved surfaces covering pedestrian walkways, parking areas, etc. From the literature and reference of SWMM manual, parameters for both the LID methods used in the study [17]. All the commercial and residential rooftops are assumed to be converted into green roofs based on the Google satellite imagery and study area survey and parking spaces, and minor roads are assumed to be converted to permeable surfaces. To determine the available area for converting the existing conditions into both scenarios, QGIS software was utilized along with Google Earth engine. An area of 0.17 km^2 covering 33% of the total study area is identified to be converted to a green roof. About 0.09 km^2 covering 17% of study area is identified as the available area that can be converted into porous pavement.

3 Results and Discussion

Rainfall intensities are taken from IDF curves developed for different return periods and as shown in Fig. 2. In SEWERGEMS, for different return periods flood peaks and surcharge locations are identified. From the SEWERGEMS results it is identified that for even a 2-h duration, design storm event of 5- and 10-year return periods, flood occurrence was identified at three and four nodes in the distribution system. The total volume of flooding was calculated as 374 m^3 and 419 m^3 for 5-, and 10-year return periods, respectively. With the inclusion of green roof and porous pavement LIDs in the model, drainage system capacity has increased in flood mitigation conditions by reducing the runoff volume flow to the outlet. Figure 2

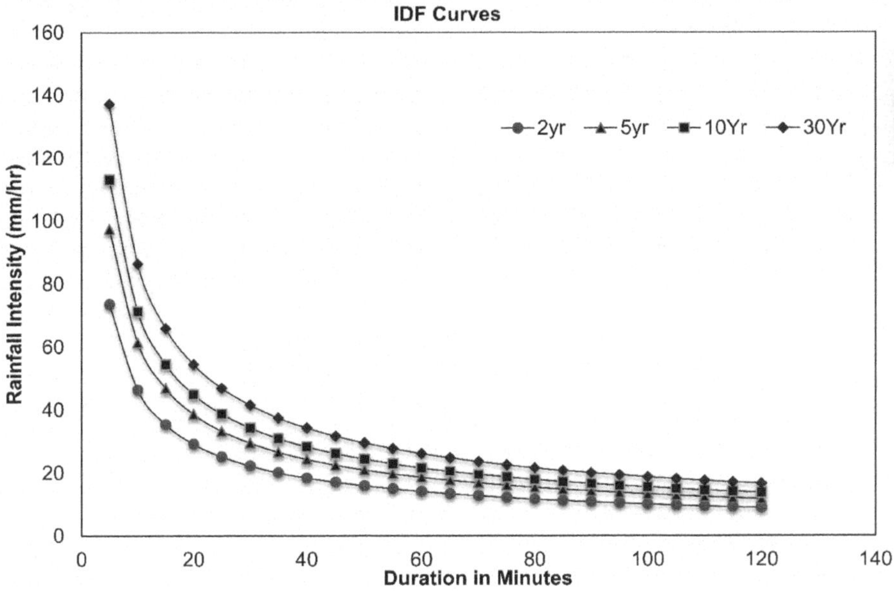

Fig. 2 Developed IDF curves based on the Indian Meteorological Department downscaling formula

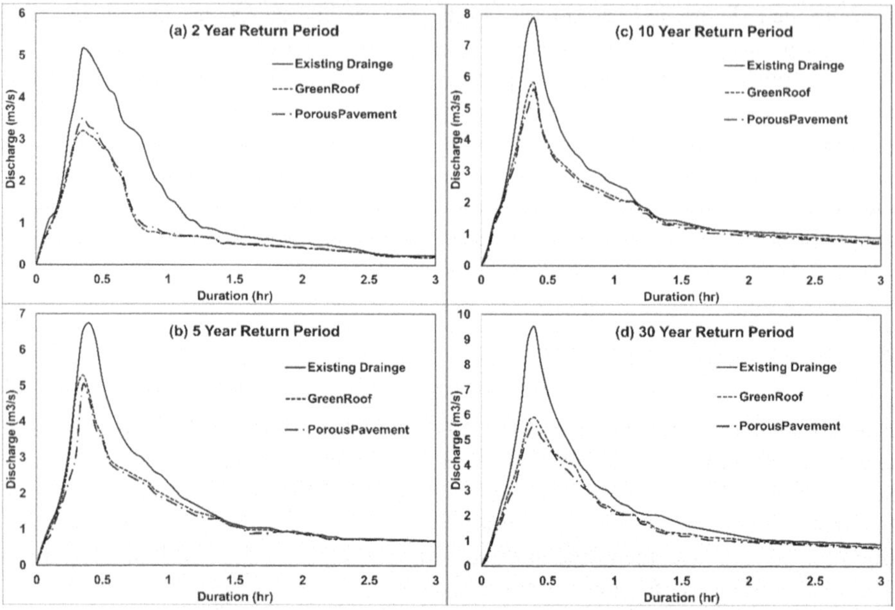

Fig. 3 Hydrographs simulated at the outlet for various return periods and implemented LIDs

shows the IDF curves developed for the study area using the Indian Meteorological Department downscaling formula. Figure 3 shows the hydrographs resulting from peak flow for the study area.

From Fig. 3, it can be identified that peak flow is reduced to a greater extent for porous pavements than green roof for all the return periods. Green roofs control the runoff at the source over the roof surface of building. Porous pavements are the conversion of parking spaces, side walking tracks, and secondary roads into permeable roads. For a 2-year return period, porous pavements indicated a better reduction in runoff rate when compared to green roof LID practice as shown in Fig. 3a. With the implementation of LID methods green roof and porous pavements, for a 2-year return period the runoff volume was reduced by 26% which implies efficient management of flooding in the study area. Flood volume reduction can be observed even for other return periods as shown in Fig. 3b, c, d. The study area has a sewage treatment plant (STP) with a capacity to treat 2.5MLD. Present STP has a capacity to treat 18,000 people, i.e., 36% of total population of the Pragathi Nagar area. With the interference of rainwater during rainy season, it cannot be used at full capacity which indicates peak flow conditions in drainage system. Calibration and validation of the model can't be performed, because of non-availability of flow gauges or flow measurement devices. The analysis conducted herein is to understand the behavior of LIDs like Green Roof and Porous pavement in effective mitigation of flood volume. To achieve flood mitigation and control, both LID and standard methods must be examined in combination.

4 Conclusions

The present study analyzed the effectiveness of green roof and porous pavements in controlling urban floods in a part of Pragathi Nagar of Hyderabad city. Results demonstrated that use of LIDs can reduce the runoff volumes mainly for small rainfall events and for high-design storm events LIDs are less effective. Green roof and porous pavement practices delay the runoff rate at the source and control the runoff volume in the study area. SewerGEMS model along with LIDs methods was effective in explaining the flooding mitigation approaches. In the study, flood magnitude was done by including the LIDs as independent approaches and further work can be done to understand the optimum combination of LIDs and parameter selection for the same.

References

1. Marlow DR, Moglia M, Cook S, Beale DJ (2013) Towards sustainable urban water management: a critical reassessment. Water Res 47:7150–7161. https://doi.org/10.1016/j.watres.2013.07.046
2. Gumma MK, Mohammad I, Nedumaran S et al (2017) Urban sprawl and adverse impacts on agricultural land: a case study on Hyderabad, India. Remote Sens (Basel) 9:1136. https://doi.org/10.3390/rs9111136
3. Biprodip M, Mukherjee P, Chaudhuri K, Das S (2023) A feasibility case study of storm and sewer drainage systems of Rajarhat Gopalpur municipality. In: Sunil K, Ghangrekar MM, Kundu A (eds) Sustainable environmental engineering and sciences. Springer Nature Singapore, Singapore, pp 1–10
4. Pawar A, Patil Y, Mane S, et al (2021) International Research Journal of Engineering and Technology (IRJET) Design of Sewer System for village using SewerGEMS
5. Arnone E, Pumo D, Francipane A et al (2018) The role of urban growth, climate change, and their interplay in altering runoff extremes. Hydrol Process 32:1755–1770. https://doi.org/10.1002/hyp.13141
6. Kabisch N, Korn H, Stadler J, Bonn A (2017) Theory and practice of urban sustainability transitions natureebased solutions to climate change adaptation in urban areas. In: Policy and practice. Springer Nature, Cham, Switzerland
7. Zoppou C (2001) Review of urban storm water models. Environ Model Softw 16:195–231. https://doi.org/10.1016/S1364-8152(00)00084-0
8. Rafiee MR, Rasouli D, Zolghadr M, Mahbod M (2022) Evaluation of EPA SWMM, ASSA and SewerGEMS models in analysis of urban flood collected by surface drainage network (Case Study: Lar New City). Water Resour Eng J 15:107–124. https://doi.org/10.30495/wej.2021.25316.2253
9. Awakimjan I, Puello LLP Timo Hartman I, Grontmij JSK Brugnachir M (2015) urban flood modelling recommendations for Ciudad Del Plata
10. Salvadore E, Bronders J, Batelaan O (2015) Hydrological modelling of urbanized catchments: a review and future directions. J Hydrol (Amst) 529:62–81
11. Krebs P, Larsen TA (1997) Guiding the development of urban drainage systems by sustainability criteria. Water Sci Technol 35:89–98. https://doi.org/10.1016/S0273-1223(97)00187-X
12. Pour SH, Wahab AKA, Shahid S et al (2020) Low impact development techniques to mitigate the impacts of climate-change-induced urban floods: current trends, issues and challenges. Sustain Cities Soc 62:102373. https://doi.org/10.1016/j.scs.2020.102373

13. Ashok RV, Bhatt CM, Patel AK, Umamahesh NV (2022) Geo-spatial analysis of October 2020 Hyderabad flood. In: Kumar DA, Narasimhan B, Kumar B, Patel AK (eds) Innovative trends in hydrological and environmental systems. Springer Nature Singapore, Singapore, pp 33–42
14. Kumar S, Agarwal A, Ganapathy A et al (2022) Impact of climate change on stormwater drainage in urban areas. Stoch Env Res Risk A 36:77–96. https://doi.org/10.1007/s00477-021-02105-x
15. Chocat B, Ashley R, Marsalek J et al (2007) Toward the sustainable management of urban storm-water. Indoor Built Environ 16:273–285. https://doi.org/10.1177/1420326X07078854
16. Barbosa AE, Fernandes JN, David LM (2012) Key issues for sustainable urban stormwater management. Water Res 46:6787–6798. https://doi.org/10.1016/j.watres.2012.05.029
17. Kourtis IM, Tsihrintzis VA, Baltas E (2018) Simulation of low impact development (LID) practices and comparison with conventional drainage solutions. MDPI AG, p 640
18. Sohn W, Kim J-H, Li M-H, Brown R (2018) The influence of climate on the effectiveness of low impact development: a systematic review. J Environ Manag 236:365–379
19. Burszta-Adamiak E, Mrowiec M (2013) Modelling of green roofs' hydrologic performance using EPA's SWMM. Water Sci Technol 68:36–42. https://doi.org/10.2166/wst.2013.219
20. Li C, Liu M, Hu Y et al (2018) Evaluating the hydrologic performance of low impact development scenarios in a micro urban catchment. Int J Environ Res Public Health 15:273. https://doi.org/10.3390/ijerph15020273
21. Matos C, Briga Sá A, Bentes I et al (2019) An approach to the implementation of low impact development measures towards an EcoCampus classification. J Environ Manag 232:654–659. https://doi.org/10.1016/j.jenvman.2018.11.085
22. Ji T, Yao Y, Dou Y et al (2022) The impact of climate change on urban transportation resilience to compound extreme events. Sustainability (Switzerland) 14:3880
23. Bibi TS, Kara KG (2023) Evaluation of climate change, urbanization, and low-impact development practices on urban flooding. Heliyon 9:e12955. https://doi.org/10.1016/j.heliyon.2023.e12955
24. Mishra V, Sadhu A (2023) Towards the effect of climate change in structural loads of urban infrastructure: a review. Sustain Cities Soc 89:104352. https://doi.org/10.1016/j.scs.2022.104352
25. Perera ATD, Hong T (2023) Vulnerability and resilience of urban energy ecosystems to extreme climate events: a systematic review and perspectives. Renew Sust Energ Rev 173:113038. https://doi.org/10.1016/j.rser.2022.113038
26. Rossman LA (2010) Storm water management model user's manual version 5.0
27. Nwaogazie IL, Sam MG, Ikebude C (2021) Improving Indian meteorological department method for 24-hourly rainfall downscaling to shorter durations for IDF modelling. Int J Hydrol 5:72–82. https://doi.org/10.15406/ijh.2021.05.00268

Impact of Seawater Intrusion on Groundwater in Kanchipuram District, Tamil Nadu Coastal Area

Manne Venkat Revanth, Soujanya Jonnalagadda, C. H. Lokesh Reddy, T. Sharj, G. Mahesh Babu, and Gokulan Ravindiran

Abstract Saltwater intrusion is a global issue that affects the coastal areas worldwide. Due to this intrusion, it will affect the growth of crops, soil fertility, and human health. The current study estimates parameters such as Water Quality Index (WQI), Human Health Risk Assessment (HHRA), Seawater Mixing Index (SMI), and Irrigation efficiency (IE) in order to examine the groundwater quality of Kanchipuram district, Tamil Nadu coastal area. A total of 60 sampling stations were considered to examine groundwater quality. For the groundwater analysis, the physicochemical parameters were taken into account. Q-GIS has been used in mapping the sampling locations and affected areas. WQI was calculated for the Kanchipuram based on the physicochemical characteristics of water namely pH, Electrical Conductivity, Total Dissolved solids, Total Hardness, Calcium, Magnesium, potassium, chlorides, fluorides, nitrates, and sodium. From the WQI results it is observed that out of 60 stations, it was observed that around 30% of groundwater is above good quality and around 42% is very poor and not suitable for consumption. The WQI states that groundwater in Kanchipuram district was moderately polluted. SMI was also calculated, it was observed that out of 60 stations SMI values of 23 stations were above 1, i.e., in 38% of stations SMI value was observed to be high. Also calculated HHRA, it was observed in 13% of stations fluoride content is greater than 1 in children. The analysis shows that the impact of fluoride content in children is higher than in adults. Irrigation Indices conclude that 65% of sampling stations have SAR values greater than 10 mEq/L and 58 stations have Na% less than 80%.

Keywords WQI · HHRA · Seawater Mixing Index · Saltwater intrusion

---

M. V. Revanth · S. Jonnalagadda (✉) · C. H. L. Reddy · T. Sharj · G. M. Babu · G. Ravindiran
Department of Civil Engineering, VNR Vignana Jyothi Institute of Engineering and Technology, Hyderabad, Telangana, India
e-mail: soujanya_j@vnrvjiet.in

P. Pathak et al. (eds.), *Advances in Environmental Sustainability, Energy and Earth Science*, https://doi.org/10.1007/978-3-031-73820-3_22

1 Introduction

Water is one of the vital natural resources in our daily life. Water is used for various purposes like domestic, industrial, and agriculture. Humans mainly depend on groundwater to fulfill their basic needs [1]. Groundwater is a vital source of water for human usage. Groundwater provides 43% of the water used for crops and all or most of the drinking water for up to 50% of the world's population. The world's population of about 2.5 billion people gets all their basic daily water needs from groundwater resources. Due to increase in population and industrialization the usage of groundwater is increasing day by day. In coastal areas, the excess usage of groundwater leads to differences in sea level which results in the intrusion of seawater into groundwater, so there is a risk of contamination of groundwater [2]. The migration of saltwater into freshwater aquifers as a result of human or natural activity is known as seawater intrusion. The seawater intrusion decreases the storage capacity of groundwater and also in some cases it leads to abandonment of wells. Areas adjacent to coastal waters are at greater risk of seawater intrusion [2]. It is a serious issue that affects coastal aquifers globally. Reductions in groundwater levels or elevations in seawater levels can result in seawater intrusion. When freshwater is rapidly pumped out of the aquifer, a cone of depression is formed, lowering its height. In response, the sea level rises proportionally for every foot of freshwater depression. Saltwater intrusion into freshwater bodies affects the soil condition, vegetation, and water quality. Higher concentrations of chlorides in water is unfit for drinking and industrial activities, higher concentrations of sodium ions in water can lead to high blood pressure in humans [3]. To assess the groundwater quality various parameters were determined. The most effective way to evaluate the quality of groundwater is WQI. We computed the Seawater Mixing Index (SMI) to figure out how groundwater is affected by seawater intrusion. Due to consumption of this polluted groundwater and also through agricultural activities like usage of excessive fertilizers and unsanitary conditions human health is at risk. To examine the fluoride content in both adults and children Human Health Risk Assessment (HHRA) was calculated. Intruded groundwater can have a negative impact on irrigation indices. To know the effect on irrigation we also calculated the Irrigation Indices (II) [4]. The present study was undertaken in the Kanchipuram district (12.8185° N, 79.6947° E) located on northern east coast of Tamil Nadu. Kanchipuram district covers a total area of 1448 Km^2, where 87.2 Km is the coastal length. It has a normal rainfall of 1200–1300 mm. The river basins of the Palar, Araniyar, and Kosasthalaiyar, and Thondiar, are utilized for irrigation and drinking purposes. This study's primary goal is to determine the groundwater quality in different blocks of Kanchipuram district and also map the results using Q-GIS.

2 Materials and Methods

2.1 *Study Area*

The Kanchipuram district's study area is near Chennai city, and is located on Tamil Nadu's northern east coast, illustrated in Fig. 1. The district has a total area of 1448 Km^2. It lies between 11° 00′ to 12° 00′ latitudes and 77° 28′ to 78° 50′ longitudes. The maximum and minimum temperatures of the Kanchipuram district are during summer 38.5 °C (101.3 °F), 29.1 °C (84.4 °F), and during winter 27.7 °C (81.9 °F), 19.0 °C (66.2 °F). The maximum rainfall of 9.8 inches occurs in October and the least amount of rainfall occurs in February with average of 0.4 inches. The average humidity conditions range from 73 to 90%. The research was conducted using 60 groundwater samples that were collected all over the Kanchipuram district. The data used in the current study was obtained from the Central Groundwater Board of India (CGWB) (Table 1).

2.2 *Water Quality Index (WQI)*

Water Quality Index (WQI) provides the overall quality of the water for any intended purpose by expressing the water's quality as an index number. For calculation of

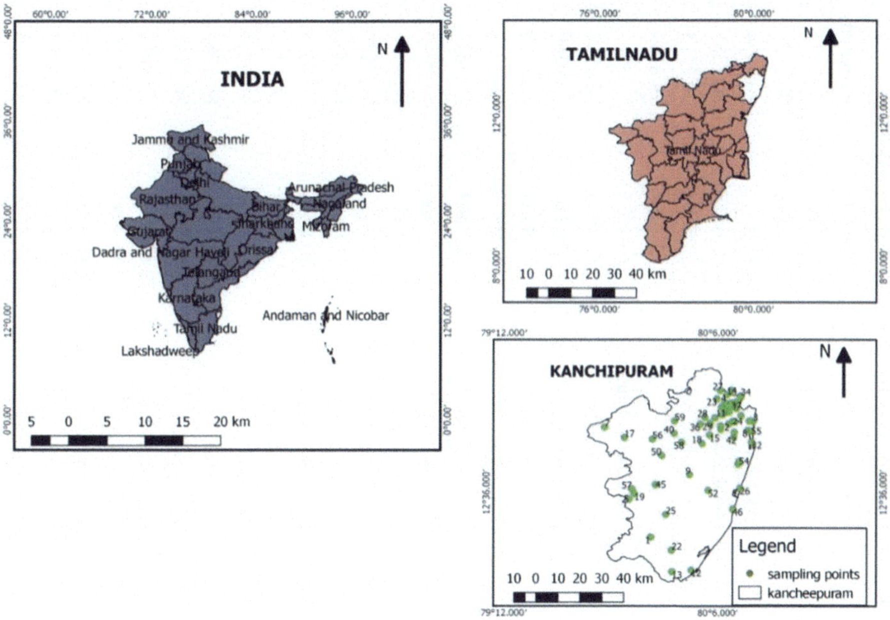

Fig. 1 Location of sampling stations of Kanchipuram district, Tamil Nadu, India

Table 1 Latitude and Longitude details of sampling locations

S.No.	Sampling location	Latitude	Longitude	S.No.	Sampling location	Latitude	Longitude
1	Acharapakkam	12.4064	79.8133	31	Mudichur	12.8943	80.0645
2	Agaram	12.8898	80.143	32	Muthukadu	12.8125	80.2433
3	Akkarai	12.902	80.2494	33	Nagalkani	12.9576	80.1261
4	Anakaputhur	12.9827	80.1286	34	Nandambakkam	13.0106	80.196
5	Arasanimangalam	12.5689	79.7233	35	Nedunkundram	12.8833	80.1106
6	Athur	12.4064	79.8133	36	Paddapai (Pz)	12.8898	80.022
7	Baluchettychatram	10.4633	77.79	37	Pallavaram	12.9814	80.156
8	Chemmancheri	12.8633	80.2261	38	Pallavaram pz	12.9725	80.1543
9	Chenglepet	12.6744	79.9769	39	Pallikaranai (Pz)	12.9263	80.1979
10	Chitlapakkam	12.9403	80.1378	40	Panruti Kandigai	12.8506	79.9119
11	Chrompet	12.95	80.1338	41	Perungulathur	12.9173	80.0856
12	Chunambedu	12.2669	79.9844	42	Ponmar	12.8338	80.1667
13	Chunamedu	12.2586	79.9028	43	Ponmar (Pz)	12.8556	80.1693
14	Girugambakkam	13.0153	80.1366	44	Porur	13.0338	80.1569
15	Guduvancheri	12.8439	80.0622	45	Rettamangalam	12.63	79.8336
16	Hastinapuram	12.9431	80.1476	46	Sadras	12.5269	80.1622
17	Kanchipuram	12.8336	79.7025	47	Shollinganallore	12.9037	80.233
18	Kattankalathur	12.8083	80.0297	48	St. Thomas Mount	13.0071	80.1976
19	Kattur-Pz	12.5708	80.9749	49	Tambaram	12.931	80.1166
20	Kilkattalai	12.9574	80.1785	50	Thirumukkudal	12.7564	79.8592
21	Kolapakkam	12.8687	80.1101	51	Thiruporur (Pz)	12.72	80.1852
22	Kolathanallur	12.3506	79.9014	52	Tirukalukundram	12.6061	80.056
23	Kundrathur	12.998	80.0936	53	Tiruneermalai	12.9628	80.1132
24	Madambakkam	12.9008	80.1615	54	Tiruporur	12.7318	80.1891
25	Madurantagam	12.5028	79.8761	55	Uthandi	12.86	80.2417
26	Mahabalipuram	12.6158	80.1925	56	Uthikadu	12.8267	79.8211

27	Mangadu	13.0342	80.1085	57	Uthiramerur	12.6117	79.7344
28	Manimangalam	12.919	80.0424	58	Vadakkupattu	12.8117	79.9425
29	Mannivakkam	12.8922	80.8019	59	Vadekkal (Palnallur)	12.9036	79.9139
30	Meenambakkam	12.9855	80.1784	60	Vengaivasal	12.9004	80.1711

WQI physicochemical parameters of water are taken into consideration [5, 6]. Based on WQI the water quality is classified as Excellent (0–25), Good (26–50), Poor (51–75), Very Poor (76–100), and Unfit (>100).

$$WQI = \frac{\sum q_i W_i}{\sum W_i} \tag{1}$$

$$q_i = 100 \times \frac{(v - v_o)}{(s_n - s_o)} \tag{2}$$

$$W_i = \frac{K}{s_n} \tag{3}$$

$$K = \frac{1}{\sum \frac{1}{s_n}} \tag{4}$$

where,

q_i—is the quality rating of each water quality parameter.

W_i– is the unit weight of each water quality parameter.

v—is the actual concentration of each water quality parameter.

v_o—is the constant value of each parameter (for pH it is 7 and for remaining all it is zero); K—is the proportionality constant.

s_n– is the standard permissible value of each water quality parameter.

The concentration of Na^+, Mg^+, Cl^-, SO_4^{2-} are used to calculate the seawater mixing index. It had been calculated using the following Eq. (10):

$$SMI = a \times \frac{C_{Na}}{T_{Na}} + b \times \frac{C_{Mg}}{T_{Mg}} + C \times \frac{C_{Cl}}{T_{Cl}} + d \times \frac{C_{SO_4}}{T_{SO_4}} \tag{5}$$

where, constants a, b, c, and d denote a relative proportion of Na^+, Mg^+, Cl^-, SO_4^{2-}, respectively, their values are (a = 0.31, b = 0.04, c = 0.57, d = 0.08), T is the regional threshold value, and C is the calculated concentration of groundwater samples. The threshold value is obtained by drawing the graph between the sampling stations vs cumulative probability.

2.3 *Human Health Risk Assessment (HHRA)*

Nitrate and fluoride are the two most pervasive contaminants found in groundwater that can be harmful to human health. Worldwide reports of nitrate contamination in shallow groundwater, particularly in agricultural areas, are on the rise [7, 8].

$$ADD_{ing} = \frac{C_W \times IR \times EF \times ED}{BW \times AT} \tag{6}$$

Table 2 HHRA Standards for the calculation of HI of fluoride

Description	IR (L/day)	EF (day/Year)	ED (Year)	BW (Kg)	AT (day)	SA (cm^2)	ET (h/day)
Adults	2	350	40	70	14,000	18,000	0.58
Children	0.78	350	4	15	1400	6600	1

$$ADD_{der} = \frac{C_W \times SA \times K_P \times ET \times EF \times ED \times 10^{-3}}{BW \times AT} \quad (7)$$

$$HQ_{ing} = \frac{ADD_{ing}}{RfD} \quad (8)$$

$$HQ_{der} = \frac{ADD_{der}}{RfD} \quad (9)$$

$$HI = HQ_{ing} + HQ_{der} \quad (10)$$

where

C_W—is the actual concentration of water quality parameter in the water (mg/L); IR—is the ingestion rate (L/day).

EF—is the exposure frequency (day/year); ED—is the exposure duration in years; BW—is the average body weight (kg); AT—is the average time (days).

ET—is the exposure time (h/day).

SA—is the average surface area of the skin exposed to water in cm^2.

K_P—is the coefficient of water for dermal activity (0.001 for Na^{2+}, F^-, NO_3^- and Cl^-).

ADD_{ing}—is the average daily exposure dose through ingestion (ing).

ADD_{der}—is the average daily exposure dose through dermal (der).

RfD—is the reference dose of water quality parameters in mg/kg/day (F^- is 0.06 and NO_3^- is 1.6); HQ_{ing}—is the Hazard Quotient of ingestion; HQ_{der}– is the Hazard Quotient of dermal (Table 2).

HI—is the Hazard Index.

2.4 *Irrigation Indices (II)*

In order to know the irrigation indices, the following were calculated Sodium Adsorption Ratio (SAR), percentage of Na, Magnesium Adsorption Ratio (MAR), Permeability Index (PI), and Kelly's Ratio (KR) [9–12].

$$SAR = \frac{Na}{\sqrt{\frac{Ca+Mg}{2}}} \quad (11)$$

$$\%Na = \frac{Na + K}{Na + k + Ca + Mg} \times 100 \quad (12)$$

$$MAR = \frac{Mg}{Ma + Ca} \times 100 \quad (13)$$

$$KR = \frac{Na^{+}}{Ca^{2+} + Mg^{2+}} \quad (14)$$

$$PI = \frac{Na + \sqrt{HCO_3}}{Ca + \mathrm{mg} + Na} \times 100 \quad (15)$$

3 Results and Discussions

The presence of higher amounts of cations and anions results in seawater intrusion. We can clearly observe that most of the sampling locations exhibit higher concentrations of anions and cations (Table 3).

The above physicochemical parameters show that TDS values vary from 266 mg/L to 5213 mg/L. As per BIS standards 10 samples are less than 500 mg/L.

The pH values vary from 6.9 to 7.9. All the 60 samples are below the desired limit (pH:6.5–8.5).

Total hardness values vary from 70 mg/L to 2780 mg/L. The desired limit is 300 mg/L. 16 samples are in the desired limit as per standards.

As per BIS standards, the desired limit for fluoride is 1.5 mg/L and nitrate is 45 mg/L. For fluoride samples 17 and 56 are not within the limits.

3.1 Water Quality Index (WQI)

In general, the water quality depends upon the values of iron, manganese, and arsenic. An increase in these values indicates a decrease in its quality. Water quality index values range from 21 to 233. Out of 60 samples, 13 samples have exceeded the desired limit of drinking water as per BSI standards.

For WQI, if the value is greater than 100 it is unfit for consumption, out of 60 stations 13 stations recorded are unfit for consumption, 12 stations are very poor and 17 stations are poor. We observed that Manimangalam (station 28) has the highest value of 233, which is greater than 100, Uthikadu has 193, Shollinganallore has 157, and many more. These are unfit for drinking because that area is covered by many dyeing factories. Whereas places like Chitlapakkam have a WQI value of 21 comes under the excellent category. This area has no risk of mixing water from the sea. From Fig. 2, it is clear that the Kanchipuram district, Tamil Nadu, was moderately polluted (Fig. 3).

Table 3 Physical and chemical characteristics of water samples from the Kanchipuram district (pH is the only unit not in mg/L)

Sample	pH	TDS	F	CI	SO_4	TH	Na	K	Ca	Mg	NO_3
1	7.81	663	1.04	142	3	390	37	1	84	44	1
2	7.4	634	0.49	192	8	350	59	2	76	39	2
3	7.48	554	0.12	121	46	310	54	17	52	44	57
4	6.9	1004	1.4	206	5	400	145	45	78	36	157
5	7.1	748	0.52	145	60	380	64	1	100	32	70
6	7.2	924	1.15	248	50	450	99	6	64	70	40
7	7	1307	0.29	347	70	450	160	5	92	53	30
8	7.45	497	0.37	114	108	190	69	40	20	30	34
9	7.9	482	0.94	191	3	190	89	8	56	12	32
10	7.42	323	0.17	60	16	165	35	1	23	23	50
11	7.09	1145	0.91	312	29	530	169	4	76	83	7
12	7	880	0.41	305	50	460	72	0	120	39	45
13	7.11	315	0.4	28	2	160	36	0	32	19	21
14	7	1102	1.13	277	70	450	163	11	112	41	1
15	7.15	884	0.69	220	10	420	90	26	110	24	99
16	7.2	768	0.55	178	35	400	89	3	76	51	9
17	7.8	2555	1.76	596	8	140	722	23	32	15	72
18	7.7	535	0.59	113	6	310	37	3	56	41	38
19	7.7	391	0.81	85	42	270	20	6	60	29	6
20	7.4	1050	0.55	305	78	460	153	4	120	39	39
21	7.3	1186	0.64	312	73	480	199	2	116	46	98
22	7.64	625	0.9	163	6	250	104	2	44	34	19
23	7.2	865	1.21	241	44	350	127	3	112	17	27
24	7.38	1274	0.78	369	44	490	214	3	104	56	26
25	7.38	837	0.36	156	5	390	78	5	136	12	38
26	7.55	288	0.59	57	54	190	17	6	24	32	3
27	7.1	1178	0.8	298	66	460	184	14	128	34	85
28	7.54	1365	0.97	277	40	330	169	268	92	24	146
29	7	660	0.38	160	56	370	37	2	76	44	35
30	7.47	672	0.63	156	64	340	87	1	80	34	27
31	7.4	1112	0.37	291	52	430	182	21	140	19	65
32	6.96	1417	0.32	540	186	380	304	15	60	56	6
33	7.12	1097	0.42	348	7	350	226	3	68	44	1
34	7.3	593	0.93	146	35	230	93	15	64	17	16
35	7.1	790	0.63	160	90	370	82	0	72	46	35
36	7.33	1229	0.48	391	48	690	126	4	132	88	49
37	7.27	1359	0.89	391	106	510	217	2	132	44	40
38	7.56	757	0.82	213	47	420	72	1	108	36	16
39	7.25	1138	0.25	369	49	580	133	3	176	34	22
40	7.3	376	0.5	82	30	70	72	0	12	10	10
41	7.32	1749	1.11	518	71	440	402	4	80	58	53
42	7.52	591	0.2	170	35	340	52	5	64	44	2

(continued)

Table 3 (continued)

Sample	pH	TDS	F	Cl	SO_4	TH	Na	K	Ca	Mg	NO_3
43	7.5	686	0.54	163	31	350	81	3	58	46	62
44	7.1	881	1.21	213	21	370	123	2	88	36	2
45	7.1	546	0.99	57	30	230	54	0	76	10	25
46	7.82	863	0.7	178	122	460	90	21	128	34	73
47	7.51	2022	1.12	696	328	340	467	102	68	41	112
48	7	1242	1.32	320	78	540	184	1	144	44	59
49	7.26	629	0.54	149	25	310	83	4	72	32	58
50	7.2	416	0.33	71	35	160	72	4	28	22	37
51	7.59	947	0.84	220	144	360	158	7	56	54	5
52	7.7	298	0.51	85	34	180	23	8	24	29	8
53	7.15	583	0.12	156	9	270	65	31	68	24	13
54	7.86	1065	1.25	213	160	280	255	9	24	54	48
55	7.47	266	0.53	64	59	140	25	8	20	22	13
56	7	1697	1.7	440	98	780	65	89	240	44	110
55	7.47	266	0.53	64	59	140	25	8	20	22	13
56	7	1697	1.7	440	98	780	65	89	240	44	110
57	7.12	5213	1.04	2382	9	2780	457	0	576	326	68
58	7.1	624	0.36	106	25	310	40	1	72	32	41
59	7.1	2002	0.77	638	155	440	398	29	64	68	25
60	7.29	1742	0.52	504	88	780	239	20	164	90	21

3.2 Seawater Mixing Index (SMI)

The threshold values obtained from the graph (Actual frequency vs Concentration) are Na =93 mg/L, Mg = 44 mg/L, Cl =440 mg/L, and SO4 = 50 mg/L. From the results obtained the SMI values vary from 0.18 to 4.92 (Fig. 4).

Generally, SMI value >1 signifies that the seawater is mixing with the groundwater. Out of total, 38% of the sampling stations namely Uthiramerur (4.92), Kanchipuram (3.21), and Muthukadu (2.06) have SMI values above 1 and are affected by seawater intrusion. Majorly seawater intrusion in that area is caused due to climatic conditions, hydraulic gradient, rate of groundwater extraction, and sea level rise. The people of Kanchipuram district use more groundwater during summer which results in a decrease in groundwater table and due to this seawater intrusion takes place (Fig. 5).

3.3 Human Health Risk Assessment (HHRA) for Fluoride

From the above HHRA, we conclude that fluoride content in children is greater than in adults (Figs. 6 and 7). For adults, the HI concentration is less than 1 in all sampling stations, as fluoride content does not have much impact on adults. For adults, the HI

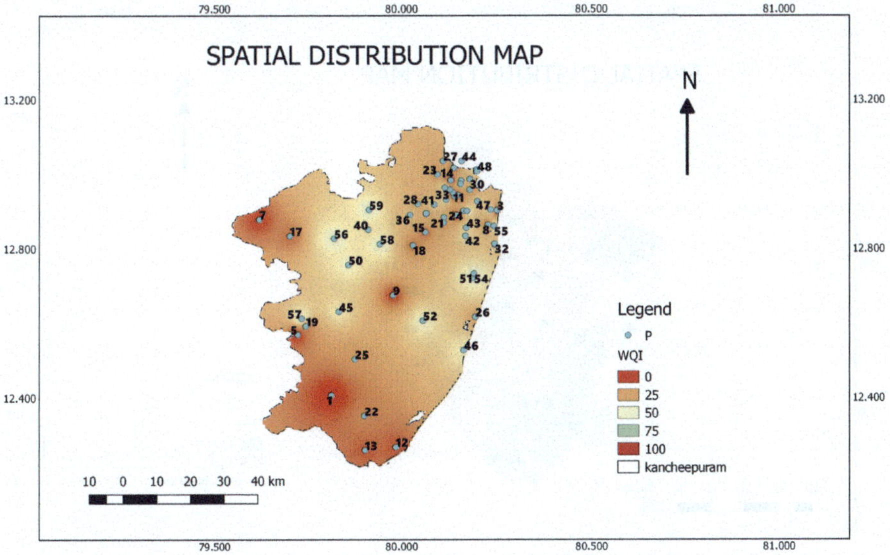

Fig. 2 Geographical distribution map of Water Quality Index (WQI) of Kanchipuram district

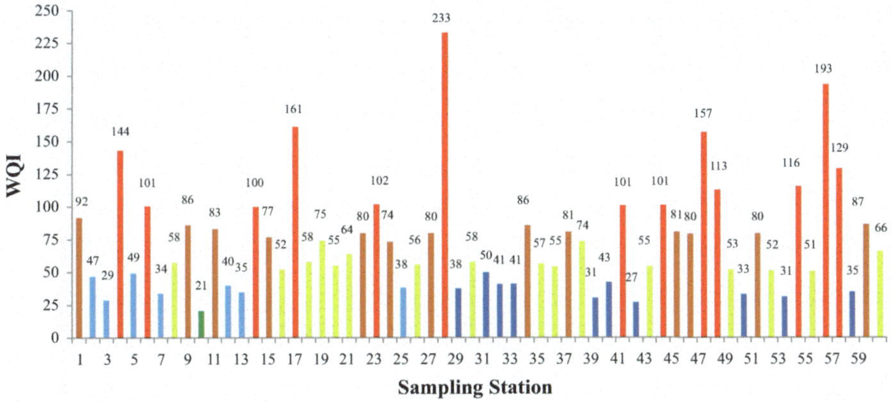

Fig. 3 WQI values of different sampling stations of Kanchipuram district

concentrations of Fluoride varied from 0.06 to 0.81 has an Average = 0.37 and for Children varied from 0.104 to 1.538 has an Average = 0.69. Kanchipuram (station 17) area has high fluoride content in children (1.54). The main reason for this is that the consumption of groundwater with high fluoride content has a greater impact on children compared to adults. Eight stations have HI value greater than 1 in children. The samples analyzed from the research area have high fluoride content in children ranging from 0.1 to 1.53. The results show that most of the samples have higher fluoride content in children than the permissible limits. Due to the excessive fluoride content present in groundwater can cause dental problems, bone damage, joint-related problems, and change in color of teeth (Fig. 8 and Table 4).

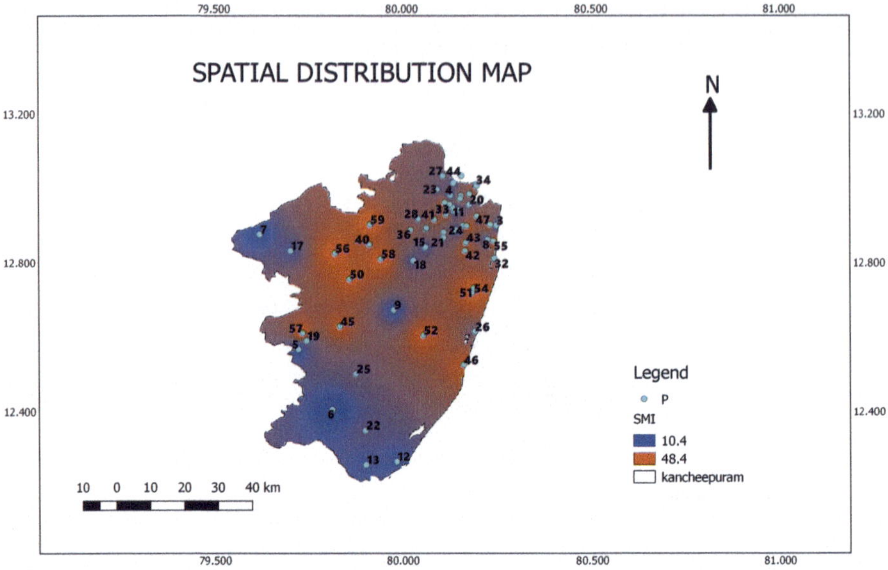

Fig. 4 Geographical distribution map of Seawater Mixing (SMI) of Kanchipuram district

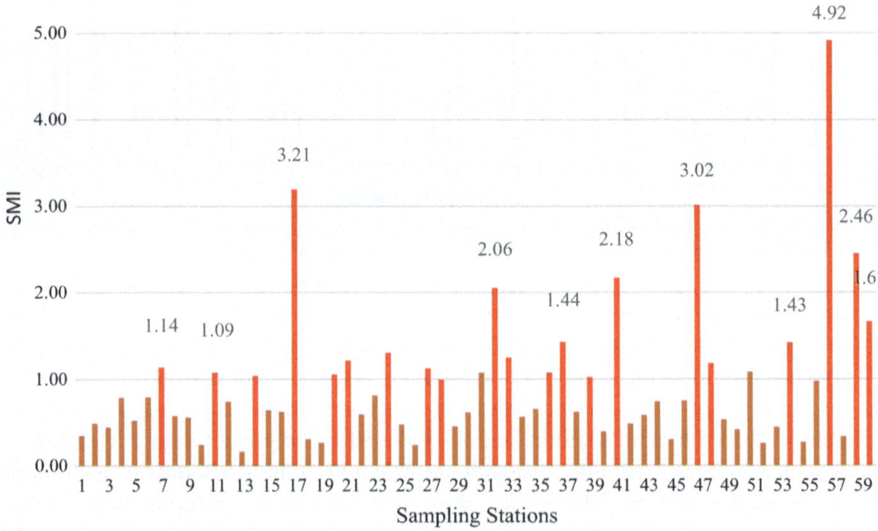

Fig. 5 SMI values of different sampling stations of Kanchipuram district

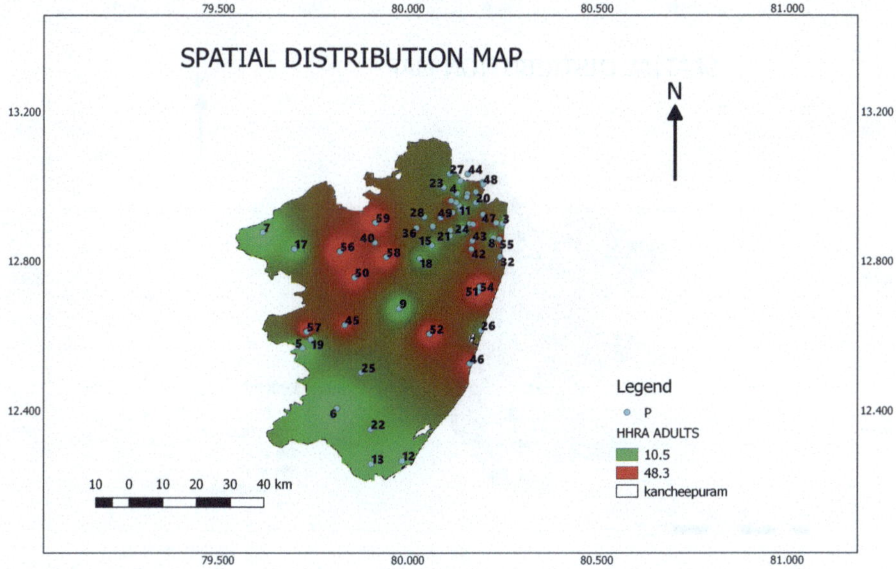

Fig. 6 Geographical distribution map of Human Health Risk Assessment (HHRA) for adults of Kanchipuram district

3.4 Irrigation Indices (II) Sodium Adsorption Ratio

One of the most important metrics for determining if water is suitable for irrigation is the Sodium Adsorption Ratio (SAR), which measures how much water can alter the permeability and structure of the soil. The ratio of sodium (Na) to calcium (Ca) and magnesium (Mg) in water is assessed using SAR. Elevated salt concentrations can cause lower infiltration and soil dispersion, which can impact plant growth. SAR values of Kanchipuram district vary from 2.99 to 148.94. In general, the SAR value should be less than 10 mEq/L. Out of 60 samples, only 20 samples are below the desired limit. These sampling stations have no effect on irrigation whereas the remaining 40 stations are unfit for irrigation (Figs. 9 and 10).

3.4.1 Kelly's Ratio

Kelly's ratio (KR) depends on sodium, magnesium, and calcium concentrations. Twenty-four stations have values greater than 1, indicating excessive sodium in the water. Elevated sodium levels in irrigation water can have negative effects on plant development, soil permeability, and soil structure. So, these stations are unsuited for irrigation (Figs. 11 and 12).

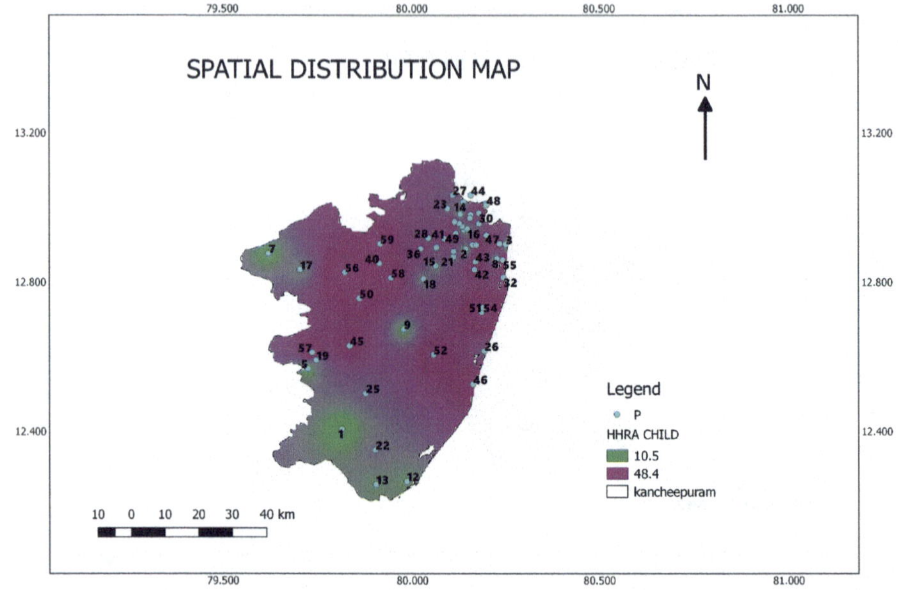

Fig. 7 Geographical distribution map of Human Health Risk Assessment (HHRA) for children of Kanchipuram district

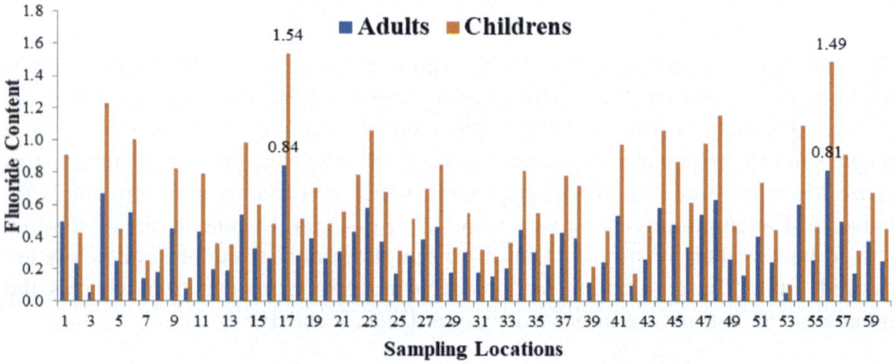

Fig. 8 Human Health Risk Assessment (HHRA) of Kanchipuram district

3.4.2 Magnesium Adsorption Ratio

The Magnesium Adsorption Ratio (MAR) is a crucial metric for determining whether water is suitable for irrigation. Understanding the possible effects of magnesium (Mg) content in relation to calcium (Ca) on crop yield and soil structure is made easier with the use of MAR. An overabundance of magnesium in irrigation water can negatively impact plant development and soil characteristics. Higher concentration of magnesium present in the groundwater affects the quality of soil and lowers the crop yield. The desired value for MAR is less than 50 (Figs. 13 and 14).

Table 4 Different parameters of HHRA for sampling stations

Station	Fluoride (Adults)					Fluoride (Children)				
	ADD_{ing}	ADDder	HQing	HQder	HI	ADDing	ADDder	HQing	HQder	HI
1	0.02971	0.00016	0.50	0.0026	0.4978	0.90133	0.00763	0.90	0.0076	0.90896
2	0.01400	0.00007	0.23	0.0012	0.2346	0.42467	0.00359	0.42	0.0036	0.42826
3	0.00343	0.00002	0.06	0.0003	0.0574	0.10400	0.00088	0.10	0.0009	0.10488
4	0.04000	0.00021	0.67	0.0035	0.6701	1.21333	0.01027	1.21	0.0103	1.22360
5	0.01486	0.00008	0.25	0.0013	0.2489	0.45067	0.00381	0.45	0.0038	0.45448
6	0.03286	0.00017	0.55	0.0029	0.5505	0.99667	0.00843	1.00	0.0084	1.00510
7	0.00829	0.00004	0.14	0.0007	0.1388	0.25133	0.00213	0.25	0.0021	0.25346
8	0.01057	0.00006	0.18	0.0009	0.01771	0.32067	0.00271	0.32	0.0027	0.32338
9	0.02686	0.00014	0.45	0.0023	0.4500	0.81467	0.00689	0.81	0.0069	0.82156
10	0.00486	0.00003	0.08	0.0004	0.0814	0.14733	0.00125	0.15	0.0012	0.14858
11	0.02600	0.00014	0.43	0.0023	0.4356	0.78867	0.00667	0.79	0.0067	0.79534
12	0.01171	0.00006	0.20	0.0010	0.1963	0.35533	0.00301	0.36	0.0030	0.35834
13	0.01143	0.00006	0.19	0.0010	0.1915	0.034667	0.00293	0.35	0.0029	0.34960
14	0.03229	0.00017	0.54	0.0028	0.5409	0.97933	0.00829	0.98	0.0083	0.98762
15	0.01971	0.00010	0.33	0.0017	0.3303	0.59800	0.00506	0.60	0.0051	0.60306
16	0.01571	0.00008	0.26	0.0014	0.2633	0.47667	0.00403	0.48	0.0040	0.48070
17	0.05029	0.00026	0.84	0.0044	0.8425	1.5233	0.01291	1.53	0.0129	1.53824
18	0.01686	0.00009	0.28	0.0015	0.2824	0.51133	0.00433	0.51	0.0043	0.51566
19	0.02314	0.00012	0.39	0.0020	0.3877	0.70200	0.00594	0.70	0.0059	0.70794
20	0.01571	0.00008	0.26	0.0014	0.2633	0.47667	0.00403	0.48	0.0040	0.48070
21	0.01829	0.00010	0.30	0.0016	0.3064	0.55467	0.00469	0.55	0.0047	0.55936
22	0.02571	0.00013	0.43	0.0022	0.4308	0.78000	0.00660	0.78	0.0066	0.78660
23	0.03457	0.00018	0.58	0.0030	0.5792	1.04867	0.00887	1.05	0.0089	1.05754
24	0.02229	0.00012	0.37	0.0019	0.3734	0.67600	0.00572	0.68	0.0057	0.68172

(continued)

Table 4 (continued)

Station	Fluoride (Adults)					Fluoride (Children)				
	ADD_{ing}	ADDder	HQing	HQder	HI	ADDing	ADDder	HQing	HQder	HI
25	0.01029	0.00005	0.17	0.0009	0.1723	0.31200	0.00264	0.31	0.0026	0.31464
26	0.01686	0.00009	0.28	0.0015	0.2824	0.51133	0.00433	0.51	0.0043	0.51566
27	0.02286	0.00012	0.38	0.0020	0.3829	0.69333	0.00587	0.69	0.0059	0.69920
28	0.02771	0.00014	0.46	0.0024	0.4643	0.84067	0.00711	0.84	0.0071	0.84778
29	0.01086	0.00006	0.18	0.0009	0.1819	0.32933	0.00279	0.33	0.0028	0.33212
30	0.01800	0.00009	0.30	0.0016	0.3016	0.54600	0.00462	0.55	0.0046	0.55062
31	0.01057	0.00006	0.18	0.0009	0.1771	0.32067	0.00271	0.32	0.0027	0.32338
32	0.00914	0.00005	0.15	0.0008	0.1532	0.27733	0.00235	0.28	0.0023	0.27968
33	0.01200	0.00006	0.20	0.0010	0.2010	0.36400	0.00308	0.36	0.0031	0.36708
34	0.02657	0.00014	0.44	0.0023	0.4452	0.80600	0.00682	0.81	0.0068	0.81282
35	0.01800	0.00009	0.30	0.0016	0.3016	0.54600	0.00462	0.55	0.0046	0.55062
36	0.01371	0.00007	0.23	0.0012	0.2298	0.41600	0.00352	0.42	0.0035	0.41952
37	0.02543	0.0013	0.42	0.0022	0.4260	0.77133	0.00653	0.77	0.0065	0.77786
38	0.02343	0.00012	0.39	0.0020	0.3925	0.71067	0.00601	0.71	0.0060	0.71668
39	0.00714	0.00004	0.12	0.0006	0.1197	0.21667	0.00183	0.22	0.0018	0.21850
40	0.01429	0.00007	0.24	0.0012	0.2393	0.43333	0.00367	0.43	0.0037	0.43700
41	0.03171	0.00017	0.53	0.0028	0.5313	0.96200	0.00814	0.96	0.0081	0.97014
42	0.00571	0.00003	0.10	0.0005	0.0957	0.17333	0.00147	0.17	0.0015	0.17480
43	0.01543	0.00008	0.26	0.0013	0.2585	0.46800	0.00396	0.47	0.0040	0.47196
44	0.03457	0.00018	0.58	0.0030	0.5792	1.04867	0.00887	1.05	0.0089	1.05754
45	0.02829	0.00015	0.47	0.0025	0.4739	0.85800	0.00726	0.86	0.0073	0.86526
46	0.02000	0.00010	0.33	0.0017	0.3351	0.60667	0.00513	0.61	0.0051	0.61180
47	0.03200	0.00017	0.53	0.0028	0.5361	0.97067	0.00821	0.97	0.0082	0.97888
48	0.03771	0.00020	0.63	0.0033	0.6319	1.14400	0.00968	114	0.0097	1.1536

49	0.01543	0.00008	0.26	0.0013	0.2585	0.46800	0.00396	0.47	0.0040	0.47196
50	0.00943	0.00005	0.16	0.0008	0.1580	0.28600	0.00242	0.29	0.0024	0.28842
51	0.02400	0.00013	0.40	0.0021	0.4021	0.72800	0.00616	0.73	0.0062	0.73416
52	0.01457	0.00008	0.24	0.0013	0.2441	0.44200	0.00374	0.44	0.0037	0.44574
53	0.00343	0.00002	0.06	0.0003	0.0574	0.10400	0.00088	0.10	0.0009	0.10488
54	0.03571	0.00019	0.60	0.0031	0.5983	1.08333	0.00917	1.08	0.0092	1.09250
55	0.01514	0.00008	0.25	0.0013	0.2537	0.45933	0.00389	0.46	0.0039	0.46322
56	0.04857	0.00025	0.81	0.0042	0.8137	1.47333	0.01247	1.47	0.0125	1.48580
57	0.02971	0.00016	0.50	0.0026	0.4978	0.90133	0.00763	0.90	0.0076	0.90896
58	0.01029	0.00005	0.17	0.0009	0.1723	0.31200	0.00264	0.31	0.0026	0.31464
59	0.02200	0.00011	0.37	0.0019	0.3686	0.66733	0.00565	0.67	0.0056	0.67298
60	0.01486	0.00008	0.25	0.0013	0.2489	0.45067	0.00381	0.45	0.0038	0.45448

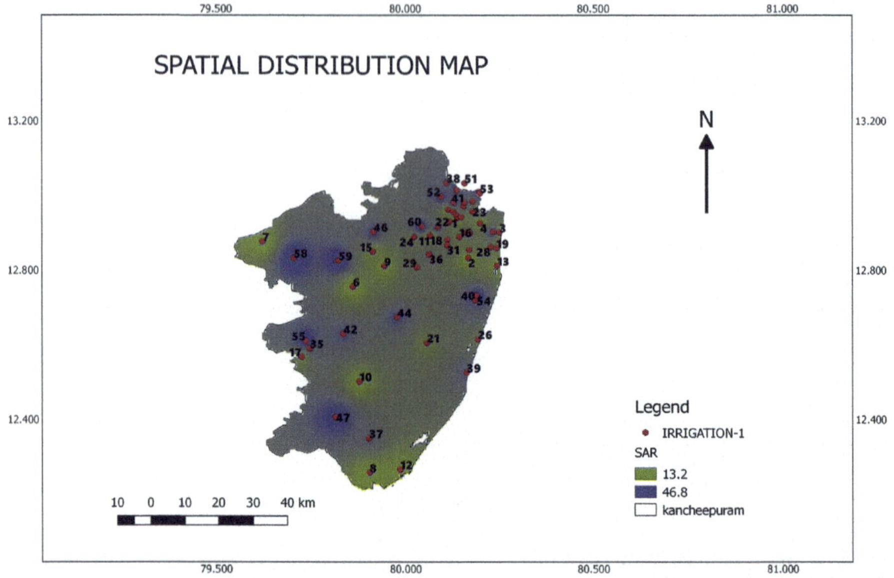

Fig. 9 Geographical distribution map of Kanchipuram district for Irrigation Indices (II) based on sodium adsorption ratio (SAR)

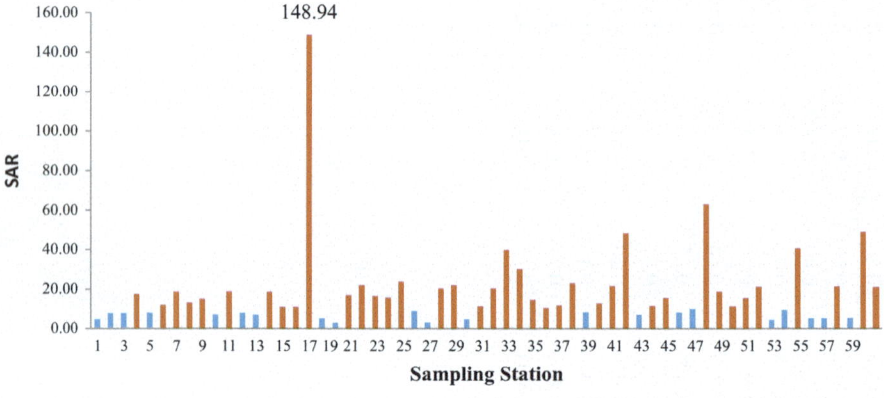

Fig. 10 Irrigation Indices (II) of Kanchipuram district based on sodium adsorption ratio (SAR)

3.4.3 Percentage of Na

Groundwater is categorized according to its percentage of Na, which ranges from excellent (<20%) to good (20–40%), acceptable (40–60%), dubious (60–80%), and inappropriate (> 80%) (Figs. 15 and 16).

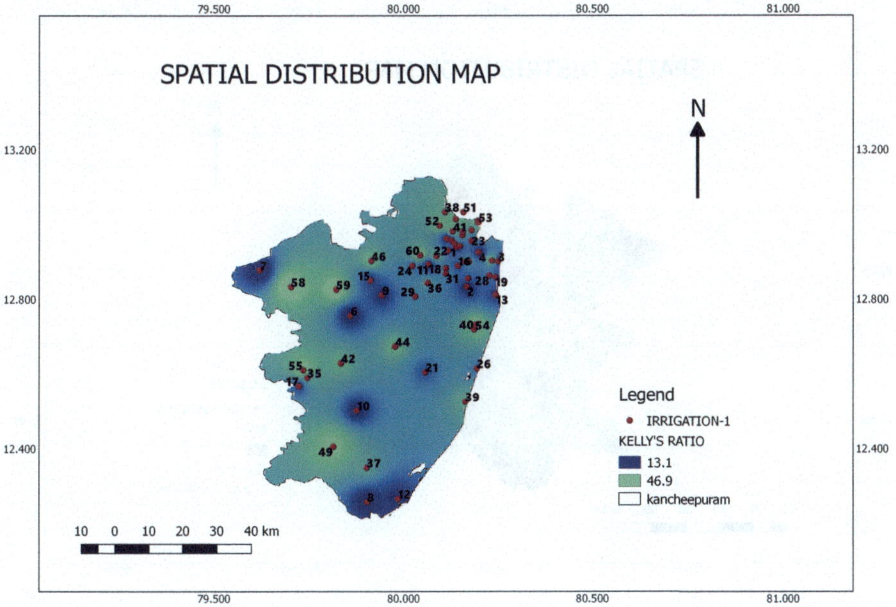

Fig. 11 Geographical distribution map of Kanchipuram district for Irrigation Indices (II) based on Kelly's ratio (KR)

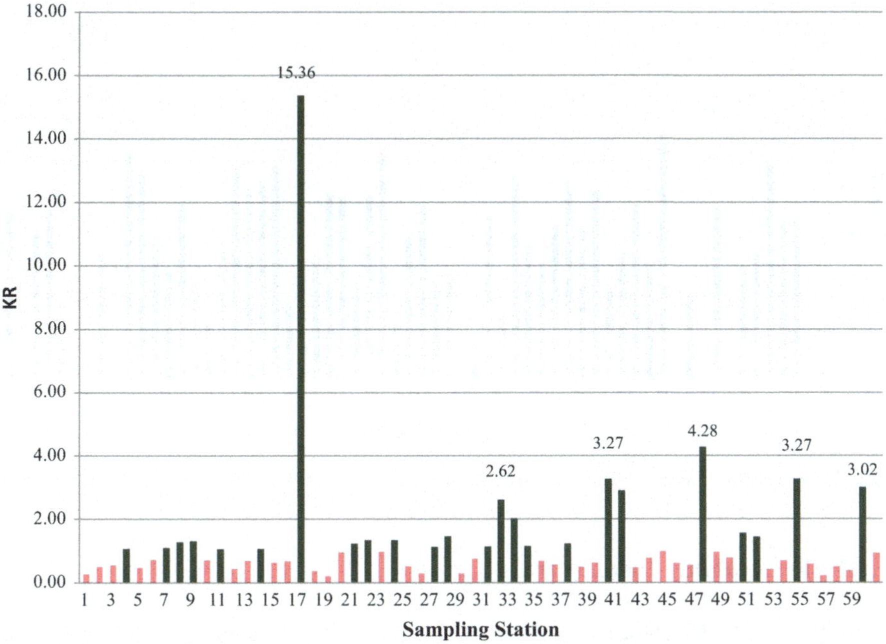

Fig. 12 Irrigation Indices (II) of Kanchipuram district based on Kelly's ratio (KR)

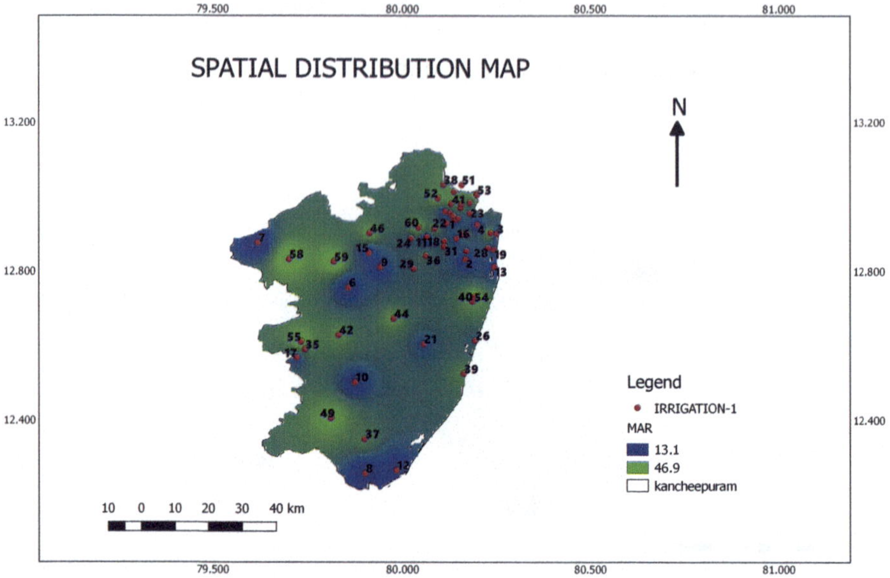

Fig. 13 Geographical distribution map of Kanchipuram district for Irrigation Indices (II) based on magnesium adsorption ratio (MAR)

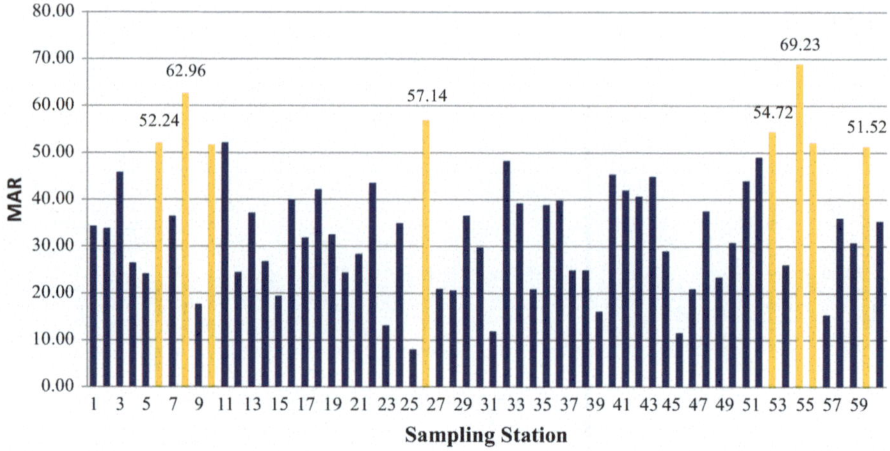

Fig. 14 Irrigation Indices (II) of Kanchipuram district based on magnesium adsorption ratio (MAR)

3.4.4 Permeability Index (PI)

From the results PI values vary from 30.13 to 98.22. The average value of PI was reported as 54.91. As per desired limits, PI values less than 25 is unsuited for irrigation (Figs. 17 and 18).

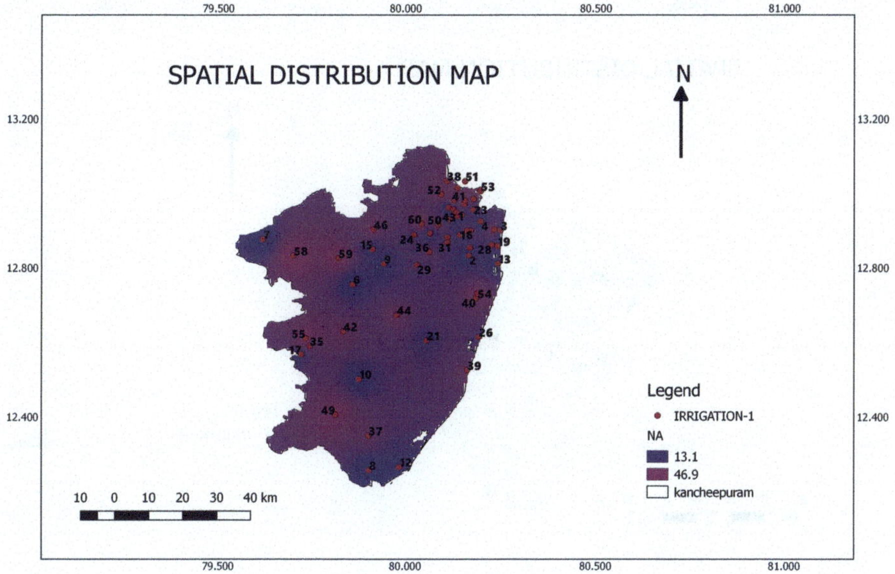

Fig. 15 Geographical distribution map of Kanchipuram district for Irrigation Indices (II) based on percentage of Na

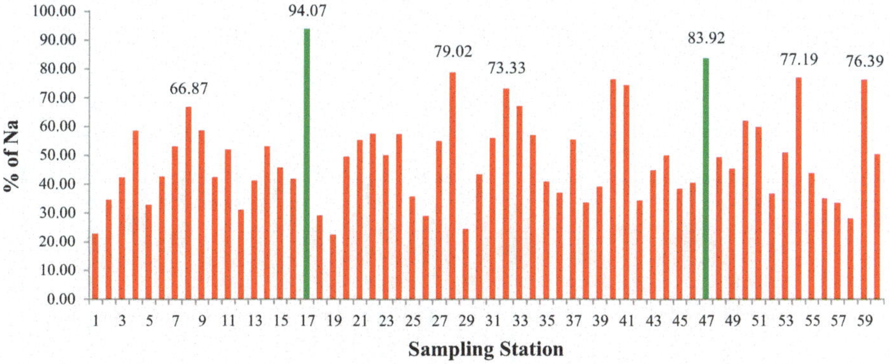

Fig. 16 Irrigation Indices (II) of Kanchipuram district based on percentage of Na

For the irrigation indices, we conclude that out of 60 stations, SAR values of 21 stations are less than 10 meq/L, so in those areas the water is fit for irrigation, and the crops do not give much yield. Due to the saltwater intrusion, these ions will mix up with the ground water. This water will carry through the irrigation fields. The water with high SAR content is supplied to soil for years, The calcium and magnesium ions in the soil may be replaced by the sodium in the water. The soil's fertility will decline as a result of this. For the percentage of Na, 98% of stations are less than 80%, this water will impact irrigation crops (Table 5).

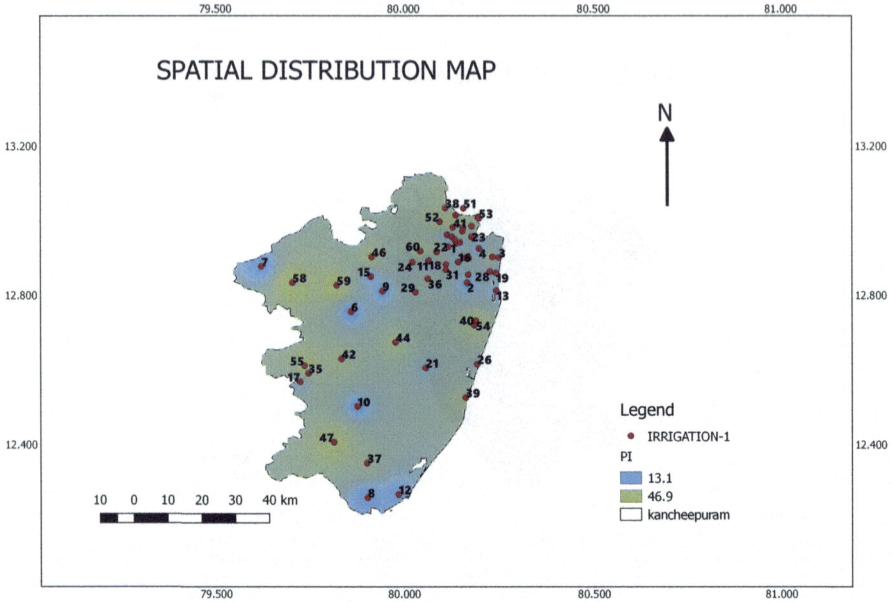

Fig. 17 Geographical distribution map of Kanchipuram district for Irrigation Indices (II) based on permeability index (PI)

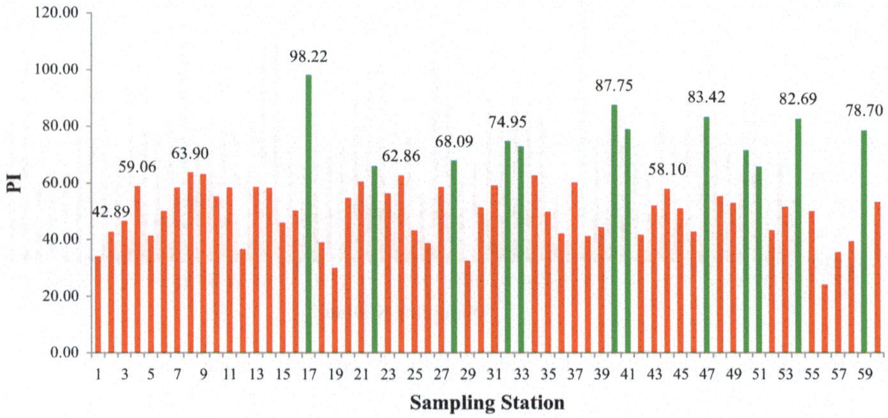

Fig. 18 Irrigation Indices (II) of Kanchipuram district based on PI

4 Conclusions

The following results were concluded based on the investigation of groundwater in Kanchipuram district. WQI values of 60 stations it was observed that around 70% of groundwater is not suitable for potable use. Proper treatment is required for the water

Table 5 Different parameters for Irrigation Indices (II) for sampling stations

S. No.	Sampling location	SAR	KR	NA	MAR	PI
1	Acharapakkam	4.63	0.29	22.89	34.38	34.30
2	Agaram	7.78	0.51	34.66	33.91	42.89
3	Akkarai	7.79	0.56	42.51	45.83	46.67
4	Anakaputhur	17.58	1.07	58.66	26.47	59.06
5	Arasanimangalam	7.88	0.48	32.99	24.24	41.48
6	Athur	12.09	0.74	42.74	52.24	50.20
7	Baluchettychatram	18.79	1.10	53.23	36.55	58.47
8	Chemmancheri	13.28	1.28	66.87	62.96	63.90
9	Chenglepet	15.26	1.31	58.79	17.65	63.18
10	Chitlapakkam	7.20	0.72	42.53	52.00	55.32
11	Chrompet	18.95	1.06	52.11	52.20	58.51
12	Chunambedu	8.08	0.45	31.17	24.53	36.63
13	Chunamedu	7.13	0.71	41.38	37.25	58.66
14	Girugambakkam	18.64	1.07	53.21	26.80	58.30
15	Guduvancheri	11.12	0.64	45.82	19.46	45.95
16	Hastinapuram	11.17	0.70	42.01	40.16	50.28
17	Kanchipuram	148.94	15.36	94.07	31.91	98.22
18	Kattankalathur	5.31	0.38	29.20	42.27	39.12
19	Kattur-Pz	3.00	0.22	22.61	32.58	30.13
20	Kilkattalai	17.16	0.96	49.68	24.53	54.91
21	Kolapakkam	22.11	1.23	55.37	28.40	60.64
22	Kolathanallur	16.65	1.33	57.61	43.59	66.14
23	Kundrathur	15.81	0.98	50.19	13.18	56.50
24	Madambakkam	23.93	1.34	57.56	35.00	62.86
25	Madurantagam	9.07	0.53	35.93	8.11	43.25
26	Mahabalipuram	3.21	0.30	29.11	57.14	38.79
27	Mangadu	20.44	1.14	55.00	20.99	58.75
28	Manimangalam	22.19	1.46	79.02	20.69	68.09
29	Mannivakkam	4.78	0.31	24.53	36.67	32.60
30	Meenambakkam	11.52	0.76	43.56	29.82	51.53
31	Mudichur	20.41	1.14	56.08	11.95	59.26
32	Muthukadu	39.92	2.62	73.33	48.28	74.95
33	Nagalkani	30.20	2.02	67.16	39.29	73.02
34	Nandambakkam	14.61	1.15	57.14	20.99	62.75
35	Nedunkundram	10.68	0.69	41.00	38.98	49.73
36	Paddapai (Pz)	12.01	0.57	37.14	40.00	42.17
37	Pallavaram	23.13	1.23	55.44	25.00	60.29
38	Pallavaram pz	8.49	0.50	33.64	25.00	41.26
39	Pallikaranai (Pz)	12.98	0.63	39.31	16.19	44.40
40	Panruti Kandigai	21.71	3.27	76.60	45.45	87.75
41	Perungulathur	48.40	2.91	74.63	42.03	79.06
42	Ponmar	7.08	0.48	34.55	40.74	41.77

(continued)

Table 5 (continued)

S. No.	Sampling location	SAR	KR	NA	MAR	PI
43	Ponmar (Pz)	11.65	0.79	44.95	44.95	52.21
44	Porur	15.62	0.99	50.20	29.03	58.10
45	Rettamangalam	8.23	0.63	38.57	11.63	51.05
46	Sadras	10.00	0.56	40.66	20.99	42.91
47	Shollinganallore	63.26	4.28	83.92	37.61	83.42
48	St. Thomas Mount	18.98	0.98	49.60	23.40	55.40
49	Tambaram	11.51	0.80	45.55	30.77	53.04
50	Thirumukkudal	15.60	1.56	62.12	44.00	71.68
51	Thiruporur (Pz)	21.30	1.44	60.00	49.09	65.92
52	Tirukalukundram	4.47	0.43	36.90	54.72	43.29
53	Tiruneermalai	9.58	0.71	51.06	26.09	51.71
54	Tiruporur	40.83	3.27	77.19	69.23	82.69
55	Uthandi	5.46	0.60	44.00	52.38	50.07
56	Uthikadu	5.45	0.23	35.16	15.49	24.28
57	Uthiramerur	21.52	0.51	33.63	36.14	35.62
58	Vadakkupattu	5.55	0.38	28.28	30.77	39.42
59	Vadekkal (Palnallur)	48.99	3.02	76.39	51.52	78.70
60	Vengaivasal	21.21	0.94	50.49	35.43	53.46

before consumption. SMI values of 60 stations it was observed that 38% of the sampling stations are affected by seawater intrusion greater than 1. There are seven sampling stations that are affected by both WQI and SMI. HHRA study concludes for Children, 13% of the sampling station is having a health impact due to excess fluoride content. Irrigation efficiency concludes that 65% of sampling stations, SAR values greater than 10 mEq/L and 2 stations have a percentage of Na greater than 80% therefore the water is unfit for irrigation. Nearly 40% of the sampling stations have a KR value less than 1, indicating that those areas are suitable for irrigation. Additionally, about 12 sampling stations have a PI value greater than 65%, making the water suitable for irrigation purposes. Nine stations having MAR value greater than 50, those stations are not recommended for irrigation purposes.

References

1. Wagh GS, Sayyed MRG, Sayadi MH (2014) Evaluating groundwater pollution using statistical analysis of hydro chemical data: a case study from southeastern part of Pune metropolitan city (India). Int J Geo Geosci 4(3):456–476
2. Neumann B, Vafeidis AT, Zimmermann J, Nicholls RJ (2015) Future coastal population growth and exposure to sea-level rise and coastal flooding – a global assessment. PLoS One 10: e0118571. https://doi.org/10.1371/journal.pone.0118571

3. Delsman JR, Hu-a-ng KRM, Vos PC, de Louw PGB, Oude Essink GHP, Stuyfzand PJ, Bierkens MFP (2014) Paleo-modeling of coastal saltwater intrusion during the Holocene: an application to The Netherlands. Hydrol Earth Syst Sci 18:3891–3905. https://doi.org/10.5194/hess-18-3891-2014
4. Dey NC, Rabbi SE (2013) Studies on the impact of BRAC WASH-1 interventions: an overview. In: In: achievements of BRAC water, sanitation and hygiene programme towards millennium development goals and beyond. BRAC, Dhaka
5. Raju NJ, Shukla UK, Ram P (2011) Hydrogeochemistry for the assessment of ground water quality in Varanasi: a fast urbanizing center in Uttar Pradesh, India. Environ Monit Assess 173: 279–300
6. Yisa J, Jimoh T (2010) Analytical studies on water quality index of river Landzu. Am J Appl Sci 7:453–458
7. Tyagi S, Singh P, Sharma B, Singh R (2014) Assessment of water quality for drinking purpose in district Pauri of Uttarakhand India. Appl Ecol Environ Sci 2(4):94–99
8. Subba Rao N (2018) Groundwater quality from a part of Prakasam District, Andhra Pradesh, India. Appl Water Sci 8:30
9. Richards LA (1954) Diagnosis and improvement of saline and alkali soils. Handbook.
10. Eaton FM (1950) Soil Sci 69:123–134
11. Raghaunth H. M.1989 Groundwater Wiley Eastern Ltd New Delhi 563
12. Doneen LD (1964) Notes on water quality in agriculture. Published as a water science and engineering paper 4001. Department of Water Science and Engineering, University of California

A Review on Groundwater and Surface Water Interaction: Insight into Measuring Methods

K. Veerendra Gopi, N. Pradeepthi, and M. Rohith Reddy

Abstract Surface water and groundwater were analyzed as individual components in the past in a hydrological cycle, though they are interconnected throughout the process. Overutilization of aquifer water and surface water has attracted attention to their recent interaction studies. Both lateral and longitudinal flows are present in this interaction. Surface water seeps into the earth, replenishing subterranean aquifers, while groundwater seeps to the surface, providing stream base flow. It is essential to comprehend and measure the interchange occurring between stream water and groundwater from the perspective of sustainable water resource management. A fundamental grasp of these interactions is crucial in developing effective strategies for disaster and risk management studies. Comprehending how surface water and groundwater connect is essential for creating numerical models that effectively simulate scenarios. The present review emphasizes various numerical coupling models and strategies for estimating fluidities in the surface water and groundwater interface. The review underlines the limitations and advantages of fully coupled and loosely coupled models in understanding interactions on a regional scale.

Keywords Groundwater and surface water interaction · Fully coupled models · Loosely coupled models

1 Introduction

Droughts are the second most impactful natural occurrences after floods that occur due to natural and unnatural means [1]. Drought events are natural phenomena that occur due to deficiency of rainfall events which do not occur for periods and the available resources are not sufficient to meet human activity needs. Droughts or floods may affect various parameters socially, economically, and damage property, individuals, or the environment [2]. Droughts were ranked first based on the severity,

K. V. Gopi (✉) · N. Pradeepthi · M. R. Reddy
Department of Civil Engineering, VNR Vignana Jyothi Institute of Engineering and Technology, Hyderabad, India

P. Pathak et al. (eds.), *Advances in Environmental Sustainability, Energy and Earth Science*, https://doi.org/10.1007/978-3-031-73820-3_23

loss of lives, areal extent, social effect, suddenness, long-term impact, and occurrence of related events [3]. Policy makers treat drought event as a rare and unplanned event in decision-making process. Evaluating the frequency and patterns of floods and droughts can yield crucial data for better management of water resources, particularly in light of ongoing climate change [4]. The main source of freshwater for the whole world, surface water, will become less dependable and predictable due to climatic instability and drought conditions [5, 6]. As a result, groundwater will become more important, and overexploitation might put groundwater under stress [7, 8].

The world's most significant freshwater resource, groundwater, is vital to irrigate agriculture and, consequently, to maintaining the security of the world's food supply. Excessive withdrawal of groundwater for irrigation in areas with slow replenishment is the primary factor contributing to depletion. Additionally, the situation may be worsened by the effects of climate change in certain regions [9, 10]. One of the main causes of increasing sea levels is the global reduction in groundwater levels, which has increased significantly since the mid-twentieth century. But the effect of changing climate on water resources becomes increasingly noticeable on a regional scale, specifically in vital agricultural regions of China, India, and the United States. Ensuring the long-term viability of food production in these regions is possible only through the stabilization of groundwater levels [11]. For different purposes, like irrigation, industrial usage, and drinking, groundwater is used in many countries. Specifically, in drought-prone areas, adequate understanding and knowledge are needed to manage available water resources properly [12]. This can be achieved only if one has strong knowledge on water resources management policies. Interaction between Surface water and groundwater is required/essential in effective decision-making process related to water resources management. Interaction studies are also necessary to identify groundwater contamination zones and depletion zones in specific areas and to implement management practices [13–15]. The present study focusses on explaining the significance of interaction studies, methods of interaction, and numerical approach methods and their significance.

2 Significance of GW–SW Interaction

For the development of such strong management policies, one of the parameters which required to be known is the amount of mass transfer from underground to surface, i.e., surface water and groundwater interactions, the two important aspects of hydrological cycle. Both hydrological components were assessed as individual entities and have been assessed individually, though the studies have evolved such that they are now assessed as single entity. Interactions between the two components are considered in two ways, i.e., mass transfer from stream to aquifer and from aquifer to stream. While the mass is being transferred between the zones there is the possibility of contamination in either of the zones or both the zones. Hence, the transition zone plays an important role in the whole process of exchange [16].

The various water resources issues can be addressed by having knowledge on interaction between the hydrological components. The simultaneous use of surface and groundwater resources, as well as the evaluation of water rights and groundwater flow accounting, creates a variety of difficulties. Important components of integrated water management include controlling floods throughout river valleys, minimizing pollution between surface and groundwater, and assessing losses and delays in water flow from surface reservoirs [17–19]. In watershed planning, groundwater flows must be considered, and the effect of riparian zones on river water quality must be evaluated, to preserve aquatic life and a diverse ecosystem. Comprehending the significance of groundwater–surface water interactions in regulating aquatic life also helps in interpreting the effects of modifications to these interactions. Recognizing the significance of active environments at the interconnection and the preservation of aquatic biodiversity makes it easier to assess the state of aquatic water quality. The construction and maintenance of wetlands also support conservation initiatives and the general health of ecosystems [20].

3 Process and Scales of Interaction

Interaction occurring between the two hydrological components has classified the surface water and subsurface water into six different types as shown in Fig. 1 [21–23]. From the saturated zone if the groundwater seeps into the river it is called a gaining stream (Fig. 1a). When the water table is underneath the bed of a river or connected along the riverbed, water seeps into the aquifer from the stream and is called losing disconnected stream or losing connected stream (Fig. 1c, d). If both conditions occur it is known as transition, losing stream (Fig. 1b). The groundwater table for parallel streams is at or below the riverbed (Fig. 1e). For varying ground elevation on any one side of the river flow through river occurs (Fig. 1f). In one way or another, the two hydrological components are related to one another and have the potential to impact human habitat and the hydrological cycle. Excessive utilization and pollution of the components can harm the system and will also make it difficult to understand the interaction.

The study of the interaction between Groundwater (GW)–Surface water (SW) is totally scale-dependent relative. Initially, interactions are classified as large-scale when the total catchment or basin is involved in the interaction process. In contrast, on a local scale, interaction occurs only in the hyporheic zone, which is primarily governed by streambed characteristics [24–26]. To better understand the process of interaction, additional classification was carried out into three scales: sediment (depth < 1 m), reach (depth < 1000 m), and catchment (depth > 1000 m) [27]. Regardless of artificial borders, these scales take into account the order of groundwater flow networks. Reach scale is correlated with the local flow system, catchment scale is correlated with the regional flow system, and sediment scale is

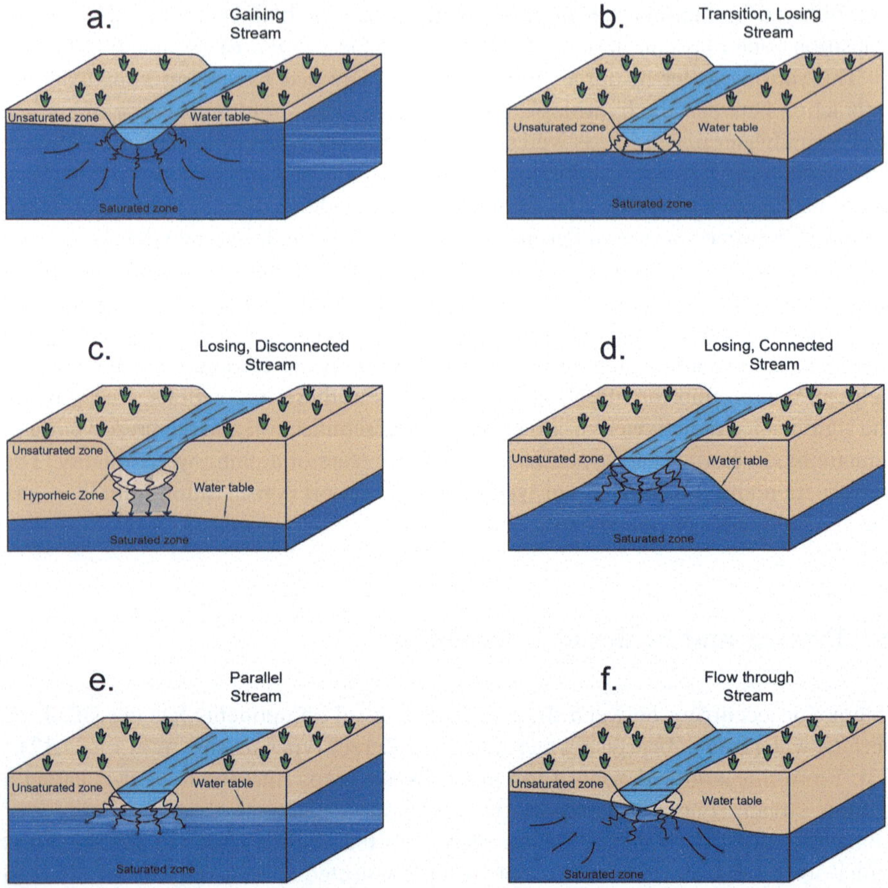

Fig. 1 Different interaction scenarios between surface water and subsurface water

correlated with the hyporheic zone. A proper scaling method should be considered to perform appropriate measurement methods or models.

4 Methods to Assess Interaction Surface Water and Groundwater

This study is regarding the various possible approaches to measure or quantify the amounts of interaction. Measuring methods are broadly classified into four categories based on scale of experimentation, boundary conditions, and flow conditions. A brief outline of each method is specified for understanding.

4.1 Field/Experimental Methods

4.1.1 Aquifer Intensity Method

A groundwater elevation map is necessary to determine the interactions between the two hydrological components using aquifer intensity approach. From the elevations of water table, increasing contours represent the exchange of water mass from aquifer to stream and decreasing contours represent the exchange of water mass from stream to aquifer. Water table elevation map gives the details concerning the transfer of water mass alone. For the assessment of quantity of water mass transferred, Darcy's law will be used [28]. Groundwater velocity, porosity, hydraulic gradient, and hydraulic conductivity need to be determined to evaluate the subsurface water flux.

4.1.2 Transition Zone Intensive Method

Banerjee D and Ganguly S [23], explained the importance of transition zone. Transition zone is a zone between a shallow aquifer and surface water in which groundwater and surface water interactions take place. The method of identifying the quantity of interaction between surface water and groundwater interaction in the transition zone is identical to the aquifer intensity method and parameters of Dracy's equation are determined. Velocity and flow observation is identified from seepage meters, hydraulic heads are identified from piezometers. These two parameters are combined to identify hydraulic conductivity in transition zone-intensive method.

4.1.3 Surface Water Intensive Method

The only reason for the physical or parameter-related changes seen in the corresponding water sources, which can supply groundwater quantities, have changed. The environmental tracer method, the heat tracer method, the hydrograph separation method, and the incremental streamflow method are the four different ways used in surface water-intensive methods. The use of stable isotopes like deuterium and oxygen-18, along with radioactive isotypes, to differentiate rainfall events from prior rainfall event flows made the environmental tracer method more effective than other methods [29].

4.2 Analytical Modelling Approach Method

In analytical modelling approach method, it assesses the flow in all three parameters that is horizontal flows and vertical flows, horizontal flows being surface water and groundwater, vertical flow being seepage, it also considers and assesses the riverbed. A vertical plane is considered, and side banks are not considered [30]. This is only true if the riverbanks are significantly smaller than the riverbed; if not, the side banks will influence the exchange process in a big way, which will make the assessment erroneous.

4.3 Semi-analytical Methods

In any of the models so developed in surface water and groundwater analysis, their accuracy would be highly influenced by its resolution and meshing process. Models so developed, sometimes do not consider unconventional boundary challenges and deviation of head, which results in the deviation of results from actual results. Hence for such scenarios, semi-analytical models are developed [31]. Semi-analytical models avail the advantages of both numerical and analytical models.

4.4 Numerical Modelling Approach Methods

The numerical modelling approach involves the development of a model which includes a mathematical tool. Mathematical/ statistical tool plays a key role in understanding the interaction behavior between GW and SW [32, 33]. The numerical models are used to replicate the real-world scenario. The numerical modelling approach is also called the regional scale approach, which works for larger catchment areas. Numerical models replicate the actual scenario by incorporating initial conditions, and boundary conditions. The type of aquifers, hydrogeological systems, and input data availability all influence the model selection. Hydrological models perform well in forecasting subsurface flow conditions, hydrological models are effective in surface water flow scenario predictions. Groundwater component forecasts done by groundwater models assess the behavior with various types, incorporating steady state, 1D, 2D, and 3D representations [34]. Numerical modelling approaches are further classified as fully coupled models and loosely coupled models.

4.4.1 Fully Coupled Models

Various variables like hydro climatic conditions, geological conditions, difference in head of water in SW and GW, and flow conditions in the aquifer affect the interaction of hydrological components. Hence, modelling the interaction process is assumed to be complex and brings the concept of coupled models. For larger scale models, best-suited model is a fully coupled model. Fully coupled model is a model in which both surface and subsurface flows are evaluated in a single software, examples of fully coupled models are Hydro Geosphere, Cathy, and MIKESHE. These models use the process-based approach in forecasting the interactions.

Cathy's model used 1D finite difference approach for surface water analysis and 3D finite element approach for subsurface interaction analysis [35, 36]. Gatto B. et al., [36], used the Cathy model to assess the pollutant transport during the interaction process between the two hydrological components. Hydraulic parameter exchange in the interaction process using Gibbs tool was explained by Bhalaji C K et al. [32] Hydro Geosphere uses control volume finite element method for assessment of both surface flow and subsurface flow conditions. Drawing out of exchange flux between subsurface and surface and writing of zoned budget are not properly specified in Hydro Geosphere code [37, 38]. Talebmorad H. et al. [38] and Abdollahi S. et al. [39] used Hydro Geosphere model to study the interaction between the surface and subsurface zones under steady state and unsteady state conditions for a larger catchment. The degree of climate change influence on the groundwater flow and soil moisture is dependent on location geological conditions were assessed using Hydro Geosphere [40]. Hydrus model uses a modified form of Richard's equation and has the capability of optimization in 1D and 2D [41]. MIKE SHE is a fully distributed physical process-based 3D model developed and wide range applicability in hydrological and hydrogeological model [41, 42].

4.4.2 Loosely Coupled Models

A loosely coupled model is a model in which one or more software tools are used in which they are coupled with each other such that the output of one software tool becomes the input for another software tool. In a loosely coupled model, surface and subsurface flows are evaluated separately and efficiently. A compelling benefit of a loosely coupled model is the ability to utilize independent modelling tools for specific hydrological processes, rather than relying on a single model to assess all hydrological processes. Loosely coupled models are developed even for regional scale, even though they are stated as local scale models. Commonly used loosely coupled models in recent times are SWAT-MODFLOW, SWAT-MODFLOW-WEAP, GSFLOW-MODFLOW, etc. And for regional scale analysis, widely used and effective loosely coupled models are FEFLOW [43], and MIKE SHE combined with MIKE11 [12]. Chunn D. et al., [44], coupled SWAT, a semi-distributed process-based model, with MODFLOW, a 3D finite difference approach model, to study the impact of changing climate on SW–GW interaction on a regional scale. Regional/catchment scale modelling was performed to understand the interactions between the hydrological components using SWAT–MODFLOW and GSFLOW–MODFLOW [45–48]. Abbas S A. et al., [49], included the WEAP model in the coupled SWAT–MODFLOW model to simulate the allocation scenarios for an urban catchment under climate change conditions.

Major drawbacks in fully coupled models are availability of observed data on a regional scale, computational capabilities available with the resource person, and accessibility of the model. Heterogeneity of the MIKE SHE model is to address selection of multiple parameters which indicates over parameterization leading to uncertainty in modelling the interaction studies [42]. In loosely coupled models for transient models with varying recharge rates, groundwater model calibration is inadequate. Full integration of dynamic model with linkage between stream and aquatic water transition zone consideration is possible. In the process of estimation of recharge rates, physiological conditions of catchment/ regional scale aquifers are simplified as diverse. Loosely coupled models cannot calculate the stream flow level along the reach of stream in the process of simulations [50]. Adjustment of boundary conditions is a more challenging factor in loosely coupled models.

5 Conclusions

Interaction between the two hydrological components is vital for the assessment/ forecasting of flooding or drought conditions. The present review addresses recent advancements and limitations in the interaction studies of regional scale models. Fully coupled models are effective in estimating the exchange flux between stream and subsurface waters. However, these models have limitations in specifying the boundary conditions and need a wide range of parameters. The advantage of loosely coupled models is that simulations from one model can be given as input to the sub-model, to study the interaction process. Individual calibration of each model can create disparity among the other in loosely coupled models. The study specified the reference of recent advancements for both loosely coupled and fully coupled models, which can help the researchers choose the appropriate methods. Further, a detailed framework is needed in the interaction studies of groundwater and surface water to address the uncertainties, as it is a complex process.

References

1. Cred U (2018) Economic losses, poverty & disasters: 1998–2017. Centre for Research on the Epidemiology of Disasters (CRED), United Nations Office for Disaster Risk Reduction (UNISDR)
2. Wilhite DA (2004) Drought and drought and water crises after crises environmental science
3. Delforge D, Wathelet V, Lanfredi Sooa C, et al (2023) EM-DAT: the emergency events database. https://doi.org/10.21203/rs.3.rs-3807553/v1
4. Sarkar S (2022) Drought and flood dynamics of Godavari basin, India: a geospatial perspective. Arab J Geosci 15:772. https://doi.org/10.1007/s12517-022-10041-5
5. Deb P, Kiem AS, Willgoose G (2019) A linked surface water-groundwater modelling approach to more realistically simulate rainfall-runoff non-stationarity in semi-arid regions. J Hydrol (Amst) 575:273–291. https://doi.org/10.1016/j.jhydrol.2019.05.039
6. Barry-Sosa A, Flint MK, Ellena JC et al (2024) Effects of surface water interactions with karst groundwater on microbial biomass, metabolism, and production. EGUsphere 2024:1–35. https://doi.org/10.5194/egusphere-2024-49

7. Kundzewicz ZW, Döll P (2009) Will groundwater ease freshwater stress under climate change? Hydrol Sci J 54:665–675. https://doi.org/10.1623/hysj.54.4.665
8. Wakode HB, Baier K, Jha R, Azzam R (2018) Impact of urbanization on groundwater recharge and urban water balance for the city of Hyderabad, India. Int Soil Water Conserv Res 6:51–62. https://doi.org/10.1016/j.iswcr.2017.10.003
9. Park J, Lee D, Kim H, Woo NC (2024) Effects of water-table changes following rainfall events on arsenic fate and transport in groundwater–surface water mixing zones. Sci Total Environ 933:173200. https://doi.org/10.1016/j.scitotenv.2024.173200
10. Xia Y, Xiao J, van der Ploeg M et al (2024) Combined effects of precipitation anomalies and dams on streamwater-groundwater interaction in the Fen River basin, China. Sci Total Environ 929:172704. https://doi.org/10.1016/j.scitotenv.2024.172704
11. Niswonger RG (2020) An agricultural water use package for MODFLOW and GSFLOW. Environ Model Softw 125:104617. https://doi.org/10.1016/j.envsoft.2019.104617
12. Haque A, Salama A, Lo K, Wu P (2021) Surface and groundwater interactions: a review of coupling strategies in detailed domain models. Hydrology 8:1–16
13. Banda K, Mulema M, Chomba I et al (2023) Investigating groundwater and surface water interactions using remote sensing, hydrochemistry, and stable isotopes in the Barotse floodplain, Zambia. Geol Ecol Landsc:1–16. https://doi.org/10.1080/24749508.2023.2202450
14. Pham QN, Le TT, Dang DN (2024) Enhancing hydrological insight: isotopic methods revealing groundwater–surface water interactions in the lower Quang Tri River group, Vietnam. Isotopes Environ Health Stud 60(3):309–330. https://doi.org/10.1080/10256016.2024.2356609
15. Zhao L, Su C, Liu W, Xie X (2024) Understanding surface water-groundwater conversion relationship and associated hydrogeochemistry evolution in the upper reaches of Luan River basin, North China. J Earth Sci 35:1010–1023. https://doi.org/10.1007/s12583-022-1629-x
16. Kalbus E, Reinstorf F, Schirmer M (2006) Hydrology and earth system sciences measuring methods for groundwater-surface water interactions: a review. Hydrol Earth Syst Sci 10(6): 873–887
17. Ramesh R, Purvaja GR (1995) Man's influence on freshwater ecosystems and water use. IAHS Publ
18. Wang Y, Chen N (2021) Recent progress in coupled surface–ground water models and their potential in watershed hydro-biogeochemical studies: a review. Watershed Ecol Environ 3:17–29. https://doi.org/10.1016/j.wsee.2021.04.001
19. Wei X, Bailey RT (2019) Assessment of system responses in intensively irrigated stream-aquifer systems using SWAT-MODFLOW. Water (Switzerland) 11:1576. https://doi.org/10.3390/w11081576
20. Khan HH, Khan A (2019) Groundwater-surface water interaction along river Kali, near Aligarh, India. HydroResearch 2:119–128. https://doi.org/10.1016/j.hydres.2019.12.001
21. Banks EW, Simmons CT, Love AJ et al (2009) Fractured bedrock and saprolite hydrogeologic controls on groundwater/surface-water interaction: a conceptual model (Australia). Hydrogeol J 17:1969–1989. https://doi.org/10.1007/s10040-009-0490-7
22. Bencala KE, Gooseff MN, Kimball BA (2011) Rethinking hyporheic flow and transient storage to advance understanding of stream-catchment connections. Water Resour Res 47. https://doi.org/10.1029/2010WR010066
23. Banerjee D, Ganguly S (2023) A review on the research advances in groundwater–surface water interaction with an overview of the phenomenon. Water (Switzerland) 15:1552
24. Larkin RG, Sharp JM Jr (1992) On the relationship between river-basin geomorphology, aquifer hydraulics, and ground-water flow direction in alluvial aquifers. GSA Bull 104:1608–1620. https://doi.org/10.1130/0016-7606(1992)104
25. Brunke M, Gonser T (1997) The ecological significance of exchange processes between rivers and groundwater. Freshw Biol 37:1–33
26. Woessner WW (2000) Stream and fluvial plain ground water interactions: rescaling hydrogeologic thought. Groundwater 38(3):423–429

27. Dahl M, Nilsson B, Langhoff JH, Refsgaard JC (2007) Review of classification systems and new multi-scale typology of groundwater-surface water interaction. J Hydrol (Amst) 344:1–16. https://doi.org/10.1016/j.jhydrol.2007.06.027
28. Kebede S, Charles K, Godfrey S et al (2021) Regional-scale interactions between groundwater and surface water under changing aridity: evidence from the river Awash Basin, Ethiopia. Hydrol Sci J 66:450–463. https://doi.org/10.1080/02626667.2021.1874613
29. Harvey FE, Lee DR, Rudolph DL, Frape SK (1997) Locating groundwater discharge in large lakes using bottom sediment electrical conductivity mapping. Water Resour Res 33:2609–2615. https://doi.org/10.1029/97WR01702
30. Nawalany M, Sinicyn G, Grodzka-Łukaszewska M, Mirosław-Światek D (2020) Groundwater-surface water interaction-analytical approach. Water (Switzerland) 12:1792. https://doi.org/10.3390/w12061792
31. Ward ND, Lough H (2011) Stream depletion from pumping a semiconfined aquifer in a two-layer leaky aquifer system. J Hydrol Eng 16:955–959. https://doi.org/10.1061/(asce)he.1943-5584.0000382
32. Bhalaji CK, Deepasri R, Kavimalar K, Eunice J (2024) Exploring surface water-groundwater interactions: a multidisciplinary approach. Int Res J Adv Eng Hub (IRJAEH) 2:944–951. https://doi.org/10.47392/irjaeh.2024.0132
33. Ahmadi H (2024) Modeling of groundwater-surface water interactions: a review of integration strategies. ISH J Hydraul Eng 30:132–146. https://doi.org/10.1080/09715010.2023.2263434
34. Ameli AA (2014) Semi-analytical methods for simulating the groundwater-surface water interface
35. Muma M, Rousseau AN, Gumiere SJ (2016) Assessment of the impact of subsurface agricultural drainage on soilwater storage and flows of a small watershed. Water (Switzerland) 8:326. https://doi.org/10.3390/w8080326
36. Gatto B, Paniconi C, Salandin P, Camporese M (2021) Numerical dispersion of solute transport in an integrated surface–subsurface hydrological model. Adv Water Resour 158:104060. https://doi.org/10.1016/j.advwatres.2021.104060
37. Brunner P, Simmons CT (2012) HydroGeoSphere: a fully integrated, physically based hydrological model. Ground Water 50:170–176
38. Ostad-Ali-Askari K, Talebmorad H, Eslamian S et al (2018) Evaluation of integrated hydro geosphere hydrologic model in modeling of large basins subject to severe withdrawal. Environ Chem Toxicol 2(1):31
39. Abdollahi S, Madadi M, Ostad-Ali-Askari K (2021) Monitoring and investigating dust phenomenon on using remote sensing science, geographical information system and statistical methods. Appl Water Sci 11:111
40. Erler AR, Frey SK, Khader O et al (2019) Evaluating climate change impacts on soil moisture and groundwater resources within a Lake-affected region. Water Resour Res 55:8142–8163. https://doi.org/10.1029/2018WR023822
41. Kumar CP (2012) Groundwater modelling software-capabilities and limitations. IOSR J Environ Sci Toxicol Food Technol 1(2):46–57
42. Ma L, He C, Bian H, Sheng L (2016) MIKE SHE modeling of ecohydrological processes: merits, applications, and challenges. Ecol Eng 96:137–149. https://doi.org/10.1016/j.ecoleng.2016.01.008
43. Hu L, Zhang M, Yang Z et al (2020) Estimating dewatering in an underground mine by using a 3D finite element model. PLoS One 15:e0239682. https://doi.org/10.1371/journal.pone.0239682
44. Chunn D, Faramarzi M, Smerdon B, Alessi DS (2019) Application of an integrated SWAT-MODFLOW model to evaluate potential impacts of climate change and water withdrawals on groundwater-surface water interactions in west-Central Alberta. Water (Switzerland) 11:110. https://doi.org/10.3390/w11010110

45. Hassan SMT, Lubczynski MW, Niswonger RG, Su Z (2014) Surface-groundwater interactions in hard rocks in Sardon catchment of western Spain: an integrated modeling approach. J Hydrol (Amst) 517:390–410. https://doi.org/10.1016/j.jhydrol.2014.05.026
46. Bailey RT, Wible TC, Arabi M et al (2016) Assessing regional-scale spatio-temporal patterns of groundwater–surface water interactions using a coupled SWAT-MODFLOW model. Hydrol Process 30:4420–4433. https://doi.org/10.1002/hyp.10933
47. Mosase E, Ahiablame L, Park S, Bailey R (2019) Modelling potential groundwater recharge in the Limpopo River basin with SWAT-MODFLOW. Groundw Sustain Dev 9:100260. https://doi.org/10.1016/j.gsd.2019.100260
48. Jafari T, Kiem AS, Javadi S et al (2021) Fully integrated numerical simulation of surface water-groundwater interactions using SWAT-MODFLOW with an improved calibration tool. J Hydrol Reg Stud 35:100822. https://doi.org/10.1016/j.ejrh.2021.100822
49. Abbas SA, Xuan Y, Bailey RT (2022) Assessing climate change impact on water resources in water demand scenarios using SWAT-MODFLOW-WEAP. Hydrology 9:164. https://doi.org/10.3390/hydrology9100164
50. Krause P, Boyle D, Markstrom S (2007) Hydrologic modeling of a groundwater dominated watershed using a loosely coupled modeling approach

Technological Advances in Leachate Leakage Detection and Monitoring in Landfill Area—A Review

H. R. Rashmi, G. V. Rathnamala, M. N. Vathsala, and R. M. Ashwini

Abstract For the betterment of public health and environmental preservation, technological advancements in leachate leakage monitoring and detection in landfill regions are becoming more and more important. The liquid that results from water interacting with waste products in landfills is known as leachate, and it may include toxic compounds and other contaminants. In order to stop groundwater pollution and environmental deterioration, leachate leak detection and monitoring are crucial. Although landfills are essential locations for disposing of trash, there are serious environmental dangers associated with them because of the possibility of leachate leaks, which can damage soil and water supplies. In order to reduce these environmental risks, leachate leakage must be effectively detected and monitored. The primary technique used in traditional leachate detection methods was manual sampling and analysis, which frequently had drawbacks in terms of accuracy, efficiency, and real-time monitoring capabilities. But as more sophisticated technologies have become available, such as remote sensing, geophysical surveys, and sensor networks, the way that leachate leakage is found and tracked has changed dramatically. The technical advancements in leachate leak detection and monitoring in landfill regions are highlighted in this abstract.

Keywords Leakage detection · Landfill areas · Remote sensing · Geophysical surveys · Machine learning

H. R. Rashmi (✉)
K S School of Engineering and Management, Bangalore, India
e-mail: rashmihr@kssem.edu.in

G. V. Rathnamala · R. M. Ashwini
GITAM School of Technology, Bangalore, India

M. N. Vathsala
SJCIT, Bangalore, India

P. Pathak et al. (eds.), *Advances in Environmental Sustainability, Energy and Earth Science*, https://doi.org/10.1007/978-3-031-73820-3_24

1 Introduction

The term "landfill leachate" refers to any liquid waste that has been released from a landfill or dump site and contains unwanted elements that seep through the trash debris. The incredibly complex geochemical processes in the waste and the underlying soil layers often leave its route of exposure and toxicity unclear and unpredictable. The rising materialistic lifestyle and planned obsolescence of items are the main causes of the global increase in the incidence of landfill garbage disposal, with or without preparation. With a population of almost 1210 million, the nation's urban population makes up over 377 million, or 31.16% of the total population. Due to migration and the loss of natural resources, towns experiencing such a rapid rate of urbanization are confronting additional challenges with regard to socio-economic and environmental prospects. The amount of waste generated per person in India has expanded exponentially, according to the Central Pollution Control Board (0.26 kg/day–0.85 kg/day). An estimated 80–90% of municipal waste is disposed of in landfills without the use of adequate management techniques or open burning, which pollutes the air, water, and soil [1]. Majority of landfills in the world are outdated and not designed to stop the poisonous leachate from contaminating the groundwater and underlying soil.

Monitoring and detecting leachate leaks are an essential part of managing landfills. Leachate, a liquid by-product of landfill degradation, is extremely toxic and contains a range of contaminants, such as ammonia, heavy metals, and other dangerous substances. Leachate's ability to leak into groundwater supplies and harm the environment is a top worry for landfill managers and owners. Leachate leakage monitoring and detection technologies have advanced, enabling landfill operators and owners to promptly recognize and address the risks associated with leachate leakage. Leachate leakage detection and monitoring have benefited from technological advancements in both hardware and software. Specialized sensors and probes used to identify and track leachate leaks are examples of hardware solutions. The placement of these hardware devices enables a thorough understanding of leachate leaks from both above and below ground. Software solutions allow landfill operators and owners to promptly detect and address any leachate leaks. These include monitoring and reporting systems [2]. By offering information on leachate levels, temperature, and other factors, these software programmes enable proactive leachate leakage control.

Technological advancements in leachate leakage detection and monitoring include automated systems that can identify and react to possible leaking in addition to hardware and software solutions. It is possible to train automated systems to notify landfill operators and owners of possible leachate leaks, giving them time to take appropriate action and avoid environmental harm [3]. Leachate leakage detection and monitoring may be done more precisely by using automated techniques to pinpoint regions that require more research. The ability to promptly identify and address any risks caused by leachate leakage has been made possible by technological advancements in leachate leakage detection and monitoring for landfill owners and operators. Landfill owners and operators can lessen the possibility of environmental harm from leachate leaks by employing automated systems, specialized hardware and software solutions, and other technical advancements. Furthermore, landfill owners and operators are now able to create more effective and efficient waste management plans because of technology advancements in leachate leakage detection and monitoring.

2 Literature Review

2.1 General

Leachate from landfills is a serious environmental issue that can affect groundwater and surface water supplies in the long run. The liquid that results from water seeping into a landfill and coming into touch with the waste materials inside is called leachate. Heavy metals, organic compounds, and other pollutants are picked up by the leachate as it passes through the landfill and can subsequently be discharged into the surrounding area. Leachate leaks from landfills may have detrimental effects on ecosystems and the local population's health as well as harm local flora and animals. Leachate leaks are most frequently caused by inadequately or unintentionally lined landfills. Leachate from unlined landfills has the ability to readily penetrate the soil and perhaps reach groundwater and surface water supplies [4]. This may cause drinking water supplies in the area to become contaminated, which might be extremely dangerous for people's health. Leachate's pollutants can also affect the health of aquatic and terrestrial ecosystems, resulting in a decline in certain habitats' productivity and biodiversity.

Poor management methods, shoddy landfill construction and maintenance, and shoddy containment system design are the primary causes of landfill leachate leakage. Inadequate leachate level monitoring, incorrect landfill closure and cover, and improper hazardous waste disposal are examples of poor management practices. A landfill's improper drainage and liner systems might be the result of poor construction and upkeep. Leachate leakage can also result from inadequate containment system design, since the system may not be able to hold leachate for extended periods of time. Leachate leaks from landfills can affect the environment in the short and long term. Leachate has the potential to short-term contaminate water, which might have detrimental effects on animal and human health. Long-term detrimental effects of pollutants in the ecosystem include lower productivity of some habitats, decreasing biodiversity, and other issues [5]. Leachate leakage can have serious consequences, thus it is critical to take precautions to lessen and avoid it. This entails better containment system design to guarantee that leachate is kept confined for extended periods of time, as well as better landfill construction and maintenance. In order to make sure that leachate levels stay below allowable bounds, frequent monitoring should also be done. Lastly, to make sure that hazardous material does not contribute to leachate leaks, adequate disposal must be guaranteed [6].

Studies often begin with characterizing the composition of landfill leachate to understand its potential contaminants and environmental impact. Analysis techniques include chemical analysis, spectroscopic methods, and biological assays to determine toxicity and degradation pathways [7]. Research explores the pathways through which leachate can contaminate groundwater and surface water, affecting ecosystems and human health [8]. Modeling studies predict the spread and persistence of contaminants from landfill sites [9]. Geosynthetic Clay Liners liners are used to enhance the barrier properties of landfill liners, reducing the risk of leachate leakage [10]. High-density polyethylene (HDPE) and other geomembranes provide impermeable barriers to contain leachate within landfill cells. These use recirculation and aeration systems to accelerate decomposition and reduce leachate production [11]. Plant-based systems are investigated for their ability to uptake and degrade contaminants from leachate [12]. Advanced monitoring technologies include remote sensing, real-time data acquisition systems, and geophysical methods to detect leaks and monitor contaminant migration. Control strategies involve active and passive systems for leachate collection, treatment, and management to minimize environmental impact [13]. Finally, it should be noted that landfill leachate leakage poses a significant risk to the ecosystem and may eventually affect groundwater and surface water supplies. Leachate leakage must be prevented and its effects must be minimized by proper landfill management, construction, and design. To make sure that leachate levels stay within allowable bounds, frequent monitoring should also be done. Lastly, to make sure that hazardous material does not contribute to leachate leaks, adequate disposal must be guaranteed.

2.2 *Importance of Detection and Monitoring*

Leachate leakage is one kind of environmental hazard that happens when contaminated fluids escape a landfill. Leachate, or liquid leachate, is created when organic materials decomposing in landfill garbage come into contact with precipitation and other liquids. Pathogens and pathogenic bacteria, as well as a variety of hazardous materials and compounds like metals, nitrates, phosphates, and volatile organic compounds, make up leachate. If leachate is not properly managed and monitored, it could have a significant negative impact on the ecosystem [14]. Leachate leak detection and monitoring are essential components of a comprehensive waste management strategy. Numerous environmental problems, including the growth of pathogenic microorganisms, a rise in air pollution, and contamination of nearby groundwater supplies, can be brought on by leachate leaks. By detecting and tracking leachate leaks, landfills are able to pinpoint potential pollution sources and mitigate the risks associated with such leaks.

Leachate leakage detection and monitoring use a number of techniques, such as visual inspections, physical sampling, and laboratory testing. Visual examinations are used to identify any overt signs of leachate leaking, such as moist patches or discolored soil. Leachate samples are taken in person, and a laboratory examines them thereafter. Through laboratory testing procedures, the composition of the leachate and the presence of any hazardous elements can be determined. Leachate leak detection and monitoring are essential for several reasons. Initially, landfills can identify potential pollution sources and address them before they pose a threat to the environment by employing this strategy. Second, it can help landfills identify any potential health risks associated with the leachate and put the appropriate safety measures in place to protect the public [15]. Finally, it might help landfills ensure that their waste management practices adhere to environmental regulations.

In conclusion, monitoring and identification of leachate leaks are essential to ensuring the safe and effective management of landfills. By detecting and tracking leachate leaks, landfills are able to pinpoint potential pollution sources and mitigate the risks associated with such leaks. Additionally, landfills can use this process to identify any health risks the leachate may pose to the general public and implement the necessary safety measures. Finally, it might help landfills ensure that their waste management practices adhere to environmental regulations.

2.3 *Overview of Landfill Leachate and Its Properties*

Leachate wastewater often contains elevated levels of organic and inorganic substances, heavy metals, and chlorinated organics [16]. It originates from landfills and carries a significant amount of dissolved contaminants that can potentially pollute soil, groundwater, and surface water, posing substantial risks to the environment, human health, and ecosystems. Overall, landfill leachate is a highly contaminated liquid that is produced by the decomposition of solid waste in landfills. It can vary significantly in composition and properties and can have a significant impact on local ecosystems if it is not properly managed. Proper management and treatment of landfill leachate is essential to protect the environment and human health.

When solid waste kept in landfills comes into touch with water or other liquids, a type of liquid waste material known as landfill leachate is created. Heavy metals, chemical compounds, and other pollutants are among the dissolved and suspended elements present in this potentially dangerous liquid. It is created at landfills as a result of surface runoff, subsurface flow, and precipitation—all of which enable water to move through the garbage and get polluted with it. Furthermore, organic stuff in the landfill, such as food waste and yard clippings, can decompose and produce landfill leachate [17].

Depending on the type of items dumped in the landfill, the leachate's composition can change significantly. Pesticides, heavy metals, organic compounds, and other hazardous elements are frequently detected pollutants in landfill leachate. Due to their potential for toxicity to both humans and animals upon exposure, the presence of these contaminants may be hazardous to the environment. Leachate production in a landfill is also influenced by the dump's features, including the kind of material used, volume of garbage held, and climate [18]. Landfill leachate has the potential to negatively influence the ecosystem, thus it is critical to manage it appropriately.

There are several ways to accomplish this, including collection and treatment. Leachate can be collected and kept out of the environment by using collection devices like sumps and lined trenches. Physical, chemical, and biological techniques can all be used to treat the leachate with the goal of lowering the quantity of toxins it contains.

In order to safeguard both the environment and public health, landfill leachate management must be done properly. This is a major environmental concern. Comprehending the origins of leachate and the possible hazards linked to it is crucial for the efficient handling of this refuse.

2.4 Characteristics of Leachate Flow and Behavior

Depending on the soil type, environmental factors, and landfill design, leachate flow and behavior might differ significantly. Leachate is generally produced when surface water, such as precipitation, seeps into a landfill and combines with organic materials to generate a liquid that may contain a range of toxins. Following that, the leachate may travel through the landfill due to capillary action and gravity, as well as via alternative channels including leachate collection systems, surface runoff, or groundwater.

Leachate is typically a complex mixture of water, organic matter, inorganic salts, heavy metals, and various contaminants from decomposing waste materials. The exact composition varies depending on the types of waste present and their stage of decomposition. The volume of leachate generated depends on factors such as rainfall, waste composition, and landfill design. Initially, leachate production tends to be high as water infiltrates through the waste. Over time, as the landfill matures and decomposes, leachate generation rates may decrease.

Leachate can vary in color, ranging from dark brown to black, due to dissolved organic matter and suspended particles. It often has a foul odor due to the presence of gases such as methane and hydrogen sulfide, which are by-products of decomposition processes. Leachate is typically acidic due to organic acids produced during waste decomposition. It may also be alkaline in certain cases, depending on the type of waste and reactions occurring within the landfill. The pH of leachate can vary widely, influencing its potential for corroding containment structures and affecting aquatic ecosystems if it enters water bodies [19].

Leachate contains various contaminants such as heavy metals (e.g., lead and cadmium), organic pollutants (e.g., polycyclic aromatic hydrocarbons), ammonia, and pathogens. These contaminants can pose environmental risks if leachate escapes from the landfill and contaminates groundwater, surface water, or soil. Leachate can migrate through the waste mass, driven by gravity and hydraulic pressure gradients within the landfill. The presence of impermeable liners and collection systems in modern landfills helps manage and control leachate flow. However, older or poorly designed landfills may allow leachate to migrate freely, potentially contaminating surrounding environments. Due to its potential environmental hazards, leachate often requires treatment before discharge or reuse. Treatment methods may include physical processes (e.g., sedimentation and filtration), chemical processes (e.g., oxidation and precipitation), and biological processes (e.g., microbial degradation). Effective leachate management involves both containment within the landfill and proper treatment to mitigate environmental impacts [6].

By understanding the characteristics of leachate flow and behavior in landfills, it is possible to better manage the contamination of groundwater and soil systems. Proper design and maintenance of landfills are essential for minimizing the risk of contamination and for minimizing the impact of leachate on the environment.

2.5 Detection Methods for Landfill Leachate Leakage

One of the main causes of groundwater pollution is landfill leachate leakage. Numerous toxic compounds and heavy metals that pose a risk to the environment and public health can be found in the leachate. Consequently, it is critical to identify and keep an eye on landfill leachate leaks as soon as feasible. Leachate leak detection can be accomplished by a variety of techniques, which fall into three categories: chemical, physical, and biological.

Leachate leaking is detected chemically by examining the leachate's chemical characteristics. Analyses of pH, dissolved oxygen, total organic carbon (TOC), and total nitrogen are included in this. The concentrations of different compounds that may be present in the leachate can be determined by examining these characteristics. Other chemical techniques include the use of chemical sensors to find volatile organic compounds (VOCs) in the leachate, such as gas chromatography and ion-selective electrodes [20].

Detecting leachate leakage through physical approaches entails making physical observations. This entails visually inspecting the landfill and its surroundings for leakage indicators, like changes in the color of the plant or soil. Other physical approaches include the use of geophysical technologies to find subsurface leachate plumes, such as seismic reflection, electrical resistivity, and ground penetrating radar.

In biological approaches, leachate is detected by the use of organisms like fungi and bacteria. Using bacteria that are sensitive to the presence of specific compounds in the leachate, biosensors are one method of doing this. Biosensors can identify particular elements, including heavy metals, chemical compounds, and poisons, in the leachate [21].

In conclusion, a variety of techniques are available to identify landfill leachate leaks. Leachate can be found in the environment using chemical, physical, or biological techniques. To safeguard the environment and public health, leachate leaks from landfills must be closely observed.

2.6 Monitoring Techniques for Landfill Leachate Leakage

Landfill leachate leakage is a major environmental problem due to the presence of toxic compounds, bacteria, and organic compounds. Groundwater pollution can be avoided by keeping an eye on landfill leachate leaking and taking action to lessen its effects.

The installation of a groundwater monitoring network is the most popular technique for keeping an eye on landfill leachate leaks. Typically, this network is made up of wells that are dug around the landfill at different depths. These wells are used to assess the degree of contamination in the groundwater and keep an eye out for any possible leachate leaks. Utilizing suitable sample methods, like grab sampling, is crucial to guaranteeing the accuracy of the data gathered. The source of the contamination and whether leachate leaking is happening can then be ascertained using the information gathered from these wells [22].

To monitor landfill leachate leakage, methods other than utilizing a groundwater monitoring network might be employed. For instance, leachate presence in the environment can be determined using remote sensing techniques. Any variations in the amount of leachate existing in the environment over time can be found using this technique. This can help identify the contamination's source and the ways in which it is impacting the environment. In certain situations, monitoring landfill leachate leaks may also require the use of chemical and biological indicators. You can utilize chemical indicators like pH, TDS, and nutrient levels to find out if leachate is present in the ecosystem. You can also employ biological indicators to find out if leachate is present, like fungi, bacteria, and plants. To give a more complete picture of the leachate leakage, these indicators can be utilized in conjunction with the data gathered from the groundwater monitoring network.

Finally, it is critical to conduct routine landfill inspections to find any possible leachate leaks. This can be achieved by regularly testing the landfill's drainage systems and visually monitoring the dump for any indications of leaks, such as discolored soil or water. In order to find any possible issues before they worsen, inspections should be carried out frequently [23]. A variety of monitoring methods can be used to identify any suspected leachate leaks and take action to lessen its effects. By doing this, you can ensure that leachate does not contaminate groundwater and that the environment is protected.

2.7 Advanced Monitoring Techniques

Geophysical surveys and remote sensing are examples of advanced approaches for monitoring landfill leachate leaks. In order to find out if there are any apparent indications of leakage, like discolored foliage or liquid-filled ponds, remote sensing uses satellites and aircraft. In order to find out whether leachate is present below the surface, geophysical investigations employ seismic refraction techniques, electrical resistivity imaging, and ground penetrating radar. These methods can be applied to map the leachate plume's shape and extent as well as to pinpoint possible contamination sources. Furthermore, real-time monitoring of the leachate's quality by automated sensors enables quicker reactions to environmental changes [24].

i. **Fiber-optic sensors**

Fiber-optic sensors are one of the most advanced monitoring techniques for landfill leachate leakage. Fiber-optic sensors use fiber-optic cables to detect changes in the environment. These cables can be used to detect changes in temperature, pressure, and pH, which can indicate the presence of leachate. The cables are placed in the landfill and connected to a monitoring system, which can then be used to detect any changes in the environment. This technology is highly sensitive and can detect even the smallest changes in the environment.

ii. **Piezoelectric sensors**

Piezoelectric sensors are another advanced monitoring technique for landfill leachate leakage. Piezoelectric sensors use piezoelectric materials to detect changes in pressure or strain. These sensors can be used to detect changes in pressure or strain in the landfill due to the presence of leachate. The sensors can also be used to detect changes in the electrical properties of the soil due to the presence of leachate.

iii. **Wireless sensor networks**

Wireless sensor networks are a relatively new technology for monitoring landfill leachate leakage. Wireless sensor networks are networks of wireless nodes that are connected to one another and to a central monitoring system. These networks can be used to detect changes in the environment, such as temperature, pressure, and pH that could indicate the presence of leachate. The nodes can be placed in strategic locations in the landfill to monitor for changes in the environment.

These advanced monitoring techniques for landfill leachate leakage can be used to detect changes in the environment that could indicate the presence of leachate. This can help landfill operators and environmental regulators detect leachate leakage before it becomes a major issue. The use of these monitoring techniques can help ensure that the landfill is properly managed and that leachate is not escaping into the environment.

2.8 Emerging Technologies in Landfill Leachate Leakage Detection and Monitoring

Leachate leaking from landfills can now be detected and monitored more easily and effectively thanks to the development of new technologies. Artificial intelligence (AI) systems, satellite photos, drones, and sensors are all useful tools for anticipating and responding to possible leaks. These devices will grow stronger as they are developed, shielding our ecosystem from the harmful effects of landfill leachate leaks.

i. **Machine learning and artificial intelligence**

AI and machine learning can be utilized to create automated systems that track leachate leakage. Leaks can be promptly found by automated systems, which can also notify waste management staff in real time so they can respond right away. This can enhance environmental safety by lowering the time and resource requirements for addressing leachate leakage [25].

In conclusion, leachate leak detection and monitoring are made possible by the potent new technologies of machine learning and artificial intelligence. Waste management firms can mitigate the health and environmental problems connected with leachate leaks by implementing automated systems and predictive models.

ii. **Internet of Things (IoT)**

The use of Internet of Things (IoT) technology has enabled the monitoring of landfill leachate leakage on a real-time basis. Sensors installed in the landfill can detect changes in the water level, temperature, and other parameters within the landfill. This data can be sent to a central server for analysis and alerting, thus enabling timely detection of any potential leachate leakage.

iii. **Big data analytics**

Big data analytics can be used to detect any potential leachate leakage by analyzing large quantities of historical data. By using advanced algorithms, patterns in the data can be identified and any deviations from the normal behavior can be detected. This can provide valuable information to take corrective actions and mitigate the effects of any potential leakage.

Overall, the use of emerging technologies such as ML/AI, IoT, and big data analytics provides an effective and efficient means of detecting and monitoring landfill leachate leakage. By using these technologies, potential leaks can be detected and prevented quickly, thus preventing any potential damage to the environment.

2.9 Challenges and Future Directions

Global landfill sites have been shown to have leachate leaks, a serious environmental hazard. Leachate leaks can seriously harm surrounding communities' health as well as the soils, surface waters, and groundwater. Effective procedures for identifying and monitoring leachate leakage are essential for landfill operators and regulators to prevent and reduce the possible repercussions of leachate leakage. Determining the cause and amount of leachate loss is a major task in the detection and monitoring of leachate leaks. Typically, leachate is created when water dissolves or removes organic and inorganic material from landfills. Leachate can be generated by a number of processes, such as surface runoff, precipitation, and water seeping into

Table 1 Challenges in landfill leachate detection

Challenges	Description
Complex composition	Due to the large variety of organic and inorganic chemicals present in leachate, analysis is intricate and varied
Variability	Leachate composition can vary over time and between dump sites, which might have an impact on detection techniques
Interference	Analytical methods may be hampered by matrix effects from soil and other materials in leachate, producing unreliable results
Low concentration	Sensitive detection techniques are necessary because certain pollutants in leachate may exist in very low amounts
Sampling difficulty	The diverse nature of leachate and the possibility of contamination during collection make it difficult to obtain representative samples
On-site analysis	Due to their isolated locations, lack of electricity, and requirement for sturdy, portable equipment, landfill sites are challenging to monitor in real time
Environmental impact	Leachate can pose serious environmental dangers, so in order to avoid pollution, quick and accurate detection techniques are required.
Regulatory compliance	Regulatory standards for detection methods must be met, which may call for certain analytical procedures and detection thresholds.
Cost and resources	Sophisticated detection techniques can be costly, and accurate analysis and result interpretation frequently require specific knowledge.

the landfill. There are several ways to release leachate into the environment, such as infiltration, surface runoff, and direct channelling to streams or other surface waters. A thorough study of the landfill environment, including the features of the waste materials, the groundwater and surface water systems, and the hydrological processes, is necessary to determine the cause and extent of leachate leakage [26].

Developing efficient monitoring plans for early detection and mitigation of leachate leakage becomes the next difficulty after the source and extent of the leaking have been determined. Active and passive techniques are frequently combined in monitoring tactics. Utilizing remote sensing technology, installing monitoring wells, and directly sampling leachate are examples of active techniques. Groundwater level indicators, geophysical techniques, and the installation of soil moisture sensors are examples of passive methods, challenges faced in landfill leachate detection are shown in Table 1.

The possibility of leachate leakage across a broad range of temporal and spatial scales complicates the development of monitoring techniques. Leachate leaks must therefore be detected and monitored over a range of spatial and temporal dimensions using appropriate monitoring systems. Monitoring plans also need to be able to recognize variations in leachate concentrations over time [27]. In order to detect potential problems related to leachate leakage, a thorough risk assessment must be conducted in conjunction with the creation of monitoring strategies. Risk assessment should take into account the possible health effects on neighboring communities as well as the possible effects of leachate leakage on soil, surface water, and groundwater. The possibility that leachate will pollute sources of drinking water should also be taken into account in the risk assessment.

In summary, a thorough understanding of the landfill environment as well as the possible sources and pathways of leachate leakage are necessary for developing successful techniques for leachate leakage detection and monitoring. Additionally, it necessitates the creation of thorough monitoring plans with the ability to identify and track leachate leaks across a variety of temporal and spatial dimensions. In order to detect potential problems related to leachate leakage, a thorough risk assessment must be conducted in conjunction with the creation of monitoring strategies.

Limited technology is currently in use to monitor and identify leachate leakage in landfill sites. Interpreting the data collected by the majority of monitoring systems now in use can be challenging because they rely on the placement of sensors and instruments inside the landfill. These systems can also be costly to install and maintain over time, and they may malfunction more frequently. It might be challenging to precisely determine the degree of any leakage because it is not always possible to detect and monitor leachate at different landfill depths. This is another drawback of the present technology.

The improvement of leachate monitoring systems' precision and dependability should be the main goal of future research. This can entail creating brand-new, cutting-edge, more dependable, and accurate tools and sensors. Research should also concentrate on creating monitoring devices that are more affordable and capable of detecting and keeping an eye on leachate at various waste depths.

3 Conclusions

Advances in technology have made it possible for landfill operators to monitor and detect leachate leakage, thereby guaranteeing proper management of the material and protection of the surrounding environment. Landfill managers can now identify, track, and respond to any possible risks to the environment and public health thanks to the use of sensors, drones, and other technologies. Relying less on human operations and more on these tools has enabled landfill operators to control leachate more proactively.

These developments have not only made it possible for landfill managers to identify and track leachate leaks more precisely, but they have also made it possible to comprehend the effects of leachate on the environment and public health. Landfill operators can now detect any problems and take the necessary precautions to mitigate them by gathering data from sensors and drones. Additionally, this data can be utilized to build new technologies and improve management tactics. Even though leachate leakage detection and monitoring have benefited from technological advancements, considerable work remains. This involves creating new technologies and tactics to increase the efficiency of current systems as well as gaining a deeper understanding of how leachate leakage affects the environment, public health, and nearby communities.

In general, technical advancements in leachate leakage monitoring and detection have helped landfill operators safeguard the general public's health and the surrounding environment. Landfill operators can identify, track, and respond to any possible dangers to the environment and public health by utilizing sensors, drones, and other technologies. To increase these systems' efficacy and gain a greater understanding of how leachate leakage affects the environment and public health, however, a great deal of effort remains.

References

1. Kumar A, Agrawal A (2020) Recent trends in solid waste management status, challenges, and potential for the future Indian cities—a review. Curr Res Environ Sustain 2:100011
2. Jia YG (2012) Three-dimensional (3D) dynamic monitoring for an underground contamination process induced by landfill leakage. Environ Forensic 13(1):68–81. https://doi.org/10.1080/15275922.2011.643336
3. Clement R, Oxarango L, Descloitres M (2011) Contribution of 3-D time-lapse ERT to the study of leachate recirculation in a landfill. Waste Manag 31(3):457–467. https://doi.org/10.1016/j.wasman.2010.09.005
4. Rashmi HR, Devatha CP (2021) Dewatering performance of sludge using coconut shell biochar modified with ferric chloride (sludge dewatering using bio-waste). Int J Environ Sci Technol 19(7):6033–6044
5. Moreira C et al (2014) Comparative analysis between biogas flow in landfill and electrical resistivity tomography in Rio Claro City, Brazil. J Geol Res 2014:1–7. https://doi.org/10.1155/2014/845906
6. Rashmi HR, Devatha CP (2020) Experimental investigation on sludge dewatering using granulated blast furnace slag as skeleton material. Environ Sci Pollut Res 27:11870–11881
7. Kjeldsen P, Barlaz MA, Rooker AP, Baun A, Ledin A (2002) Present and long-term composition of MSW landfill leachate: a review. Crit Rev Environ Sci Technol 32(4):297–336
8. Tchobanoglous G, Theisen H, Vigil SA (1993) Integrated solid waste management: engineering principles and management issues. McGraw-Hill
9. Tafuri AN, Braga BC, Gazzola J, Zuliani DQ (2018) Mathematical modelling and simulation of landfill leachate migration. Comput Geosci 118:36–46
10. Daniel DE, Koerner RM, Koerner GR (2019) Handbook of Geosynthetic engineering. Springer
11. Christensen TH, Kjeldsen P, Bjerg PL, Jensen DL, Christensen JB, Baun A (1992) Biogeochemistry of landfill leachate plumes. Appl Geochem 7(3):251–267
12. Mendez MO, Maier RM (2018) Phytostabilization of mine tailings in arid and semiarid environments—an emerging remediation technology. Environ Health Perspect 116(3):278–283
13. Mara DD, Pearson HW, Oragui JI (2010) Wastewater treatment in constructed wetlands with horizontal sub-surface flow. IWA Publishing
14. Casado I, Mahjoub H, Lovera R, Fernandez J, Casas A (2015) Use of electrical tomography methods to determinate the extension and main migration routes of uncontrolled landfill leachates in fractured areas. Sci Total Environ 506:546–553. https://doi.org/10.1016/j.scitotenv.2014.11.068
15. Xiang R, Lei GY, Xu Y, Zhou Q, Liu YQ, Dong L, Liu JC, Huang QF (2020) Aging behaviors of HDPE geomembrane in landfill environment and its impact on pollution risk of surrounding groundwater. Res Environ Sci 33:978–986
16. Bashir MJ, Aziz HA, Yusoff MS, Aziz SQ (2012) Color and chemical oxygen demand removal from mature semi-aerobic landfill leachate using anion-exchange resin: An equilibrium and kinetic study. Environ Eng Sci 29(5):297–305
17. Deng RJ, Jin CS, Ren BZ, Hou BL, Hursthouse AS (2017) The potential for the treatment of antimony-containing wastewater by iron-based adsorbents. Water 9:794

18. Christensen TH, Kjeldsen P, Bjerg PL, Jensen DL, Christensen JB, Baun A, Albrechtsen H-J, Heron G (2021) Biogeochemistry of landfill leachate plumes. Appl Geochem 16(7–8):659–718
19. Kjeldsen P, Barlaz MA, Rooker AP (2022) Present and long term composition of MSW landfill leachate: a review. Crit Rev Environ Sci Technol 32:297–336
20. Zhao YL, Zhang CS, Wang YX, Lin H (2021) Shear-related roughness classification and strength model of natural rock joint based on fuzzy comprehensive evaluation. Int J Rock Mech Min Sci 137:104550
21. Xie Q, Ren BZ, Hursthouse AS, Shi XY (2022) Effects of mining activities on the distribution, controlling factors, and sources of metals in soils from the Xikuangshan south mine, Hunan Province. Integr Environ Assess Manag 18:748–756
22. Pandey LMS, Shukla SK (2019) An insight into waste management in Australia with a focus on landfill technology and liner leak detection. J Clean Prod 225:1147–1154
23. Cheng YX, Liu HS, Zhao ZY (2004) Investigation of urban landfill contamination using geophysical methods. Chin J Eng Geophys 1:26–30
24. Jernberg J, Pellinen J, Rantalainen A-L (2013) Qualitative nontarget analysis of landfill leachate using gas chromatography time-of-flight mass spectrometry. Talanta 103:384–391. https://doi.org/10.1016/j.talanta.2012.10.084
25. Baghanam AH, Nourani V, Aslani H, Taghipour H (2020) Spatiotemporal variation of water pollution near landfill site: application of clustering methods to assess the admissibility of LWPI. J Hydrol 591:125581
26. Gautam P, Kumar S (2021) Characterisation of hazardous waste landfill leachate and its reliance on landfill age and seasonal variation: a statistical approach. J Environ Chem Eng 9(4):105496. https://doi.org/10.1016/j.jece.2021.105496
27. Ma Y, Liu Z-H, Xi BD, Li W-T, Xu YQ, Zhao HZ, Chen ZQ, He XS, Xing B (2021) Molecular structure and evolution characteristics of dissolved organic matter in groundwater near landfill: implications of the identification of leachate leakage. Sci Total Environ 787:147649. https://doi.org/10.1016/j.scitotenv.2021.147649

Part IV
Smart Agricultural Solutions for Sustainability

Examining Agrovoltaic System: Impacts on Energy Yield and Crop Productivity

Abitha Neethirasu, Ishwarya Perumalsamy, Kamelash Kannan, Revanth Mani, Revathy Subbiah Rajaram, and Kirubakaran Victor

Abstract This study investigates the performance of agrovoltaic systems by analyzing module efficiency, energy yield, microclimate conditions, and crop productivity. A field experiment was conducted to compare the parameters between agrovoltaic systems and traditional agricultural practices. Module efficiency and energy yield were measured to assess the solar panels' performance under co-located crop cultivation. Microclimatic conditions, including temperature, humidity, and wind speed, were monitored to understand their influence on both crop growth and solar module efficiency. Crop yield was evaluated to determine the agricultural productivity of the agrovoltaic system compared to conventional farming practices. The study outcome provides a deep insight into the feasibility and benefits of integrating agriculture and solar energy production, highlighting opportunities for maximizing land use efficiency and promoting sustainable farming practices.

Keywords Agrovoltaics · Module efficiency · Co-production · Energy Yield · Microclimate · Temperature

1 Introduction

The escalating energy demand in India, driven by exponential population growth and increasing industrial investments, emphasizes the country's urgent need for sustainable energy solutions. As of May 2024, India recorded a maximum power demand of 250 GW [1], with an installed capacity of 238 GW from conventional power plants and 180 GW from new and renewable energy sources. Solar PV contributes significantly to this capacity, accounting for 84.27 GW as of May 2024 and is expected to grow further in the coming years [2, 3]. To promote residential solar PV installations, the Indian government has introduced Central

A. Neethirasu · I. Perumalsamy · K. Kannan · R. Mani · R. S. Rajaram (✉) · K. Victor
Centre for Rural Energy, The Gandhigram Rural Institute-Deemed to be University, Gandhigram, Tamil Nadu, India

P. Pathak et al. (eds.), *Advances in Environmental Sustainability, Energy and Earth Science*, https://doi.org/10.1007/978-3-031-73820-3_25

Financial Assistance Schemes, offering subsidies of up to 30% of the total installation charges for domestic and residential installations and up to 70% for special category states. These incentives have been instrumental in achieving a Compound Annual Growth Rate (CAGR) of around 46% in rooftop-based solar installations. In addition, the Solar Park Scheme aims to establish 50 solar parks across the country, each with an individual capacity of 500 MW by 2026. The PM-KUSUM scheme focuses on integrating 30.8 GW of decentralized solar power plants in the agricultural sector. These initiatives are part of India's broader strategy to enhance its renewable energy capacity and meet the growing energy demand sustainably. The optimized renewable energy utilization thereby ensures the effective climate change mitigation through soft computing techniques [4, 5]. The National Institute of Solar Energy has estimated India's solar capacity as 748 GW. This vast potential signifies a significant opportunity for clean energy generation which can further contribute substantially to achieving the UN's Sustainable Development Goal No. 7 (Affordable and Clean Energy) and Goal No. 11 (Sustainable Cities and Communities). By harnessing solar energy for power production, we can reduce fossil fuel usage and thereby reduce the carbon footprint, thus also achieving Goal No. 13 (Climate Action). Though there are many opportunities, solar PV installation is not without its challenges. Rooftop PV installations can contribute only to small-scale energy production due to factors such as roof suitability, structural integrity, shading, and roof orientation. On the other hand, ground-mounted PV systems face significant challenges related to land availability, acquisition, and land levelling.

In a densely populated country like India, land is in high demand for agriculture, housing developments, and other infrastructure projects, making it challenging to allocate land for solar power plants. Consequently, agricultural land often becomes a target for solar power projects, which can lead to conflicts over land use and impact food production (SEIA). Addressing these challenges requires careful planning and innovative solutions to balance energy needs with sustainable land use practices. The land requirement for generating 60 GW of power is estimated to be 1.2 lakh hectares or 1200 km^2 [6]. This surge in installed capacity underscores the increasing utilization of land for solar photovoltaics. Notably, India, being agriculturally rich with 60% of its land dedicated to cultivation [7, 8], faces the challenge of balancing energy needs with agricultural requirements. In addition, the effective land utilization paves a way for effective last mile electrical connectivity for achieving rural electrification [9–11].

Green energy sources emerge as viable solutions in response to the escalating energy demand and the imperative to curtail carbon emissions. Given India's geographical positioning, photovoltaics stands out as the most suitable mode of energy generation. However, photovoltaic technology has its drawbacks, such as low conversion efficiency, high initial cost, and weather-dependent energy production. [12, 13] have reported that the increase in temperature will result in the reduction of the production voltage of the PV systems. India is a tropical country with hot summers, mild winters, and varying monsoons; the consistency in the power production of the PV system is non-linear.

Furthermore, the erection of a solar PV power plant demands a larger area than the footprint of the panels themselves. For instance, a typical 255 W solar panel occupies 1.64m^2 of land, and a 10 MW solar power plant requires 20.24 hectares or 0.20 [14]. In the United States, it is estimated that 1 GW of power generation necessitates 10–50 km^2 of land [15]. Given that agricultural land is often flat and well-maintained, it becomes a preferred choice for installing solar photovoltaics. However, this practice significantly reduces the available area for cultivation, a trend that is expected to persist and intensify.

A holistic approach to sustainability must encompass agriculture, recognizing its pivotal role alongside the surging energy demand and population growth [16]. This study advocates for the integrated implementation of ground-mounted solar photovoltaics and agriculture, aiming to generate energy efficiently while concurrently supporting food production. This dual-purpose approach seeks to strike a balance between meeting the escalating energy needs and ensuring the sustainability of agriculture in the face of a growing population.

2 Literature Review

Agrovoltaics, integrating agricultural activities with photovoltaic systems, has developed as a significant approach to mitigate the dual bottlenecks of escalating energy demand and the need for sustainable food production. This innovative concept involves harnessing solar energy for electricity generation while concurrently utilizing the same land for agricultural purposes. The concept of agro-photovoltaic systems is the integration of solar photovoltaics and agricultural production aimed at synergizing the outcome of renewable energy and agro-food production was tried [17]. The study indicates that integrating agriculture with solar photovoltaics can boost land yield by up to 70%. Additional shadings offered benefits for crop production, improvements in water productivity, and increased economic values. The agrovoltaics experiments conducted and studied by [18] highlight the potential threat for land resources as both food and energy demands escalate. Their findings demonstrate an upsurge in economic value of up to 30% and with a further boost in PV power between 40 and 70 GW for lettuce cultivation in agrovoltaic systems.

The agrovoltaics-based system simulated for maize crops resulted in increased productivity under the shadow of PV modules [19]. This study suggests that increased production can be achieved even under drought and climate change by integrating agriculture with solar photovoltaics. Agrovoltaic scenarios were found to be more advantageous than conventional maize production for biogas and electric energy, with the land equivalent ratio (LER) increasing with module density and being greater with sun position following compared to stationary (or) fixed module mounting system.

The agronomic impacts for EVL7 member states for wheat production during winter under photovoltaic modules were explored by [20]. Their results indicated the greatest capability in Europe for the cultivation of winter wheat through agrovoltaics around the Mediterranean province. Agrovoltaic structures were also found to limit the losses in grain yield and protect the crop from heat stresses and drought. A sustainable solution which emphasizes the integration of commercially available photovoltaic panels, with an efficiency of 12–15%, with agriculture to meet the increasing demands for energy and food [21]. They highlighted the potential for solar photovoltaic power plants to coexist with agriculture and bioenergy crops on one piece of land, providing a sustainable solution.

Agrovoltaics-based farming systems enhance microclimatic conditions, lowering operating temperatures and enhancing the efficiency of solar photovoltaic systems. The study conducted by [22, 23] showed that agrovoltaic-based PV systems generated increased power than the ones without crops. The increase in effectiveness of the solar PV modules was observed to be around 3.05% and 3.2% on bright and gloomy days, correspondingly. The implementation of agrovoltaics promotes more electricity productivity due to the crop's trans-evaporation process, exclusively important for the solar module surface temperature. The optimization of surface temperature of module without compromising the efficiency of the system is carried out through hybrid PV-T system [23, 24].

The environmental impacts and energy production efficiency of agrovoltaic systems are evaluated by [25, 26]. Their findings indicate that agrovoltaic systems might decrease water utilization by 14–29%, offering a dual-use solution for land in terms of electrical energy and crop productivity while minimizing freshwater needs. The effects of different levels of shading on PV in a kiwi fruit orchard, its growth, yield, and water productivity in an agrovoltaic system were demonstrated by [27, 28]. Their experiment concluded that PV shading increased the efficiency of leaf light use and crop water productivity by 19%, with a slight effect on kiwi fruit growth and yield.

The crops need sunlight for photosynthesis, whereas the same sunlight is required for power production through photovoltaic models. The paradigm is proposed in which crops and photovoltaic modules contest for solar irradiance, emphasizing the importance of design constraints [16, 17]. For instance, mechanical cultivation practices necessitate a solar module mount height of at least 4 m, with adequate gaps between strings to allow cultivators to traverse the field without causing damage to the solar photovoltaics. The study also delved into eco-physiological constraints, highlighting those plants in Agrovoltaics receive lower radiation compared to those in open lands. The author underscored the significance of selecting shade-tolerant plant species, as a decrease in irradiance below 75% of natural levels can significantly reduce crop yields, although shading can also confer benefits such as reduced water requirements and mitigated nitrogen stresses [29]. The study concluded by suggesting that Agrovoltaics might enhance the productivity of the land by up to 60% to 70% using simulation models, with the incorporation of solar photovoltaics with mirror backs further increasing light availability through multiple reflections. The effectiveness of solar photovoltaics ensured for their solid state behavior in the field without huge deterioration in the efficiency [11, 28, 30].

The literature on Agrovoltaics explores the intricate interplay between crops and photovoltaic modules [31, 32], shedding light on design constraints, eco-physiological considerations, and the potential synergies between renewable energy concepts and principles to promote green agriculture practices. This paper aims to investigate the performance of agrovoltaic systems by studying factors such as module efficiency, energy yield, microclimate conditions, and crop productivity.

3 Methodology

The agrovoltaic study is carried out according to the following methodology. The proposed methodology as shown in Fig. 1 provides a structured approach to experimenting, from initial feasibility assessment to data collection and analysis, allowing for a comprehensive evaluation of the integrated solar PV and crop cultivation system.

3.1 Feasibility Study

3.1.1 Site Selection

The potential sites for the experiment, considering factors such as sunlight exposure, land availability, accessibility, and local climate conditions were identified. The site selected is at the location 10.2793°N 77.9346°E, refer Fig. 2. There was a building located on the north side of the side. The site was selected in such a way that the shadow of the building had no effect on the site.

3.1.2 Resource Assessment

Prior to the initiation of the project, a comprehensive evaluation had been completed concerning the solar irradiance levels at the chosen sites. This evaluation utilized historical data or on-site measurements. Factors such as solar insolation, shading, and terrain effects had been meticulously assessed during this process. By utilizing PVsyst synthetic data generation, it was determined that the average annual horizontal solar irradiance amounted to 1909.8 W/m^2.

Fig. 1 Block diagram of methodology

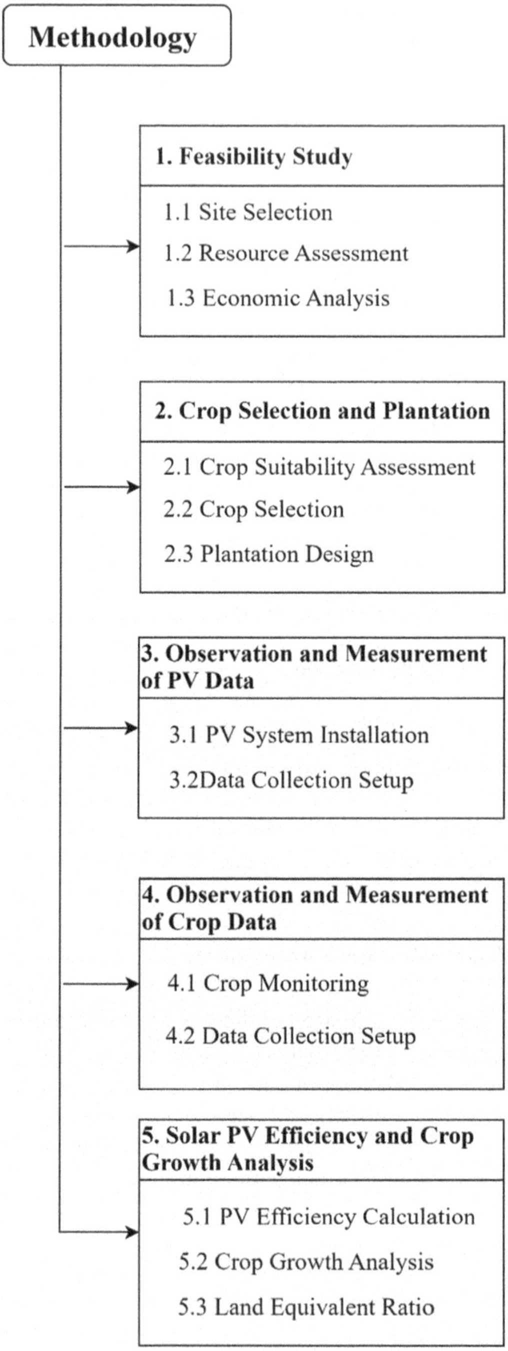

Fig. 2 Satellite image of the site

3.2 *Crop Selection and Plantation*

3.2.1 Crop Suitability Assessment

To determine the suitability of various crops for co-cultivation with solar PV panels, we have considered factors such as shade tolerance, water requirements, and growth characteristics. These considerations were crucial in assessing which crops were most compatible with the circumstances formed by the existence of solar panels. For instance, crops with high shade tolerance and moderate water requirements were preferable, as they could thrive despite reduced sunlight availability and potentially limited irrigation access.

3.2.2 Crop Selection

To choose appropriate crops that could thrive under the partial shading conditions created by the solar panels, we have considered crops with high economic value, fast growth rates, and compatibility with local soil and climate conditions. These factors have been crucial in determining which crops were most suitable for cultivation alongside solar panels. For instance, crops such as tomatoes, peppers, and lettuce would have been selected for their ability to tolerate partial shade while offering high economic returns and rapid growth rates. In this study, fenugreek (Trigonella foenum-graecum) has been chosen for its resilience to reduced sunlight, enhanced economic value, and adaptability to local soil and climate conditions. Moreover, fenugreek plants are short plants which are very much suitable for the pre-installed PV modules in the site.

3.2.3 Plantation Design

The crops were planted in an area of about 2 square meters. The crops were strategically arranged within the 5 ridges and 6 furrows layout to optimize space utilization and sunlight exposure. 4 PV modules were used for this system. Each module was positioned with a south orientation at a tilt angle of 15° to maximize sunlight exposure for both crops and PV modules. Adequate spacing between rows of solar modules allows for maintenance access and airflow, which can benefit both the PV modules and the plants.

3.3 Observation and Measurement of PV Data

3.3.1 PV System Installation

The solar PV panels at the site were installed according to the predetermined layout and configuration as shown in Figs. 3 and 4.

3.3.2 Data Collection Setup

Daily data collection was conducted to closely monitor the performance of the PV system, refer Fig. 5. Meco data logging equipment was utilized to monitor and log

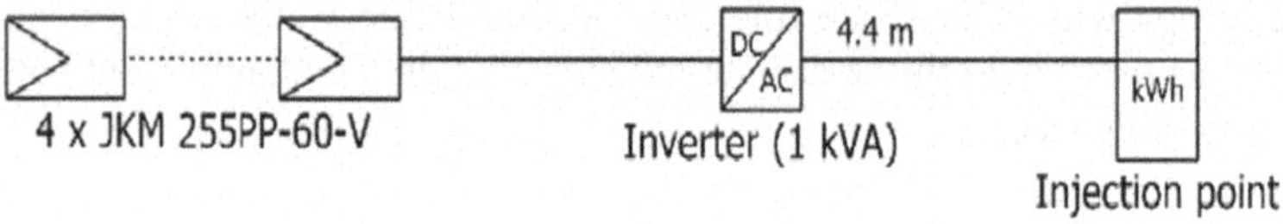

PV module	JKM 255PP-60-V
Inverter	Zeverlution 1000S
String	4 x JKM 255PP-60-V

Fig. 3 Line diagram of panels with crop

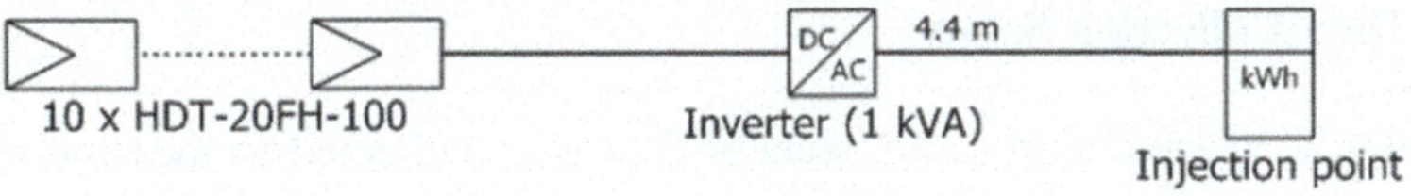

PV module	HDT-20FH-100
Inverter	Zeverlution 1000S
String	10 x HDT-20FH-100

Fig. 4 Line diagram of panel without crop

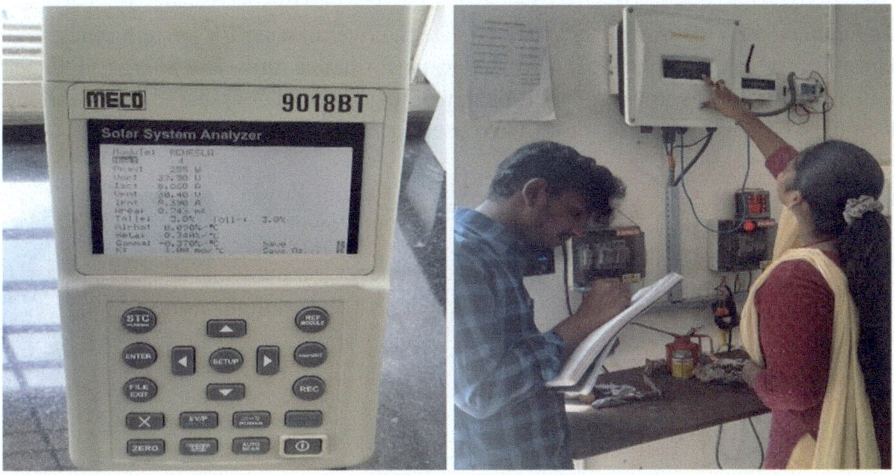

Fig. 5 PV data collection

the key parameters of the PV system, solar irradiance, panel temperature, electrical output, and system performance metrics.

3.4 *Observation and Measurement of Crop Data*

3.4.1 Crop Monitoring

Regular monitoring was done to track the growth and development of selected crops throughout the experiment. This data included observations on sunlight exposure, crop health, and crop yield.

3.4.2 Data Collection Setup

In addition to the planting setup, weekly data were collected to measure relevant crop parameters such as height, canopy coverage, leaf area index, biomass accumulation, and crop yield. Conduct periodic measurements and observations of crop data at predefined intervals, correlating crop growth dynamics with environmental variables and solar PV performance.

3.5 *Solar PV Efficiency and Crop Growth Analysis*

3.5.1 PV Efficiency Calculation

The collected PV data were used to calculate the overall solar PV system efficiency, considering factors such as capacity factor, performance ratio, and energy yield per unit area [33–35]. Moreover, the efficiency of the system was assessed using various formulas. The following formulas were used:

$$\text{Power Efficiency}(\%) = \frac{\text{Useful Output Power}}{\text{Input Power}} \times 100 \tag{1}$$

$$\text{Performance Ratio}(\%) = \left\{ \frac{\text{Total energy output}}{(\text{Installed capacity} \times \text{Total sunlight hours})} \right\} \times 100 \tag{2}$$

3.5.2 Crop Growth Analysis

Analyzing the collected crop data to evaluate the influence of co-cultivation with solar PV panels on crop growth, yield, and quality parameters and to compare the performance of crops grown under shaded and unshaded conditions using different methods like.

3.5.3 Land Equivalent Ratio

The Land Equivalent Ratio (LER) was also considered in the assessment of the system's efficiency. It was calculated by the following formula:

$$\text{LER} = (\text{Relative yield of intercrop}/\text{Relative yield of monoculture}) \tag{3}$$

If the Land Equivalent Ratio obtained is above 1, it will indicate increased land utilization. This analysis further highlights the synergistic benefits of combining

(a) (b)

Fig. 6 (**a**) Crops beneath solar panels, (**b**) Crops in open condition

crop cultivation with photovoltaic energy generation in the agrovoltaic system as shown in Fig. 6a,b.

4 Results and Discussion

The study was conducted according to the methodology above for 60 days. The observations from the PV data and the crop data are discussed below.

4.1 Crop Growth and Microclimate

4.1.1 Crop Yield

Crop growth analysis reveals that shoot length, root length, number of leaves, and quantity of seeds germinated under the APV systems were 21%, 40%, 33%, and 28% higher than conventionally grown crops as shown in Fig. 7. The fenugreek crops cultivated under agrovoltaics demonstrated better yields than those grown in open conditions, indicating that the presence of solar panels did significantly increase crop productivity.

4.1.2 Microclimate Variations

Analysis of microclimate data revealed slight differences between the two fields. While soil moisture levels tended to be higher in the agrovoltaics field due to reduced evaporation under the solar panels, temperature and humidity levels showed marginal variations between the two settings. The shade provided by PV modules in agrovoltaic systems contributed to reduced evaporation rates, maintaining moisture

Fig. 7 Crop yield in open condition

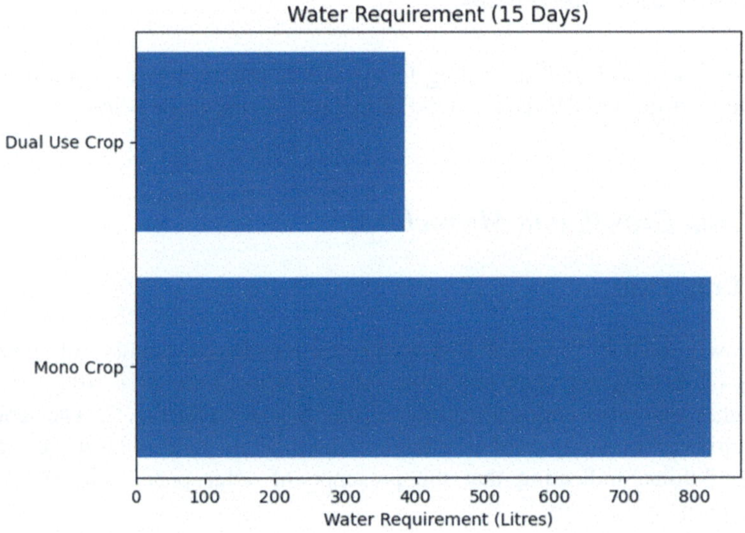

Fig. 8 Water requirement (15 Days)

levels throughout the day and resulting in 53.5% lower water consumption compared to the mono crop control group (Fig. 8).

Effect of Manure Addition: The application of biogas slurry as manure resulted in improved plant growth and vigor in both fields, as evidenced by enhanced flowering and overall biomass production.

4.2 *Power Generation*

The introduction of agrovoltaics resulted in a significant reduction of 34.76% in the working temperature of the modules, showcasing improved thermal performance under the co-production of energy and crops as shown in Fig. 9a,b. The negative correlation observed between the working temperature of the PV modules and their performance highlights the importance of temperature control for optimal module functionality, refer Fig. 10a,b.

The decreased module temperature led to a notable 54.57% increase in the conversion efficiency of the APV modules than that of conventional solar PV, contributing to enhanced overall energy generation. Performance of Agrovoltaics: The solar power plant integrated with agrovoltaics exhibited comparable or slightly improved power generation efficiency compared to the conventionally mounted plant. This suggests that incorporating agricultural activities beneath solar panels does not necessarily compromise energy production.

4.3 *Advantages of APV Systems*

4.3.1 Agrovoltaics and Crop Productivity

The comparable crop yields observed in both agrovoltaics and open-field conditions are promising for the feasibility of integrating agricultural activities with solar energy generation.

4.3.2 Environmental Benefits and Sustainability

The utilization of biogas slurry as organic manure not only enhances soil fertility but also aligns with sustainable agricultural practices. This approach reduces dependency on synthetic fertilizers, mitigating environmental pollution and promoting soil health [36].

4.3.3 Economic Viability

The comparable or slightly improved power generation efficiency of the agrovoltaics system underscores its economic viability. By maximizing land use efficiency, agrovoltaics present an opportunity to enhance agricultural productivity while simultaneously generating renewable energy.

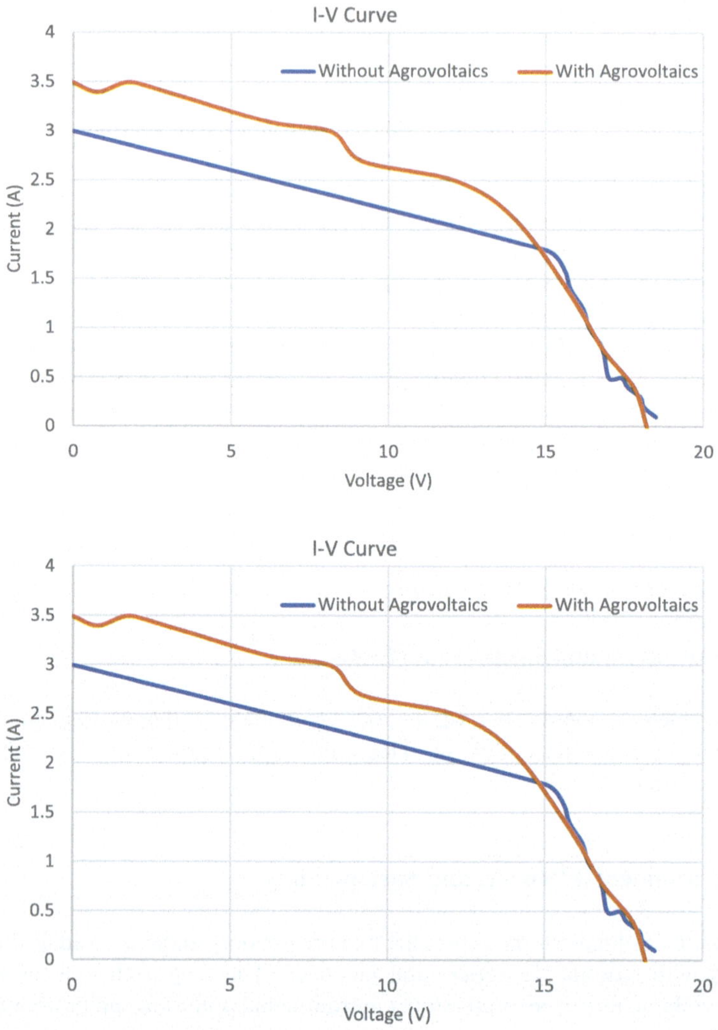

Fig. 9 (**a**) I-V curve, (**b**) P-V curve

5 Conclusions

The findings of this study demonstrate the feasibility and potential benefits of integrating agricultural activities with solar energy generation through agrovoltaics. By optimizing land use and promoting sustainable practices, agrovoltaics is a promising technology to address the dual bottleneck of food security and renewable energy production. The proposed APV systems contribute to increasing land use productivity and increase 54.57% of PV module efficiency. Utilizing solar energy through the agrovoltaic system aids in distributed energy production, lessening

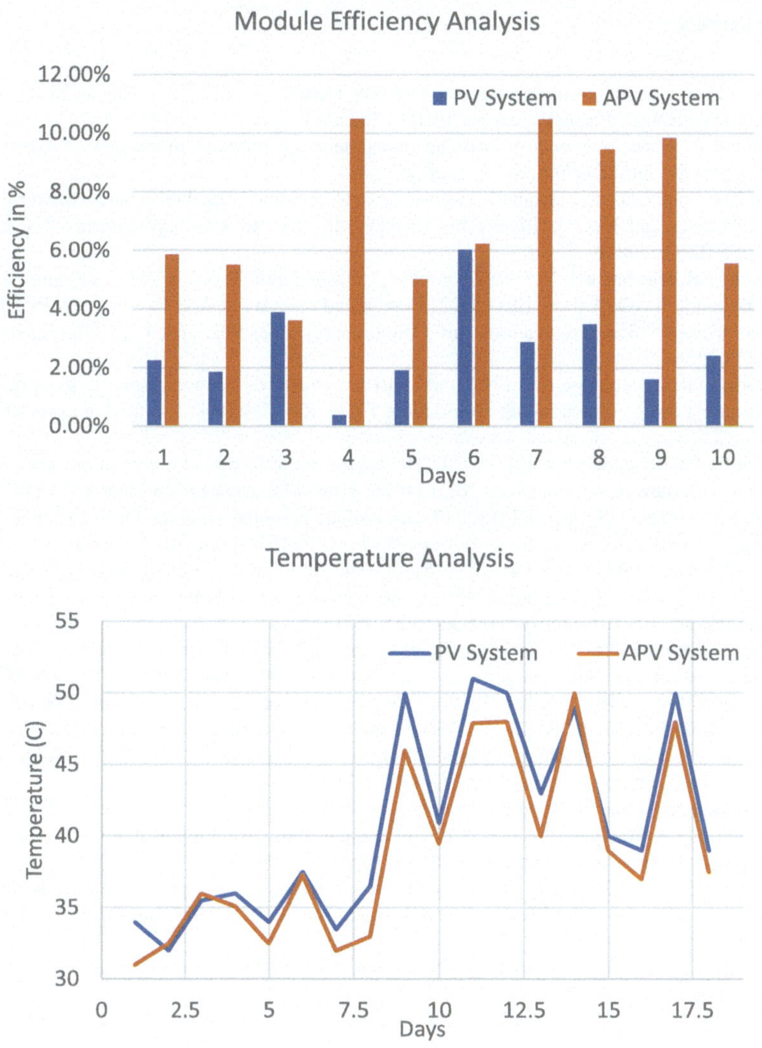

Fig. 10 (**a**) Module efficiency analysis, (**b**) Temperature analysis

requirement on conventional fuels and curbing greenhouse gas emission. This eco-friendly energy approach plays a pivotal role in fighting climate change and facilitating the shift to a net zero economy.

Through methodical planning and integration of these agrovoltaic systems we can enhance crop yields, reduction in water consumption, mitigate climate change impacts, and contribute to the transition toward a greener future. Beyond energy and agricultural advantages, the integration of panels with agriculture can provide additional environmental benefits. These may include soil conversion, local biodiversity enhancement, and carbon sequestration, thus contributing to overall ecosystem health and resilience.

References

1. India's Power Sector Achieves Record 250 GW Demand met on 30th May 2024. https://pib.gov.in/PressReleaseIframePage.aspx?PRID=2022257
2. Manohar A: Renewable energy in India: Investment opportunities in the pow... https://www.investindia.gov.in/sector/renewable-energy
3. Roof Top Solar Capacity installations growing at Compound annual growth rate of around 46%: Union Power and New & Renewable Energy Minister. https://pib.gov.in/PressReleasePage.aspx?PRID=1989806
4. Revathy SR, Kirubakaran V, Rajeshwaran M, Balasundaram T, Sekar VSC, Alghamdi S, Rajab BS, Babalghith AO, Anbese EM (2022) Design and analysis of ANFIS—based MPPT method for solar photovoltaic applications. Int J Photoenergy 2022:9625564. https://doi.org/10.1155/2022/9625564
5. Revathy SR, Kirubakaran V (2020) A critical review of artificial neural networks based maximum power point tracking techniques. J Crit Rev 7:400–412. https://doi.org/10.31838/jcr.07.12.75
6. Rösch C, Fakharizadehshirazi E (2024) The spatial socio-technical potential of agrivoltaics in Germany. Renew Sust Energ Rev 202:114706. https://doi.org/10.1016/j.rser.2024.114706
7. Malu PR, Sharma US, Pearce JM (2017) Agrivoltaic potential on grape farms in India. Sustain Energy Technol Assess 23:104–110. https://doi.org/10.1016/j.seta.2017.08.004
8. Santra P, Meena HM, Yadav OP (2021) Spatial and temporal variation of photosynthetic photon flux density within agrivoltaic system in hot arid region of India. Biosyst Eng 209:74–93. https://doi.org/10.1016/j.biosystemseng.2021.06.017
9. Singh A, Strating AT, Romero Herrera NA, van Dijk HW, Keyson DV (2017) Towards an ethnography of electrification in rural India: social relations and values in household energy exchanges. Energy Res Soc Sci 30:103–115. https://doi.org/10.1016/j.erss.2017.06.031
10. Nadimuthu LPR, Victor K (2021) Performance analysis and optimization of solar-powered E-rickshaw for environmental sustainability in rural transportation. Environ Sci Pollut Res 28:34278–34289. https://doi.org/10.1007/s11356-021-12894-x
11. Chindarkar N, Goyal N (2023) Did it increase energy consumption? A difference-in-differences evaluation of a rural electrification policy in Gujarat, India using night-time lights data. Energy Policy 183:113814. https://doi.org/10.1016/j.enpol.2023.113814
12. Rajaram RS, Victor K (2020) Validation of shadow effects on solar photovoltaic modules based on module positioning. Int J Innov Technol Explor Eng 9:1017–1022. https://doi.org/10.35940/ijitee.C7989.019320
13. Rajaram RS, Babu PK, Victor K (2024) Fuzzy logic-based power optimizer for solar photovoltaic power systems. Pertanika J Sci Technol 32:93–110
14. Shiva Kumar B, Sudhakar K (2015) Performance evaluation of 10 MW grid connected solar photovoltaic power plant in India. Energy Rep 1:184–192. https://doi.org/10.1016/j.egyr.2015.10.001
15. van de Ven D-J, Capellan-Peréz I, Arto I, Cazcarro I, de Castro C, Patel P, Gonzalez-Eguino M (2021) The potential land requirements and related land use change emissions of solar energy. Sci Rep 11:2907. https://doi.org/10.1038/s41598-021-82042-5
16. Lalith Pankaj Raj Nadimuthu GN, Karthik VM, Mohanraj M, Kirubakaran V (2019) Fast thermal degradation of biomass using scrapped solar cell with special focus on photovoltaic (PV) waste disposal. In: Waste valorisation and recycling. Springer, Singapore, pp 349–361. https://doi.org/10.1007/978-981-13-2784-1_33
17. Weselek A, Ehmann A, Zikeli S, Lewandowski I, Schindele S, Högy P (2019) Agrophotovoltaic systems: applications, challenges, and opportunities. A review. Agron Sustain Dev 39:35. https://doi.org/10.1007/s13593-019-0581-3
18. Dinesh H, Pearce JM (2016) The potential of agrivoltaic systems. Renew Sust Energ Rev 54:299–308. https://doi.org/10.1016/j.rser.2015.10.024

19. Willockx B, Herteleer B, Cappelle J (2020) Theoretical potential of agrovoltaic systems in Europe: a preliminary study with winter wheat. 2020 47th IEEE Photovoltaic Specialists Conference (PVSC). 996–1001
20. Teng JW, Soh CB, Devihosur SC, Tay RH, Jusuf SK (2022) Effects of agrivoltaic systems on the surrounding rooftop microclimate. Sustainability 14(12):7089. https://doi.org/10.3390/su14127089
21. Ott E, Kabus C, Baxter B, Hannon B, Celik I (2020) Environmental analysis of agrivoltaic systems. In: Reference module in earth systems and environmental sciences. https://doi.org/10.1016/B978-0-12-819727-1.00012-1
22. Agostini A, Colauzzi M, Amaducci S (2021) Innovative agrivoltaic systems to produce sustainable energy: an economic and environmental assessment. Appl Energy 281:116102. https://doi.org/10.1016/j.apenergy.2020.116102
23. Lalith PRGN, Prabakaran G, Murugaiyan A, Kirubakaran V (2018) Hybrid photovoltaic-thermal systems: innovative CHP approach. In: Proceedings of the 4th International Conference on Electrical Energy Systems. ICEES 2018. https://doi.org/10.1109/ICEES.2018.8442352
24. Nadimuthu LPR, Victor K (2021) Optimization of energy-intensive process in Ayurvedic medicine manufacturing unit—a case study. Process Integr Optim Sustain 5(4):975–992. https://doi.org/10.1007/s41660-021-00194-3
25. Jiang S, Tang D, Zhao L, Liang C, Cui N, Gong D, Wang Y, Feng Y, Hu X, Peng Y (2022) Effects of different photovoltaic shading levels on kiwifruit growth, yield and water productivity under "agrivoltaic" system in Southwest China. Agric Water Manag 269:107675. https://doi.org/10.1016/j.agwat.2022.107675
26. Dupraz C, Marrou H, Talbot G, Dufour L, Nogier A, Ferard Y (2011) Combining solar photovoltaic panels and food crops for optimising land use: towards new agrivoltaic schemes. Renew Energy 36:2725–2732. https://doi.org/10.1016/j.renene.2011.03.005
27. Gorjian S, Bousi E, Özdemir ÖE, Trommsdorff M, Kumar NM, Anand A, Kant K, Chopra SS (2022) Progress and challenges of crop production and electricity generation in agrivoltaic systems using semi-transparent photovoltaic technology. Renew Sust Energ Rev 158:112126. https://doi.org/10.1016/j.rser.2022.112126
28. Sirnik I, Sluijsmans J, Oudes D, Stremke S (2023) Circularity and landscape experience of agrivoltaics: a systematic review of literature and built systems. Renew Sust Energ Rev 178:113250. https://doi.org/10.1016/j.rser.2023.113250
29. Mahto R, Sharma D, John R, Putcha C (2021) Agrivoltaics: a climate-smart agriculture approach for indian farmers. Land 10(11):1277. https://doi.org/10.3390/land10111277
30. Lalith Pankaj Raj N, Kirubakaran V (2021) Energy efficiency enhancement and climate change mitigations of SMEs through grid interactive solar photovoltaic system. Int J Photoenergy 2021:6651717. https://doi.org/10.1155/2021/6651717
31. Lee S, Lee J, Jeong Y, Kim D, Seo B, Seo Y, Kim T, Choi W (2023) Agrivoltaic system designing for sustainability and smart farming: agronomic aspects and design criteria with safety assessment. Appl Energy 341:121130. https://doi.org/10.1016/j.apenergy.2023.121130
32. Gomez-Casanovas N, Mwebaze P, Khanna M, Branham B, Time A, DeLucia EH, Bernacchi CJ, Knapp AK, Hoque MJ, Du X, Blanc-Betes E, Barron-Gafford GA, Peng B, Guan K, Macknick J, Miao R, Miljkovic N (2023) Knowns, uncertainties, and challenges in agrivoltaics to sustainably intensify energy and food production. Cell Rep Phys Sci 4:101518. https://doi.org/10.1016/j.xcrp.2023.101518
33. Chopade SS, Pankaj Raj GNL, Prabakaran G (2018.): Special Issue 33) Bench marking of grid tied solar roof top photovoltaic system: a case comparison. Int J Eng Technol 7(2.33):553. https://doi.org/10.14419/ijet.v7i2.33.14832
34. Selvaraj D, Nadimuthu L, Victor K (2021) Monitoring and simulation of parabolic trough collector powered vapor absorption refrigeration system for rural cold storage. Therm Sci 26:298–298. https://doi.org/10.2298/TSCI201110298S

35. Jamil I, Lucheng H, Habib S, Aurangzeb M, Ali A, Ahmed EM (2022) Performance ratio analysis based on energy production for large-scale solar plant. IEEE Access 10:5715–5735. https://doi.org/10.1109/ACCESS.2022.3141755
36. Shravanth Vasisht M, Srinivasan J, Ramasesha SK (2016) Performance of solar photovoltaic installations: effect of seasonal variations. Sol Energy 131:39–46. https://doi.org/10.1016/j.solener.2016.02.013

Fate and Risk Assessment of In vitro Hydroponic Cropping System under Dye and Saline Medium

Srisesh Sriram and C. Ramprasad

Abstract The major global issues are the scarcity in availability of freshwater resources, diminishing environmental pollution, constant increase in the population density, and diminishing fertile land resources for crop cultivation. The availability of land for agriculture in urban and peri-urban is impassable due to dense population. Hydroponic systems are a one-stop solution for dealing with water pollution, water scarcity, and feeding a growing population. In hydroponics, crops are cultivated without soil under controlled environment with constant supply of nutrients and for faster growth rate. In the study, Hydroponic Cropping System (HCS) under saline and inorganic dye concentrations was carried out. The species used for the study was *Mentha genus*, enclosed in a chamber under controlled environment. *Mentha* sp. were spiked with salinity of 0 to 3000 mg/L and dye concentrations (methylene blue) of 0 to 15 mg/L. The study found that under saline circumstances, the plants grew unhindered; however under dye concentrations, they displayed hindrance in their growth. At a concentration of 15 mg/L of dye the plant leaf turned into pale yellow colour and gradually the wilting stage is reached. In addition, the risk assessment of pollutants on plant growth was evaluated. The study concludes that possible health risk due to salinity was very less. As clearly indicated, the health risk index (HRI) and targeted hazardous quotient (THQ) values under saline conditions are less than 1 implies less risk associated, but with dye contaminant the risk assessment was on higher side with HRI and THQ values are more than 1. Hence the study concludes that hydroponic crop production is possible in saline water, but suitable pre-treatment needs for dye-containing wastewater for crop growth.

Keywords Hydroponic cropping systems · Salinity · Inorganic dye · Risk assessment · Hazard quotient · Mentha species · Methylene blue

---

S. Sriram · C. Ramprasad (✉)
School of Civil Engineering, Centre for Advanced Research in Environment (CARE), SASTRA Deemed to be University, Thanjavur, Tamil Nadu, India
e-mail: ramprasad@civil.sastra.edu

P. Pathak et al. (eds.), *Advances in Environmental Sustainability, Energy and Earth Science*, https://doi.org/10.1007/978-3-031-73820-3_26

1 Introduction

The urban population in India is increasing in an exponential rate leading to more wastewater generation, denser cities with less open space for cultivation, and reduced food supply. The reports say that 38,354 MLD of wastewater generated from class A cities, and only 30% is treated (11,786 MLD), among that only 5% is reused [1, 2]. Additionally, the import of food products from various developed countries has increased several folds as well as the food distribution among the population was biased, it further increases the urban and rural poverty [3, 4]. There is a need to tackle the above scenarios of wastewater treatment, high crop growth, and lack of space for agriculture. Hydroponic Cropping System (HCS) is one such technology, which grows crops in a solution of water or wastewater in an enclosed and controlled environment, thus not restricted by the climate, location, better utilization of vertical spaces, and thus utilizing only 10% of irrigation water when compared to conventional agriculture providing a sustainable alternative to it in feeding the growing global population [5–7].

Hydroponics is a subset of horticulture and has historical references such as 'The Hanging gardens of Babylon' which was an effective way to utilize minimal water for the crop growth. There are various hydroponic techniques such as Static Cone Culture, Continuous-flow Solution Culture, Deep Water Culture, and Aeroponics, to name a few. Further, the use of substrates (growing support materials) for the enhancement of growth of the crops such as rock wool, expanded clay aggregate, coconut coir, rice husk, etc., depending on the growing technique utilized. The success of these methods depends on the proper implementation of the same in an effective manner on a nation-wide scale being the example of Israel where when only a small fraction of farmers utilize the technique, in Israel more than 3/4th of the farmers utilize the same due to which even after being in a drought-prone region they can cultivate crops at an enormous scale all-year round [8].

Hydroponics provides advantages such as water efficiency, quicker growth, and year-round growing, but it also has disadvantages such as high start-up expenses, continuing upkeep, and the requirement for expertise. Overall, hydroponics is a resource-efficient form of plant growing with benefits and drawbacks that vary depending on the grower's aims and ability. Several authors have reported the usability of treated wastewater in hydroponics for the growth of economically important plants [9–11]. Nie et al. reported the uptake of triclosan in celery under hydroponic culture and the metabolism in the nutrient-rich wastewater [10]. Al-Karaki reported that barley fodder production was higher under the hydroponic condition with treated wastewater [12]. Similarly, Lira et al. have done a study with lettuce (*Lactuca sativa*) cultivated in hydroponics in a brackish water and shown significant growth and in anaerobically treated domestic wastewater [9, 13]. Salinity is key concern for the plant growth in hydroponic culture, it retards the plant biomass growth and decreases the photosynthesis process due to excessive salt accumulation [14, 15]. Additionally, researchers have extensive study on the impact of salinity and nitrogen stress on the primary metabolism of *Stamnagathi plants*, revealing ecotype-specific responses to stress conditions [16]. The selection of plant species is an important criteria, varied plant species exhibit varied behaviour. Halophytes such as *Suaeda fruticosa* and *Halocnemum strobilaceum* show greater tolerance to hypersaline conditions [17], *Reichardia picroides* shows moderate salt tolerance, while *Taraxacum officinale* is severely affected by high salinity levels [18]. The works are very scanty in terms of hydroponics under dye stress conditions. Hence, the use of hydroponics can be viewed as a promising solution for feeding the global growing population as it is cost-effective, efficient, and sustainable alternative for an ever-growing population which leads to food scarcity due to limited space for growing crops. This can be further helpful for the protection from diseases transmitted through soil or animals in open fields with fewer use of chemicals for pest control, thus reducing food hazards and risks [4, 19].

Plant development is hampered by salinity in water and it leads to nutritional imbalances, osmotic stress, and harmful ion build-up. Additionally, dye-based growing media might be deficient in vital nutrients and contain hazardous compounds, thus impacting plant health. Risks include pH imbalances and root suffocation. Hence, there is an urgent need to address these concerns carefully by regulating salinity with proper fertilizer solutions and use of acceptable growing media. Watering and drainage are critical parameters, especially in soil-based systems. Hence, the study aims for growing of crop effectively in a hydroponics system under saline and dye environment. The risk assessment and hazard quotient due to toxins are estimated for the hydroponic cropping system.

2 Materials and Methods

2.1 Mint Species

Mint crops, belonging to the '*Mentha Lamiaceae*' are widely cultivated, grow in moderate climates rich in vitamin A, C, and K, and are low in calories with potential health benefits. The mint crops are suited for hydroponic cultivation for several reasons such as high rapid growth rate of 4 to 6 weeks giving high yield including controlled nutrient availability and water supply (Fig. 1).

At initial stages, the plants were grown in the controlled environment for a period of one month at a room temperature of 28–32 °C and grown in a Hoagland nutrient media [20].

2.2 Deep Water Culture (DWC)

The plants were grown in the DWC system containing the nutrient solution that has all the essential minerals and nutrients required for plant growth. The various benefits of Deep Water Culture (DWC) over other types of hydroponics such as

Fig. 1 Spear mint species grown hydroponically

increased oxygen availability to the roots, promote rapid growth with efficient nutrient uptake due to direct root contact with the nutrient solution, thus ensuring high water conservation as the system recirculates and reuses the nutrient solution [21].

2.3 Pollutants and Analysis Method

The hydroponics culture was induced with salinity and dye with varied concentrations. The salinity was introduced into the hydroponics system using sodium chloride (AR grade, Merck, India), and the most common dye pollutants such as Methylene Blue (MB) were from Loba Chemicals, India. The concentration of sodium chloride varied from 0 mg/L to 3000 mg/L, whereas the dye concentrations varied from 0 to 15 mg/L. The salinity of the water sample was measured by electrical conductivity probe and gravimetric method by measuring total solids, total suspended and total dissolved solids as per the APHA standard operating procedure. The dye concentrations were measured using the UV-Vis spectrophotometer at wavelengths of 660 nm.

2.4 Risk Assessment

The risk assessment of the pollutants on the plant species was evaluated using the hazard quotient (HQ), carcinogenic and non-carcinogenic health risk index (HRI). The uptake of the pollutant by the plants depends on many factors such as type of plant species, root-zone growth, properties of growth medium, and bioavailability of the metal in water phase. The daily uptake rate (D_{UR}) of the pollutant across the mint variety was calculated using the below equation:

$$D_{UR} = \frac{C_m}{T\ x\ BM_p} \quad (1)$$

where the average daily uptake rate of heavy metals is represented by D_{UR}, i represents that normalized by the dry biomass of lettuce (μg/day); concentration of heavy metals in the lettuce tissue is represented by C_m (μg/g); the total growth time was represented in T (days); and the dry plant biomass (g) was represented by BM_p.

The daily intake rate is the most vital parameter to determine the health risk of the pollutant and its exposure pathways. The plant or vegetable grown in hydroponics is ingested by humans or animals, hence daily intake rate (D_{IR}) in mg/kg/day is determined using the below formula:

$$D_{IR} = \frac{C_m\ x\ C_f\ x\ IR}{BW} \quad (2)$$

where the pollutant concentration in the hydroponics system is represented by C_m (g/kg); C_f is the conversion factor that converts fresh plant weight to dry weight (conversion factor considered in our study is 0.065); the daily intake of hydroponic mint species is represented by IR (g/day) and assumed to be 0.05 kg/day consumed by an adult; and the average body weight is taken as BW and assumed to be 70 kg for this study.

The health risk index can be calculated using the ratio of daily intake rate (D_{IR}) with the reference dose (R_{fD}). The expression for health risk due to the pollutant for a non-carcinogenic effect is

$$HRI = \frac{Daily\ Intake\ Rate\ (D_{IR})}{Reference\ Dose\ (R_{fD})} \quad (3)$$

HRI less than 1 indicates that the exposed population is safe from the health risk that comes from chemical consumption [7, 11, 22].

The targeted hazard quotient (THQ) is a ratio of pollutant concentration to oral reference dose, weighted by the amount and duration of exposure, intake rate, and body weight. The THQ parameter is a dimensionless indicator that expresses a level of concern but does not quantify risk. THQ < 1 implies that pollutant exposure through crop consumption is unlikely to result in apparent health impacts. THQ scores between 1 and 5 indicate that the exposed population faces a specific level of health risk concern. The THQ is computed by the formula:

$$THQ = \frac{EF\ x\ ED\ x\ IR\ x\ C\ x\ 10^{-3}}{R_{fD}\ x\ BW\ x\ TA} \quad (4)$$

where *EF* indicates the exposure frequency (days per year); exposure duration is expressed by *ED* (years, equivalent to an average lifetime); the vegetable ingestion rate is expressed by *IR* (g per person per day); concentration of metal in mint crop is represented by the term C (mg per kg); the average exposure time for non-carcinogens is cited as TA in days per year.

The average body weight (BW) of a human is different from region to region, and in the present study the average body weight of an adult was assumed to be 70 kg. The literature supports that average daily mint consumption by an adult was assumed to be 50 g annually. Furthermore, exposure frequency and duration were considered to be 104 days per year and 70 years of life expectancy, respectively [9, 11].

3 Results and Discussion

3.1 Plant Growth Monitoring

The growth of plants in the hydroponic system in saline environment was monitored at different phases and its observations were tabulated in Table 1. It was observed that the growth of mint crop showed very healthy growth up to 15 days and started to turn yellowish. Similarly, it was observed that the number of leaves and plants also declined its growth level after 15 days. In the dye environment, the plant growth showed a similar trend of results with declined growth after certain days (Table 2). As researchers stated that, the reason for such phenomenon is due to over concentration, reduced amount of nutrients, and dissolved oxygen levels [23, 24].

Table 1 Growth characteristics of the mint species in a saline environment

Days	Leaf colour	Average no of fresh leaves in plants	Total no of plants	Stem height (cm)
3	Green	2	6	4
6	Green	3	6	4
9	Green	5	6	6.5
12	Green	5	5	8.1
15	Greenish-yellow	6	5	8.4
18	Greenish-yellow	4	3	8.5
21	Pale yellow	3	2	8.5

Table 2 Growth characteristics of the mint species in a dye environment

Days	Leaf colour	Average no of fresh leaves in plants	Total no of plants	Stem height (cm)
3	Green	2	6	4
6	Green	3	6	4
9	Greenish-yellow	4	5	5.2
12	Pale yellow	4	4	5.8
15	Yellow	3	3	6.0

3.2 *Effect of Salinity*

The salinity in the hydroponics culture was introduced by adding sodium chloride (AR grade) into the distilled water. After the plant attained certain growth level, salinity was introduced to plants along with a nutrient medium. The salinity level in terms of total solids, total suspended solids, and total dissolved solids dropped gradually and reached a saturation level after 18 days. The total solids level from 4500 mg /L reduced to less than 2000 mg /L, while suspended solids reduced to 120 mg /L from initial concentration of 220 mg /L. The suspended solids concentration was very lesser value and it may be due to impurities present in the salts used for nutrient media, saline solution, and also due to open space experimental setup. The total dissolved solids (TDS) concentrations reduced from 3500 mg /L to nearly to 1000 mg /L (Fig. 2).

It implies that the plants are actively removing the saline pollution from the water and treating the water. Similarly, trend was observed by Puccinelli et al. [25] and Yolcu et al. [26] on the cultivated Sea beet and Wild Swiss chard beet species grown in the saline drainage water. The authors conclude that high salinity causes osmotic stress to the plants, affecting its growth in stages.

3.3 *Effect of Dye*

The methylene blue dye is one of the basic and most common dye forms. In the present study, the initial concentration of 15 mg /L has dropped as the time progressed. The concentration of methylene blue after 12 days started to reach saturation level and reached a concentration of 2 mg /L (Fig. 3). Similar trend was

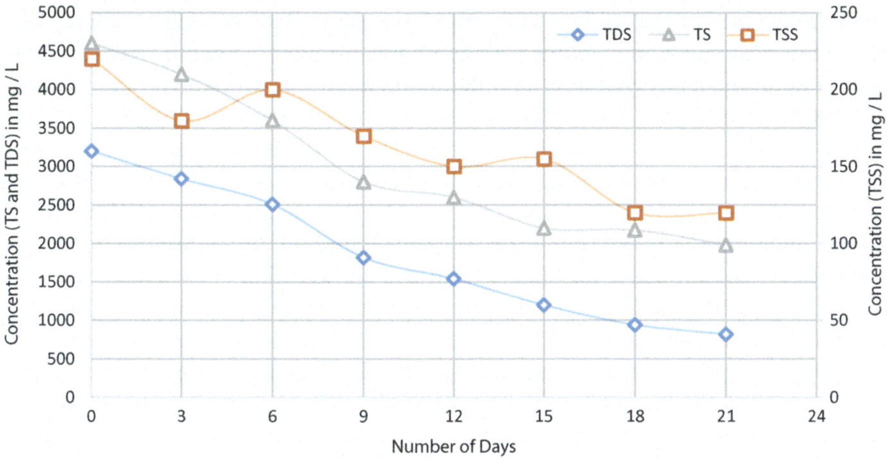

Fig. 2 Concentration of salinity in the hydroponics culture

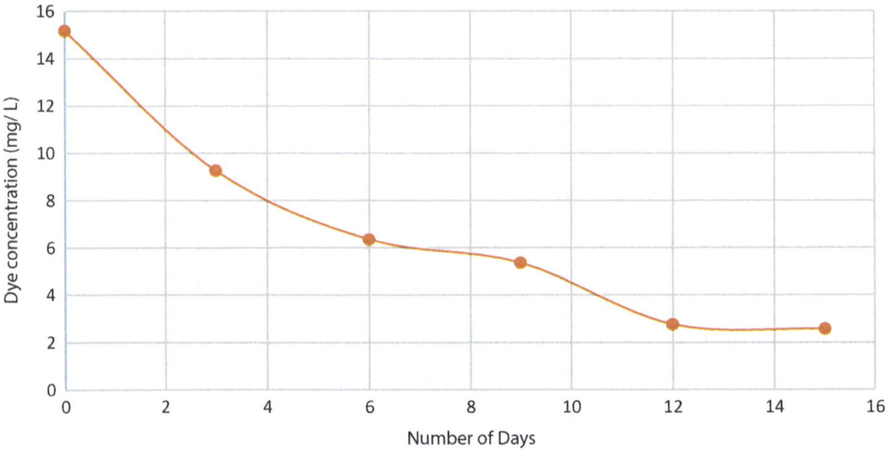

Fig. 3 Concentration of dye in the hydroponics culture

observed for the Tolfenpyrad (water-soluble insecticides) in romaine and edible amaranth grown in hydroponic conditions [27]. The pollutant concentration was reduced from the solvent and translocated into plant tissue such as stems, leaves, and roots. Garcia et al. [28] concluded that hydroponically grown lettuces absorb carbamazepine and diclofenac nearly to 90% and translocate them into plant tissues. The authors state that the translocation of pollutants onto plant root, stem, or leaf depends on the log K_{ow} value.

3.4 Risk Assessment

The ultimate aim of the study was to determine the pollutants such as salinity and dye and its accumulation in edible parts of mint constitutes a potential risk to human health due to their incorporation into the food chain. The worst exposure scenario was put-in-forth for the risk assessment evaluation considering the highest concentration of salt and dye uptake by mint species. Daily human uptake of salinity (in terms of total dissolved solids) was observed to be 0.708 mg / day, daily intake rate was estimated to be 2.76E-3 mg/kg/day. The human risk index (HRI) and total hazard quotient (HQ) for the salinity were found to be 5.26E-5 and 0.155, respectively. Additionally, the studies with dye showed that daily uptake rate was 5.25E-3 mg per day and daily intake rate was found to be 1.46E-6 mg/kg/day. The health risk index and hazard quotient were found to be 5.85E-5 and 1.26, respectively. It was cited by many researchers that HRI and HQ lesser than 1 represent less potential for risk to human [29, 30]. The risk assessment of various pesticides, PAEs, and heavy metals in four vegetable species grown hydroponically was studied by [31]. The study concludes that hazard index for the lead pollution on the plant species was in the range of 0.22 to 0.64, which implies exposure to these hazardous chemicals through consumption of the studied vegetables may not pose significant health risks. Hence, the study with salinity proved that there is no risk to plant species and the humans, whereas with the dye sample HQ was 0.8 and increased to 1.12 after the sixth day. It implies that dye showed a slight risk to plant as well as humans. Although the study concludes that consumption of mint grown in saline environment is safer, in dye it's not safe. It's a preliminary study result, a detailed investigation on the mechanism or fate of dye in plant species needs to be explored.

4 Conclusions

The results of the research show the mint crop's extraordinary resistance to salinity, demonstrating not only its ability to withstand high salt concentrations but also its capacity to actively lower them in its growth environment. With soil salinity still a persistent problem in many regions, threatening crop yields and food security, mint showed an impressive tolerance for salinity by thriving in an environment with salinity levels as high as 3000 mg/L. It actively contributed to lowering the level of salt in the medium, resulting in a notable reduction of roughly 72.7% after the period of 21 days. This finding has broad ramifications because it implies that mint crops may act as phytoremediation agents, improving soil quality and reducing salinity while supporting more environmentally friendly farming methods. Mint crops further demonstrate the value of their improvement of soil conditions by lowering electrical conductivity. It is important to note, though, that although mint crops can withstand salinity, when they were moved from the saline medium to a 15 mg / L methyl dye solution, their susceptibility to specific stresses was revealed. When this specific stressor was applied to the crop, its absorbance significantly decreased. This observation emphasizes how crucial it is to comprehend the crop's unique sensitivities when thinking about using it in various agricultural and remediation contexts.

Acknowledgement The authors would like to thank Prof. T. R. Rajagopalan research fund (Grant no: SASTRA-TRR-SOCE-1-12022024), an intra-mural start-up fund at SASTRA Deemed University, for their financial support.

References

1. Philip L, Ramprasad C, Krithika D (2019) Sustainable wastewater management through decentralized systems: case studies. In: Water scarcity and ways to reduce the impact: management strategies and technologies for zero liquid discharge and future smart cities, pp 15–45
2. Yadav RK, Sahoo S, Yadav AK, Patil SA (2023) Green wall system coupled with slow sand filtration for efficient greywater management at households. Npj Clean Water 6(1):73

3. Ravichandran MK, Philip L (2022) Fate of carbamazepine and its effect on physiological characteristics of wetland plant species in the hydroponic system. Sci Total Environ 846: 157337
4. Romeo D, Vea EB, Thomsen M (2018) Environmental impacts of urban hydroponics in Europe: a case study in Lyon. Procedia Cirp 69:540–545
5. Chadwick JJ, Witteveen A, Zhang P, Lynch I (2023) Hydroponics and alternative forms of agriculture: opportunities from nanotechnology. In: Nano-enabled sustainable and precision agriculture. Academic Press, pp 259–272
6. Yasmin H, Naeem S, Bakhtawar M, Jabeen Z, Nosheen A, Naz R et al (2020) Halotolerant rhizobacteria pseudomonas pseudoalcaligenes and Bacillus subtilis mediate systemic tolerance in hydroponically grown soybean (Glycine max L.) against salinity stress. PLoS One 15(4): e0231348
7. Carvalho RDSC, Bastos RG, Souza CF (2018) Influence of the use of wastewater on nutrient absorption and production of lettuce grown in a hydroponic system. Agric Water Manag 203: 311–321
8. Jung HY, Kim JK (2020) Complete reutilisation of mixed mackerel and brown seaweed wastewater as a high-quality biofertiliser in open-flow lettuce hydroponics. J Clean Prod 247: 119081
9. Lee E, Rout PR, Bae J (2021) The applicability of anaerobically treated domestic wastewater as a nutrient medium in hydroponic lettuce cultivation: nitrogen toxicity and health risk assessment. Sci Total Environ 780:146482
10. Nie E, Chen Y, Zhou X, Xu L, Zhang S, Li QX et al (2022) Uptake and metabolism of 14C-triclosan in celery under hydroponic system. Sci Total Environ 846:157377
11. Zhu Z, Yogev U, Keesman KJ, Rachmilevitch S, Gross A (2023) Integrated hydroponics systems with anaerobic supernatant and aquaculture effluent in desert regions: nutrient recovery and benefit analysis. Sci Total Environ 904:166867
12. Al-Karaki GN (2011) Utilization of treated sewage wastewater for green forage production in a hydroponic system. Emir J Food Agric 23:80–94
13. Lira RM, Silva ÊF, Silva GF, Soares HR, Willadino LG (2018) Growth, water consumption and mineral composition of watercress under hydroponic system with brackish water. Hortic Bras 36:13–19
14. Sambo P, Nicoletto C, Giro A, Pii Y, Valentinuzzi F, Mimmo T et al (2019) Hydroponic solutions for soilless production systems: issues and opportunities in a smart agriculture perspective. Front Plant Sci 10:923
15. Treftz C, Omaye ST (2016) Hydroponics: potential for augmenting sustainable food production in non-arable regions. Nutr Food Sci 46(5):672–684
16. Chatzigianni M, Savvas D, Papadopoulou EA, Aliferis KA, Ntatsi G (2023) Combined effect of salt stress and nitrogen level on the primary metabolism of two contrasting hydroponically grown Cichorium spinosum L. Ecotypes. Biomolecules 13(4):607
17. Barbafieri M, Bretzel F, Scartazza A, Di Baccio D, Rosellini I, Grifoni M et al (2023) Response to hypersalinity of four halophytes growing in hydroponic floating systems: prospects in the phytomanagement of high saline wastewaters and extreme environments. Plan Theory 12(9): 1737
18. Alexopoulos AA, Assimakopoulou A, Panagopoulos P, Bakea M, Vidalis N, Karapanos IC, Petropoulos SA (2021) Impact of salinity on the growth and chemical composition of two underutilized wild edible greens: Taraxacum officinale and Reichardia picroides. Horticulturae 7(7):160
19. Tola E, Al-Gaadi KA, Madugundu R, Patil VC, Sygrimis N (2023) Impact of water salinity levels on the spectral behavior and yield of tomatoes in hydroponics. J King Saud Univ Sci 35(2):102515
20. Li H, Cheng Z (2015) Hoagland nutrient solution promotes the growth of cucumber seedlings under light-emitting diode light. Acta Agriculturae Scandinavica, Section B—Soil Plant Sci 65(1):74–82

21. Holmes SC, Wells DE, Pickens JM, Kemble JM (2019) Selection of heat tolerant lettuce (Lactuca sativa L.) cultivars grown in deep water culture and their marketability. Horticulturae 5(3):50
22. León-Cañedo JA, Alarcón-Silvas SG, Fierro-Sañudo JF, de Oca GARM, Partida-Ruvalcaba L, Díaz-Valdés T, Páez-Osuna F (2019) Mercury and other trace metals in lettuce (Lactuca sativa) grown with two low-salinity shrimp effluents: accumulation and human health risk assessment. Sci Total Environ 650:2535–2544
23. Park JH (2020) Contrasting effects of Cr (III) and Cr (VI) on lettuce grown in hydroponics and soil: chromium and manganese speciation. Environ Pollut 266:115073
24. Sowmya RS, Warke VG, Mahajan GB, Annapure US (2022) Quality and shelf-life assessment of pesto prepared using herbs cultivated by hydroponics. Int J Gastron Food Sci 30:100608
25. Puccinelli M, Galati D, Carmassi G, Rossi L, Pardossi A, Incrocci L (2023) Leaf production and quality of sea beet (Beta vulgaris subsp. maritima) grown with saline drainage water from recirculating hydroponic or aquaculture systems. Sci Hortic 322:112416
26. Yolcu S, Alavilli H, Ganesh P, Panigrahy M, Song K (2021) Salt and drought stress responses in cultivated beets (Beta vulgaris L.) and wild beet (Beta maritima L.). Plan Theory 10(9):1843
27. Chen S, Ye Y, Liao F, Wu S, Zhang K (2024) Insight into the uptake, translocation, metabolism, dissipation and risk assessment of tolfenpyrad in romaine and edible amaranth grown in hydroponic conditions. Food Chem 437:137896
28. García MG, Fernández-López C, Pedrero-Salcedo F, Alarcón JJ (2018) Absorption of carbamazepine and diclofenac in hydroponically cultivated lettuces and human health risk assessment. Agric Water Manag 206:42–47
29. Goumenou M, Tsatsakis A (2019) Proposing new approaches for the risk characterisation of single chemicals and chemical mixtures: the source related Hazard Quotient (HQS) and Hazard Index (HIS) and the adversity specific Hazard Index (HIA). Toxicol Rep 6:632–636
30. Mary R, Nasir R, Alam A, Tariq A, Nawaz R, Javied S et al (2023) Exploring hazard quotient, cancer risk, and health risks of toxic metals of the Mehmood Booti and Lakhodair landfill groundwaters, Pakistan. Environ Nanotechnol Monit Manag 20:100838
31. Chen S, Yao C, Zhou J, Ma H, Jin J, Song W, Kai Z (2024) Occurrence and risk assessment of pesticides, phthalates, and heavy metal residues in vegetables from hydroponic and conventional cultivation. Food Secur 13(8):1151

Empowering Women in Agri-Startups for Sustainable Food Security

Ruqaiya Bano and Mariya Hasnat

Abstract The contribution of women to the advancement of agriculture has grown significantly. Women play a crucial role in guaranteeing food security and conserving local agro-biodiversity. Agriculture sector employs 80% of all economically active women; they comprise 33% of the agricultural labor force and 48% of self-employed farmers. Despite playing a significant role in various aspects of agriculture and food production, yet their contributions are often undervalued and underrepresented. As women are the primary participants in all stages of agricultural production, including pre-harvest, post-harvest processing, packaging, and marketing, it is crucial to implement gender-specific interventions to enhance agricultural productivity. Empowering women in agri-startups not only promotes gender equality but also has the potential to significantly enhance food security. This paper explores the crucial role of women in agri-startups for achieving sustainable food security, challenges faced by women in agri-startups, the benefits of their empowerment, and strategies for fostering their leadership in this sector. By addressing these issues, we can create a more inclusive and sustainable agricultural landscape, ultimately leading to improved food security for communities worldwide.

Keywords Women entrepreneurship · Agri-startups · Food security · Sustainable agriculture

R. Bano (✉)
Department of Agriculture, Integral Institute of Agricultural Science and Technology, Integral University, Lucknow, UP, India
e-mail: ruqaiyab@iul.ac.in

M. Hasnat
Department of Environmental Science, Integral University, Lucknow, UP, India

P. Pathak et al. (eds.), *Advances in Environmental Sustainability, Energy and Earth Science*, https://doi.org/10.1007/978-3-031-73820-3_27

1 Introduction

Entrepreneurship plays a crucial role in the social and economic integration of any country [1]. In rural areas, the entrepreneurial approach provides solutions to address the personal and family problems faced by rural residents, while also generating new prospects for entrepreneurial development. For women, entrepreneurship is a path out of poverty and of self-realization toward equality. Rural women's entrepreneurship can contribute to economic growth in developing countries and clearly represents an untapped potential [2, 3]. Women participation in entrepreneurial activities not only contributes to their family income but also plays a crucial role in fostering economic development and social well-being of the society [4]. India is the second-largest agricultural landholders with around 50% of Indians are making their livelihood from agriculture so the role of agri startups is so crucial in creating opportunities for people [5]. In developing countries, the entrepreneurship of women has the potential to contribute to economic growth, representing an untapped opportunity. Despite the increasing number of rural women running their own businesses, their socio-economic contributions and entrepreneurial potential often go unrecognized and underutilized. They are concentrated in informal, micro-size, low productivity, and low-return activities. It is essential to implement policies, services, and business environments that are responsive to gender and enable the startup and advancement of women's businesses. This support can help create decent and productive employment, promote gender equality, alleviate poverty, and build stronger economies and societies.

Women entrepreneurs play a substantial role in the development of the small business sector and drive enterprise in transition economies [6, 7]. The World Bank reports that India has the potential to boost its GDP growth by 1.5 percentage points by integrating 50% of women into the workforce. According to McKinsey Global Institute (MGI), India could increase its GDP by USD 0.7 trillion by engaging 68 million more women into India's workforce by 2025.

In recent years, there has been a growing recognition of the crucial role that women play in agriculture and food security. Women make up a significant portion of the agricultural workforce globally, yet they often face numerous challenges that limit their full potential in this sector [3]. Empowering women in agri-startups is not only a matter of gender equality but also a key strategy for achieving sustainable food security [8]. The study aimed to examine the barriers hindering women's participation in agri-startups and to explore the role of women in sustainable agriculture and food security. The strategies for empowering women in agri-startups to enhance food security have also been proposed.

2 Agri-Startups in India

Agri-startups in India are transforming the traditional agricultural landscape by introducing innovative solutions and leveraging technology to address various challenges faced by farmers. These startups are playing a crucial role in enhancing productivity, improving supply chain efficiency, and ensuring better market access for farmers. They are increasingly adopting technologies such as artificial intelligence, machine learning, the Internet of Things, and blockchain to offer precision farming solutions, crop monitoring, and predictive analytics. Many agri-startups focus on streamlining the supply chain from farm to fork, reducing wastage, and ensuring timely delivery of fresh produce. They employ digital platforms to connect farmers directly with consumers, retailers, and restaurants. Also, there is a growing emphasis on sustainable farming practices [9]. Startups are promoting organic farming, soil health monitoring, water conservation techniques, and the use of renewable energy sources [10]. Agri-fintech startups are providing farmers with access to credit, insurance, and financial planning tools, while digital marketplaces help farmers get better prices for their produce by connecting them with a larger pool of buyers [11, 12]. Innovations in biotechnology are also helping in the development of high-yield, pest-resistant crop varieties, including advancements in seed technology and bio-fertilizers. Several notable agri-startups are making significant strides in the industry, refer Fig. 1.

The impact of agri-startups on agriculture is profound. Precision farming and real-time monitoring have led to better crop management, resulting in higher yields and reduced input costs [13]. Digital platforms have democratized market access, allowing farmers to reach a wider audience and get better prices for their produce. Supply chain innovations and predictive analytics help in reducing post-harvest losses and ensuring that produce reaches consumers in optimal condition [14, 15]. Agri-fintech startups are making financial services more accessible to farmers, providing them with the necessary capital to invest in better farming practices. Emphasis on organic farming, soil health, and water conservation is

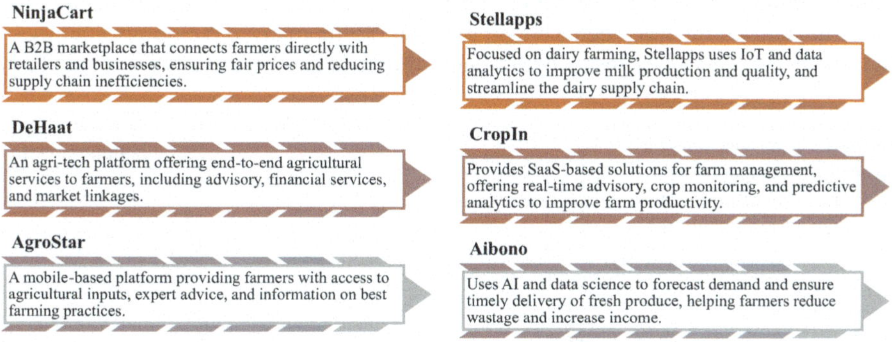

Fig. 1 Agri-startups in India

promoting sustainable agricultural practices, which are crucial for long-term food security.

Despite the promising advancements, agri-startups in India face several challenges, including infrastructure gaps, limited access to quality infrastructure in rural areas, and the need for continuous education and training programs to help farmers understand and utilize new technologies effectively. Navigating the regulatory landscape can be challenging for startups, especially in sectors such as biotechnology and financial services. Ensuring that solutions are scalable and can be tailored to the diverse needs of farmers across different regions is crucial for long-term success. To overcome these challenges, collaboration between startups, government agencies, and other stakeholders is essential. Initiatives such as the government's Startup India program and various state-level schemes aimed at promoting agri-innovation can provide the necessary support and resources for these startups to thrive. Agri-startups in India are revolutionizing the agricultural sector by introducing technology-driven solutions that address key challenges faced by farmers. Their continued growth and success are vital for achieving sustainable and inclusive agricultural development in the country.

3 Women in Agriculture: The Backbone of Food Production

Agriculture forms the lifeline of India's economy, employing nearly half of its workforce and contributing significantly to its GDP. Within this sector, women play an indispensable role, serving as the backbone of food production and rural development. Women have always been the backbone of agricultural production, contributing significantly to food production, processing, and marketing. From sowing seeds to harvesting crops, they have been involved in every step of the farming process. According to the Census of India 2011, women constitute about 33% of the agricultural labor force, but various studies and anecdotal evidence suggest this figure could be much higher, especially in rural areas where entire families engage in farming activities. Year-wise percentage of total female employment in agriculture has been shown in Table 1. In many developing countries, women make up the majority of the agricultural labor force, yet they often lack

Table 1 Year-wise data on women's participation in agriculture (% of total female employment in agriculture)

Year	Participation rate (%)
2018	58.8
2019	59.2
2020	58.7
2021	59.3
2022	58.9
2023	58.5

Source: World Bank

access to resources, technology, and markets. This limits their productivity and perpetuates the cycle of poverty and food insecurity.

According to the National Institution for Transforming India (NITI Aayog), there has been a notable increase in women agripreneurs. Initiatives such as the Mahila Kisan Sashaktikaran Pariyojana (MKSP) under the National Rural Livelihood Mission (NRLM) aim to empower women by providing training and resources for sustainable agricultural practices and entrepreneurship. Various schemes and policies have been introduced to promote women in agriculture. The Agricultural Technology Management Agency (ATMA) Scheme provides capacity-building opportunities and encourages women to take up leadership roles in Farmer Producer Organizations (FPOs). The number of female operational holders, as reported in the Agriculture Census 2015–2016, was about 11.72% of the total operational holders. This indicates a significant presence of women managing agricultural land [16]. A study by the Indian Council of Agricultural Research (ICAR) in 2020 showed that women constituted around 30% of the total rural entrepreneurs in agriculture-related enterprises [17]. Women agripreneurs play a crucial role in the rural economy by not only contributing to agricultural production but also creating job opportunities and driving innovation in the sector. For instance, women-led enterprises in areas such as organic farming, food processing, and agro-tourism have been on the rise [17].

4 Empowering Women in Agri-Startups

Empowering women in agri-startups in India is a transformative initiative that addresses various socio-economic barriers, fostering greater participation and success for women in the agricultural sector [18]. This initiative is crucial in a country where agriculture remains a significant source of livelihood, yet women often face restricted access to resources, training, and markets. By providing women with access to land, credit, technology, and inputs, their productivity and economic empowerment can be significantly enhanced. Furthermore, investing in training and capacity-building programs equips women with essential skills in sustainable farming practices, agribusiness management, and entrepreneurship, enabling them to take on leadership roles and drive innovation in agri-startups. Facilitating market access ensures that women farmers and agripreneurs receive fair prices for their produce and can engage in value addition and agro-processing. The integration of technology and innovation tailored to the needs of women farmers, such as mobile applications for market information and farm management, further enhances their efficiency and productivity. Enacting gender-responsive policies that support women's rights, access to finance, and extension services creates an enabling environment for their success. Thus, empowering women in agri-startups not only promotes gender equality but also contributes to the overall development of the agricultural sector and the rural economy in India [18].

Agri-startups can serve as catalysts for empowering women in agriculture by addressing these challenges and providing tailored solutions to meet the specific needs of women farmers and agripreneurs. These startups can create inclusive business models that prioritize the participation of women, provide mentorship and networking opportunities, and offer access to innovative technologies and market linkages. By engaging women as active participants in the design and implementation of agricultural innovations, agri-startups can ensure that their solutions are gender-responsive and effectively address the needs of women in agriculture.

5 Challenges Faced by Women in Agri-Startups

Women in agricultural startups may face several barriers that can impede their entry, participation, and success in the industry. These barriers can vary across regions and contexts, but some common challenges include:

5.1 Limited Access to Resources

Women often have limited access to crucial resources such as land, credit, and technology. This can make challenging for them to start and develop agricultural enterprises. Women may find it difficult to use land as collateral for loans due to cultural norms and laws related to land ownership. Financial institutions may exhibit bias toward male entrepreneurs which makes it challenging for women to secure funding for their business. Women may lack access to modern technology, which is essential for improving agricultural productivity and efficiency, due to financial constraints and insufficient training opportunities [19, 20].

5.2 Gender Stereotypes and Bias

Traditional gender roles and stereotypes can affect how women are perceived in the agricultural sector. Biases may exist that assume certain roles are more suitable for men, hindering women's participation in decision-making and leadership positions. Women may be perceived as less capable of managing the physical demands of farming or the strategic aspects of agribusiness. This can lead to a lack of support from male colleagues and stakeholders, making it challenging for women to assert themselves and achieve their entrepreneurial goals [19, 21].

5.3 *Education and Training Gaps*

In some cases, women may have limited access to education and training opportunities, which are essential for acquiring the skills and knowledge needed to succeed in agribusiness. Unequal access to education and training opportunities can impede women from acquiring the necessary skills and knowledge for success in agribusiness. Women might not have the same opportunities as men to attend agricultural workshops, or extension services, leaving them less informed about the latest farming techniques, market trends, and business management practices [20, 21].

5.4 *Lack of Networking Opportunities*

Networking is crucial in the business world, and women may face challenges in building professional networks due to cultural, social, or systemic factors. Women might be excluded from formal and informal networking events dominated by men, limiting their ability to form valuable business connections and partnerships. Additionally, social norms in some cultures may restrict women's participation in public life, further isolating them from potential mentors, investors, and collaborators.

5.5 *Unequal Access to Markets*

Women entrepreneurs may encounter difficulties in accessing agricultural markets. This could be due to discriminatory practices, lack of market information, or challenges in transportation and logistics. For example, women might not have the same access to market information as men, making it harder for them to identify profitable opportunities and negotiate fair prices. Logistical challenges such as inadequate transportation infrastructure can disproportionately affect women, who may have less mobility and fewer resources to overcome these barriers.

5.6 *Limited Representation in Leadership Roles*

The underrepresentation of women in leadership positions within agricultural startups and related organizations can contribute to a lack of diverse perspectives and hinder the development of supportive policies. Women's perspectives are essential for creating inclusive policies and practices that address the unique challenges they face in the agricultural sector. Without sufficient representation in leadership roles, the development and implementation of such policies may be inadequate, perpetuating the existing barriers.

5.7 Inadequate Financial Support

Limited access to finance and credit is a significant barrier for women entrepreneurs in agriculture. They may face challenges in securing loans or investment capital, which can affect the growth and sustainability of their startups. Financial institutions may have stricter lending criteria for women, or women may lack the necessary collateral to secure loans. This financial constraint limits their ability to invest in necessary resources, technologies, and expansion opportunities [22, 23].

5.8 Work–Life Balance

Work responsibilities with traditional gender roles, including caregiving and household duties, can be a significant challenge for women in agribusiness. The dual burden of managing a business and fulfilling family obligations can lead to physical and emotional stress, reducing their productivity and effectiveness in their entrepreneurial endeavors. This challenge is exacerbated in societies where women are expected to prioritize household duties over professional ambitions [24, 25].

5.9 Lack of Mentorship and Role Models

The absence of female mentors and role models in the agricultural sector can make it difficult for aspiring women entrepreneurs to find guidance and support. Mentors and role models provide valuable advice, encouragement, and networking opportunities, helping new entrepreneurs navigate the complexities of the agribusiness landscape. The scarcity of such figures can leave women feeling isolated and unsure of how to overcome the challenges they face.

5.10 Policy and Regulatory Barriers

Some policies and regulations may inadvertently disadvantage women in agribusiness. Land tenure laws or inheritance practices may disproportionately affect women's access to and control over land. Policies that do not consider the unique needs and circumstances of women can create additional hurdles, such as complex bureaucratic processes or requirements that are difficult for women to meet due to existing gender inequalities [26, 27].

Addressing these barriers requires a multifaceted approach involving policy changes, increased access to education and resources, efforts to challenge gender stereotypes, and the promotion of inclusivity in the agricultural sector. Supporting women in agribusiness not only benefits individual entrepreneurs but also contributes to the overall development and sustainability of the agricultural industry.

6 Strategies for Empowering Women in Agri-Startups

Empowering women in agri-startups is a multifaceted endeavor that involves addressing various barriers that hinder women's participation and success in the agricultural sector. This can be achieved through several key strategies:

6.1 Access to Resources

Implement policies and initiatives that ensure equitable access to resources, including land, credit, and technology, for women in agribusiness. Providing women with access to land, credit, inputs, and technology is essential for enhancing their productivity and economic empowerment. Agri-startups can play a crucial role in facilitating access to these resources and providing tailored support to women farmers [19, 20].

6.2 Training and Capacity Building

Investing in training and capacity-building programs for women in agriculture can enhance their skills in sustainable farming practices, agribusiness management, and entrepreneurship. These programs can cover a range of topics from modern farming techniques to financial literacy and business management. By equipping women with these skills, they can take on leadership roles within agri-startups and drive innovation in the sector. Training programs should be designed to be accessible to women, considering their time constraints and educational backgrounds [20, 21].

6.3 Market Access

Facilitating market access for women farmers and agripreneurs is vital for ensuring that they receive fair prices for their produce and have opportunities to engage in value addition and agro-processing. Agri-startups can serve as platforms for connecting women farmers to markets and creating market linkages for their products. Additionally, cooperatives and farmer groups can be instrumental in negotiating better prices and accessing larger markets.

6.4 Networking

Establishing mentorship programs and networking opportunities is a crucial strategy for empowering women entrepreneurs in agri-startups. By connecting women with experienced mentors and industry professionals, these programs provide invaluable guidance, support, and knowledge sharing. Mentorship can help women entrepreneurs navigate the complexities of the agricultural sector, offering insights into best practices, innovative techniques, and effective business strategies. Additionally, networking opportunities create platforms for women to build relationships with key stakeholders, including investors, suppliers, and potential customers. These connections can open doors to new markets, partnerships, and funding sources, thereby enhancing the growth and sustainability of their ventures. Through these networks, women entrepreneurs can also gain visibility, share their success stories, and advocate for policies that support gender equality in agriculture [15].

6.5 Financial Support

Developing financial mechanisms and support systems that address the specific needs of women in agri-startups is crucial for fostering their success and growth. Women often face unique challenges in accessing financial resources, such as collateral requirements and gender bias in lending practices. To address these barriers, it is essential to create targeted financial products, including low-interest loans, microfinance options, and grant programs specifically designed for women entrepreneurs in agriculture. Establishing investment funds and venture capital opportunities dedicated to supporting women-led agri-startups can provide the necessary capital for scaling their businesses. Financial literacy programs and advisory services can further equip women with the knowledge and skills to effectively manage their finances, make informed investment decisions, and navigate the financial landscape. By implementing these comprehensive financial support systems, women can empower to overcome economic constraints, drive innovation, and contribute significantly to the agricultural sector's development and sustainability [28].

6.6 Technology and Innovation

Using technology and innovation can significantly enhance the productivity and efficiency of women in agriculture. Agri-startups can develop and promote technologies tailored to the needs of women farmers, such as mobile applications for market information, weather forecasting, and farm management. These technologies can provide women with real-time information and tools that improve their decision-making and farming practices. Additionally, innovations in agricultural equipment designed to be more user-friendly for women can also contribute to increased productivity [29].

6.7 Gender-Inclusive Education and Training

Promoting gender-inclusive education and training programs is essential for empowering women with the skills and knowledge necessary for success in agri-startups. These programs must be designed to address the specific learning needs and challenges faced by women in agriculture. This includes offering training in sustainable farming practices, agribusiness management, financial literacy, and technological proficiency. Gender-inclusive education ensures that women have equal opportunities to access these resources and develop their capabilities. By creating an inclusive learning environment, these programs also encourage peer support and networking, which are crucial for personal and professional growth. Gender-inclusive education and training not only enhance the productivity and efficiency of women in agri-startups but also contribute to the broader goal of gender equality in the agricultural sector.

6.8 Advocacy and Awareness

Conducting advocacy campaigns to challenge gender stereotypes and biases is crucial for fostering a more inclusive environment for women in agriculture. These campaigns should highlight the contributions of women in agriculture and the importance of gender equality for sustainable development. Engaging men and community leaders in these efforts is also important to change societal attitudes and support gender equality. Advocacy efforts can also promote policies and practices that support women's rights and participation in the agricultural sector.

6.9 Policy Support

Enacting policies that promote gender equality in agriculture and provide support for women in agri-startups is essential for creating an enabling environment. This includes measures to address land rights, access to finance, and gender-responsive extension services. Governments and policymakers should work toward removing legal and institutional barriers that limit women's participation in agriculture. Supportive policies can also include subsidies, tax incentives, and other forms of assistance to encourage women's involvement in agri-startups [29].

7 Conclusions

This paper highlighted the importance of empowering women in agri-startups as a strategy to enhance sustainable food security in India. By addressing gender disparities and leveraging the potential of agri-entrepreneurship, the agricultural sector can not only become more resilient but also contribute significantly to the overall development of the nation. Embracing a holistic approach to empower women in agri-entrepreneurship is not just a matter of equality but a strategic imperative for achieving sustainable food security. By fostering an enabling environment that addresses the diverse challenges faced by women in the sector, we can unlock their full potential as drivers of innovation, resilience, and inclusivity in agri-startups, contributing to a more sustainable and secure global food system. Policymakers, industry stakeholders, and civil society must collaborate to create an enabling environment that fosters the empowerment of women in agribusiness.

References

1. Stam E, Spigel B (2017) Entrepreneurial ecosystems. Blackburn, Thousand Oaks, CA
2. Akehurst G, Simarro E, Mas-Tur A (2012) Women entrepreneurship in small service firms: motivations, barriers and performance. Serv Ind J 32(15):2489–2505
3. Ghouse S, McElwee G, Meaton J, Durrah O (2017) Barriers to rural women entrepreneurs in Oman. Int J Entrep Behav Res 23(6):998–1016
4. Sajjad M, Kaleem N, Chani MI, Ahmed M (2020) Worldwide role of women entrepreneurs in economic development. Asia Pac J Innov Entrep 14(2):151–160
5. Verma RK, Sahoo AK, Rakshit S (2017) Opportunities in agripreneurship in India: need challenges and future prospects. Rashtriya Krishi 13(1):71–72
6. Lerner M, Brush C, Hisrich R (1997) Israeli women entrepreneurs: an examination of factors affecting performance. J Bus Ventur 12(4):315–339
7. Naser K, Rashid Mohammed W, Nuseibeh R (2009) Factors that affect women entrepreneurs: evidence from an emerging economy. Int J Organ Anal 17(3):225–247
8. Maleki Fard F, Ghanbari Movahed R, Pakravan Charvadeh M (2021) Investigating the effect of rural women's entrepreneurship on households' sustainable food security (case study: Khorramabad township). J Rural Res 12(2):274–289
9. Youngberg G, Harwood R (1989) Sustainable farming systems: needs and opportunities. Am J Altern Agric 4(3–4):100–100
10. Bügel PCF (2020) Sustainability in agriculture: a qualitative study of how agtech startups in Norway contribute (Master's thesis). Norwegian University of Life Sciences, Ås
11. Chaudhary S, Suri PK (2024) Agri-tech: experiential learning from the Agri-tech growth leaders. Tech Anal Strat Manag 36(7):1524–1537
12. More A, Aslekar A (2022) Role of ICT & Fintech in Indian agriculture. In: 2022 international conference on Decision Aid Sciences and Applications (DASA). IEEE, pp 900–904
13. Gawande V, Saikanth DRK, Sumithra BS, Aravind SA, Swamy GN, Chowdhury M, Singh BV (2023) Potential of precision farming technologies for eco-friendly agriculture. Int J Plant Soil Sci 35(19):101–112
14. Bustos CA, Moors EH (2018) Reducing post-harvest food losses through innovative collaboration: insights from the Colombian and Mexican avocado supply chains. J Clean Prod 199: 1020–1034

15. Kelley DJ, Brush CG, Greene PG, Litovsky Y (2013) Global entrepreneurship monitor 2012 Women's report. Global Entrepreneurship Research Association (GERA)
16. PIB (2021) Women farmers in the country. https://www.pib.gov.in/PressReleaseIframePage.aspx?PRID=1780263
17. Government of India (2023) Agricultural statistics at a glance 2022. Department of Agriculture, Cooperation & Farmers Welfare, Ministry of Agriculture & Farmers Welfare. https://desagri.gov.in/wp-content/uploads/2023/05/Agricultural-Statistics-at-a-Glance-2022.pdf
18. Devi HN, Halim RA, Deka N, Naresh H (2023) Economic analysis of women Agri startups in Manipur. Econ Aff 68(04):1911–1919
19. Ajadi AA, Oladele OI, Ikegami K, Tsuruta T (2015) Rural women's farmers access to productive resources: the moderating effect of culture among Nupe and Yoruba in Nigeria. Agric Food Secur 4(1):1–9
20. Azumah FD, Onzaberigu NJ, Adongo AA (2023) Gender, agriculture and sustainable livelihood among rural farmers in northern Ghana. Econ Chang Restruct 56(5):3257–3279
21. Shekhar M, Singh N (2023) Women farmers, constraints and policy around them to harness maximum benefit. In: Shekhar M, Singh N (eds) Women farmers: unheard being heard. Springer Nature Singapore, Singapore, pp 37–53
22. Makena P, Kubaison ST, Njati CI (2014) Challenges facing women entrepreneurs in accessing business finance in Kenya: case of Ruiru township, Kiambu County. J Bus Manag 16(4):83–91
23. Ghosh PK, Ghosh SK, Chowdhury S (2018) Factors hindering women entrepreneurs' access to institutional finance-an empirical study. J Small Bus Entrep 30(4):279–291
24. Lakshmi N, Prasanth VS (2018) A study on work-life balance in working women. Int J Adv Multidiscip Sci Res 1(7):76–88
25. Sundaresan S (2014) Work-life balance–implications for working women. OIDA Int J Sustain Dev 7(7):93–102
26. Crain WM, Crain NV (2020) Regulatory barriers to women in business. Bus Public Adm Stud 14(2):15–21
27. Adnane S (2021) Women's economic participation and the impact of regulatory barriers. World Bank
28. Panda SM (2016) Constraints faced by women farmers in access to agricultural credit: a case study of India. Agric Econ Res Rev 29(2):267–274
29. Akinwale YO, Grobler WC (2023) Research and technology innovation, food security and economic growth in Nigeria: implications for agripreneurs and policymakers. Afr J Food Agric Nutr Dev 23(4):23220–23237

Uncovering the Impact of Forest Resource Extraction on Mammalian Communities in Selected Wildlife Sanctuaries of Uttar Pradesh, India

Azram Tahoor, Azra Musavi, and Jamal Ahmad Khan

Abstract The removal of forest resources, driven by increased human demand for timber, fuelwood, and other natural commodities, has significant repercussions on forest ecosystems and their inhabitants. Katarniaghat (400 sq.km.) and Kaimoor (500 sq.km.) Wildlife Sanctuaries situated in Utttar Pradesh were selected to carry out the study. Depending on the reconnaissance survey regarding the anthropogenic disturbance variables such as human trails, grazing cover, weed cover, lopping density, and fire, the protected area was categorized into high, medium, and low disturbed sites. For data collections, line transect method (direct sighting) and pellet group method (indirect evidence) were used to record mammals. Circular plot of 10 m radius was used to evaluate and quantify habitat variables. The data were analyzed using Kruskal-Wallis test using SPSS comparing density, abundance, area preference, mean group size, and encounter rate of the species. The low disturbed areas of Katarniaghat and Kaimoor showed higher values for overall mean pellet group density (13.26 ± 2.01 and 17.84 ± 1.6), abundance (0.007 ± 0.001 and 0.008 ± 0.001), area preference (0.52 ± 0.07 and 0.25 ± 0.01), encounter rate (2.9 ± 0.43 and 1.26 ± 0.22), and mean group size (2.9 ± 0.43 and 2.6 ± 0.5) of mammals, respectively. However, some species showed affinity toward human disturbance, for example, Nilgai, Jackal, and Langur, which showed their coexistence with human interference. The present study aims to contribute to a better understanding of the complex interactions between forest resource extraction and mammalian communities, offering insights to inform sustainable management practices, conservation efforts, and policy formulation in forested landscapes.

A. Tahoor (✉)
Department of Environmental Science, Integral University, Lucknow, India
e-mail: azram@iul.ac.in

A. Musavi
Advanced Centre of Women's Studies, Aligarh Muslim University, Aligarh, India

J. A. Khan
Department of Wildlife Sciences, Aligarh Muslim University, Aligarh, India

P. Pathak et al. (eds.), *Advances in Environmental Sustainability, Energy and Earth Science*, https://doi.org/10.1007/978-3-031-73820-3_28

Keywords Mammalian community · Katarniaghat · Kaimoor · Line transect · Pellet group method

1 Introduction

In recent years, wildlife conservation has become a significant concern, largely due to the settlement of residents and their reliance on forest resources for sustenance and livelihoods. The dependence on forest resources, including illegal grazing of livestock, collection of fuelwood, logging, and non-timber forest extraction, brings detrimental effects such as habitat loss, which in turn impacts wildlife populations. As human populations grow and expand, the pressure on these ecosystems intensifies, leading to habitat loss, fragmentation, and degradation, all of which threaten the survival of numerous species. Additionally, the socio-economic dynamics of residents play a significant role in shaping their relationship with wildlife and natural resources. Poverty, lack of alternative livelihood options, inadequate access to education and healthcare, and limited awareness about conservation issues can further contribute to unsustainable resource use practices and human-wildlife conflicts. Consequently, the quality of the forest habitat has a notably positive influence on the species richness of wildlife [1]. The positive association between forest habitat quality and wildlife species richness underscores the importance of conserving and restoring natural habitats to sustain biodiversity and ecosystem functioning. The investigation of numerous studies explored the detrimental effects of forest resource extraction on mammals [2–9].

Human activities determine the magnitude and reasons behind the dispersal of wild species [2], owing to ecological distribution in habitat utilization between wildlife and domestic [1]. The wildlife population is influenced by the intensity of human use, which can limit food supplies [6, 10]. However, ecological factors such as vegetation also play role in shaping wildlife populations affected by human activities [11, 12]. Few studies have shown a positive correlation between wildlife and human-induced disturbance, demonstrating that primates and elephants can tolerate logging and survive in logged forests [13, 14], and ungulates prefer open habitats resulting from logging and grazing [15].

The study aims to understand how the level of reliance on forest resources by local communities affects mammal populations within protected areas. The study compared the density, abundance, area preference, encounter rate, and mean group size of mammals across three sites categorized by the magnitude of human disturbance. The present study can have needful measures for conservation policies and strategies, helping to inform efforts to balance the needs of both human communities and wildlife within protected areas. Therefore, by examining the relationship between forest resource dependence and wildlife, this study aims to enhance understanding of how the magnitude of human activities influences mammal density, thereby contributing to conservation biology.

2 Study Area

To carry out the study, two wildlife sanctuaries were selected from Uttar Pradesh, India.

2.1 Katarniaghat Wildlife Sanctuary

The sanctuary is situated in the Bahraich district of Uttar Pradesh, India, with GPS location 28°06'N & 28°24'N latitude and 81° 02É 81° 19′É longitude (Fig. 1), having an area of 400 sq.km. The sanctuary lies on the alluvium soil of upper Gangetic plain. The climate of this area are summer, winter, and monsoon. The rivers present here are: Girwa, Kauriayala, Ghaghra, and Saryu. The forest type present in the sanctuary is tropical moist deciduous type. There are three types of habitats present in the sanctuary: Sal forest, Miscellaneous forest, and Grasslands [3]. Major vegetation present in the area are *Shorea robusta, Aegle marmelos, Schleichera oleosa, Adina cordifolia, Tectona grandis, Saccharum munja,*

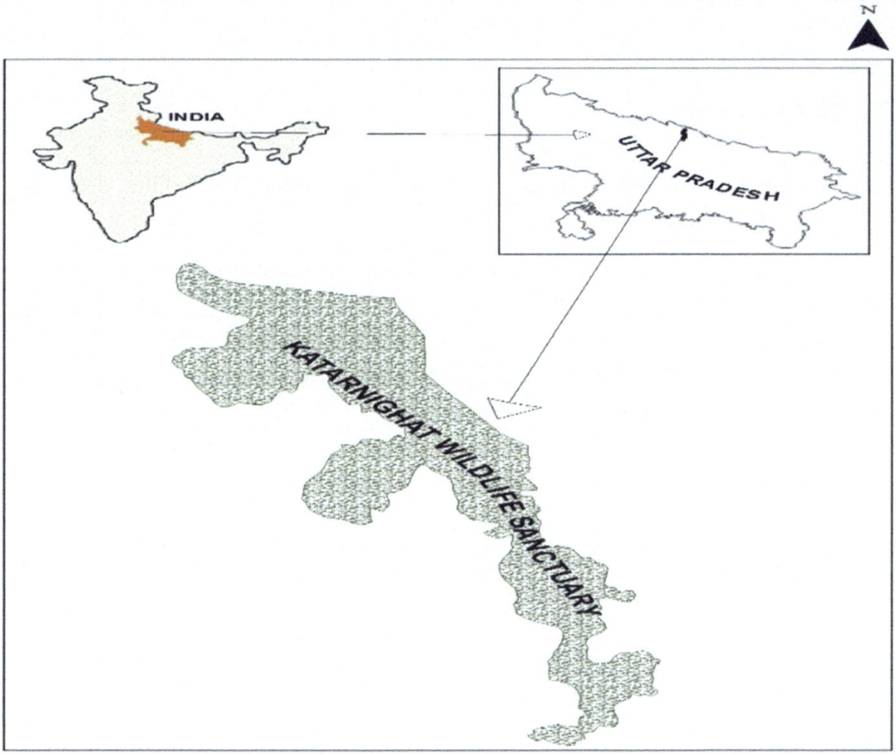

Fig. 1 Katarniaghat Wildlife Sanctuary, Inset (Uttar Pradesh)

Phragmites karka, and *Calamus tenuis.* The sanctuary is accompanied by wide variety of faunal diversity. Inside the sanctuary, six villages are present whereas more than hundred villages are present at the peripheries [16].

2.2 *Kaimoor Wildlife Sanctuary*

The sanctuary is situated in Mirzapur and Sonbhadra district of Uttar Pradesh, India, with GPS location 82°20'15″E to 24°52'00"N and 82°08'23″E to 24°27'51"N (Fig. 2), having an area of 500 sq.km. The sanctuary lies on red clay and is an undulating hilly terrain. The whole area has three seasons: summer, winter, and monsoon. The rivers present here are Son and Belan. The area by dry deciduous type of forest (Champion's and Seth Classification, 1968) and is divided into three major habitats: Sal forest, Bamboo forest, Scrub forest, and Deciduous forest. The floral species present in the sanctuary are: *Madhuca indica, Buchanania lanzan, Diospyros melanoxylon, Nyctanthes arbor-tristis,* and *Cynodon dactylon.* A wide variety of mammals are also present in the sanctuary. Eighteen villages are inside the sanctuary having human and livestock populations of 30,350 and 22,004, respectively [4].

3 Methodology

The methodology section is divided into the demarcation of sampling sites, data collection, and its analysis.

Fig. 2 Kaimoor Wildlife Sanctuary, Inset (Uttar Pradesh)

3.1 Sampling Sites

To delineate the sampling sites, a reconnaissance survey was conducted to identify areas affected by human-induced disturbances. Indicators such as human trails, the presence of cow dung, cut or lopped trees, and weed cover were used to assess the extent of disturbance. Based on discussions with residents, local guides, forest personnel, and personal observations, three sites were chosen in both the sanctuaries. These sites were compared when characterized as high disturbed (experiencing significant anthropogenic influence), medium disturbed, and low disturbed (with minimal or no human interference). The designations of high, medium, and low accurately represent varying degrees of anthropogenic disturbance.

In the three characterized sites, transects were established, with circular plots (10 m radius) placed at an interval of 200 m. In Katarniaghat, 9 transects, each 1 km in length, with 54 circular plots were evenly distributed. In Kaimoor, 6 transects, each 2 km in length, with 66 circular plots were laid out evenly across the three selected sites. Disturbance indicators were measured once the sites were selected.

Each stratified site was discovered to have four specific indications of human disturbance. The lopping of tree species was assessed on a scale of 0 to 4, with no lopping and the complete tree being chopped down. Human trails were evaluated on a similar scale, based on the quantity and width of trails. Weed and grazing cover were also scored on a scale of 0 to 4, signifying the percentage of land affected by grazing. The values of these disturbance variables were added for each circular plot to produce a composite index value, which was then averaged across stratified sites. These averaged indices made it easier to categorize areas as highly, moderately, or low disturbed.

Based on the average values of the distinct human disturbance variables, the areas were identified as high, medium, and low disturbed areas within both sanctuaries. Because low disturbed regions in both sanctuaries showed a low level of disturbance, places with no disturbance were not chosen for the study. Data collection commenced promptly after sample site selection.

3.2 Data Collection

To gather data on mammal abundance and density across the sample sites, two distinct methods were employed. These methods include the line transect for direct observation and the pellet group count for indirect evidence [17–19]. The pellet group count method undergone refinements, becoming widely favored [19], and is effective for species not readily observable during transect walks, resulting in low density counts [18]. Rain-related pellet deterioration, increased dung beetle activity, and pellet shifting present difficulties during rainy seasons. In this procedure, animal density and abundance were estimated using pellet groups of different species in the circular plot used for floral assessment.

Transect lines were drawn inside each categorized site in both protected areas to allow for direct observation. Every transect was observed by the same person, guaranteeing consistency. Excessive 500-meter gaps between transects were used to offset possible biases in data collecting. Transect monitoring in Katarniaghat and Kaimoor covered a total of 108 km and 144 km, respectively (each repeated four times). Transects' lengths varied within each protected area because of logistical limitations, but their number and length were the same for all three classified locations. Both species and individual counts were noted during each transect walk.

3.3 Data Analysis

The data gathered from both methods were utilized to estimate the density, abundance, area preference, encounter rate, and group size of mammals using the following formulas:

Density = Number of pellets/Plot area.

where plot area = πr^2.

Abundance = Total number of pellets present in the plot/Total number of plots sampled.

Area preference = (Pellets present in each area/Total pellets present in the three categorized sites) * 100.

Encounter rate = Sightings of a species/transect length.

Group size = Number of individuals present in a group/transect length.

To assess the significant differences among the three categorized sites, the Kruskal-Wallis test, a rank-based non-parametric test, was employed. Statistical Package for Social Sciences (SPSS) version 17.0 was utilized for data analysis.

4 Results

4.1 Density Estimation

Table 1 displays extensive information on the mean pellet group densities and matching Chi-square values of recorded mammal species in sampled plots of Katarniaghat Wildlife Sanctuary. The Chital had the highest pellet group density in low disturbed areas (7.6 ± 1.4), followed by medium and high disturbed parts of the sanctuary (3.53 ± 0.96 and 2.06 ± 1.04, respectively). Similarly, the Nilgai had the highest mean pellet group density in low disturbance areas, followed by high and medium disturbed areas. Pellet group densities were highest in low disturbed portions of the sanctuary (3.24 ± 0.9 for Hare, 3.24 ± 0.9 for Elephant, and 7.6 ± 1.3 for Wild Boar). Overall, Chital had the highest mean pellet group density, followed by Wild Boar (4.4 ± 0.6 and 3.24 ± 0.5). The overall average pellet group density of mammals was highest in the low disturbed area of the sanctuary (13.26 ± 2.01). The results of the Kruskal-Wallis test indicated a significant

Table 1 Mean pellet group density of mammalian species

Sanctuary name	Species	High	Medium	Low	Overall	Chi-square (degree of freedom), Significance level
Katarniaghat	Chital	2.06 ± 1.04	3.53 ± 0.96	7.6 ± 1.4	4.4 ± 0.6	16.6 (2), p < 0.01
	Barking deer	0	0	2.9 ± 0.8	0.9 ± 0.3	20.57 (2), p < 0.01
	Nilgai	1.17 ± 0.6	0.58 ± 0.4	4.4 ± 1.06	2.06 ± 0.43	14.9 (2), p < 0.05
	Hare	0	0.3 ± 0.29	3.24 ± 0.9	1.17 ± 0.3	19.1 (2), p < 0.05
	Elephant	2.06 ± 0.75	0.8 ± 0.5	3.83 ± 1	2.25 ± 0.4	7.1 (2), p < 0.05
	Wild boar	1.5 ± 0.64	0.6 ± 0.4	7.6 ± 1.3	3.24 ± 0.5	34.5 (2), p < 0.05
	Overall	5.3 ± 1.8	5 ± 1.12	13.26 ± 2.01	7.8 ± 1	18.1 (2), p < 0.01
Kaimoor	Jackal	2.17 ± 0.7	0.24 ± 0.24	0.24 ± 0.24	0.9 ± 0.26	13.9 (2), p < 0.05
	Chital	0	0	2.9 ± 0.8	1 ± 0.3	24.7 (2), p < 0.01
	Sambar	0	0	2.9 ± 0.8	0.9 ± 0.27	24.68 (2), p < 0.01
	Nilgai	7.5 ± 1.1	5.5 ± 1.6	6.7 ± 1.1	6.6 ± 0.75	6.3 (2), p < 0.05
	Blackbuck	1.5 ± 0.6	1.7 ± 0.6	0	1.04 ± 0.3	7.86 (2), p < 0.05
	Chinkara	0	1 ± 0.5	0	0.32 ± 0.1	8.06 (2), p > 0.05
	Wild boar	0.96 ± 0.67	0.5 ± 0.5	1.44 ± 0.82	0.96 ± 0.38	3.2 (2), p > 0.05
	Hyena	0.96 ± 0.67	0.01 ± 0.01	0	0.38 ± 0.22	4.1 (2), p > 005
	Overall	17.4 ± 1.8	10.7 ± 2	17.84 ± 1.6	15.31 ± 1.03	16.25 (2), p < 0.01

Source: Tahoor A (2017)

difference in the mean pellet group density of mammals recorded in high, medium, and low disturbed areas of the sanctuary, as depicted in Table 1.

The Kaimoor Wildlife Sanctuary's high disturbed areas had the highest jackal scat density, which was followed by medium and low disturbed areas (2.17 ± 0.7, 0.24 ± 0.24, and 0.24 ± 0.24, respectively). Similar results were observed for Nilgai. Wild Boar, on the other hand, had the highest mean pellet group density in low disturbed parts of the sanctuary (1.44 ± 0.82, 0.96 ± 0.67, and 0.5 ± 0.5, respectively), followed by high and medium disturbed areas. Meanwhile, Blackbuck showed the highest mean pellet group density in the areas facing medium human disturbance. Overall, Nilgai had the highest mean pellet density in the sanctuary (6.6 ± 0.75), whereas Chinkara had the lowest value (0.32 ± 0.1). The sanctuary's low impacted region had the highest mean pellet group density of mammals (7.8 ± 1.6). The Kruskal-Wallis test found a significant difference in pellet group densities of recorded mammals across the sanctuary's high, medium, and low disturbed regions.

4.2 Abundance Estimation

Table 2 depicts the mean abundance of reported pellets in sampled plots of Katarniaghat Wildlife Sanctuary. The sanctuary's least affected sections had the

Table 2 Mean abundance estimation of mammalian species [20]

Sanctuary name	Species	High	Medium	Low	Overall	Chi-square (degree of freedom), Significance level
Katarniaghat	Chital	0.001 ± 0.0	0.002 ± 0.0	0.004 ± 0.0	0.002 ± 0	16.6 (2), p < 0.05
	Barking deer	0	0	0.002 ± 0.0	0.001 ± 0.0	20.57 (2) p < 0.01
	Nilgai	0.001 ± 0.0	0.003 ± 0.0	0.002 ± 0.0	0.001 ± 0.0	14.9 (2), p < 0.05
	Hare	0	0.002 ± 0.0	0.001 ± 0.0	0.001 ± 0.0	19.1 (2), p < 0.01
	Elephant	0.001 ± 0.00	0.005 ± 0.0	0.002 ± 0.0	0.001 ± 0.0	7.1 (2), p < 0.01
	Wild boar	0.001 ± 0.0	0.003 ± 0.0	0.04 ± 0.0	0.001 ± 0.0	34.5 (2), p < 0.01
	Overall	0.003 ± 0.001	0.002 ± 0.001	0.007 ± 0.001	0.004 ± 0.001	18.1 (2), p < 0.01
Kaimoor	Chital	0	0	0.001 ± 0.0	0.0 ± 0.0	24.7 (2), p < 0.01
	Sambar	0.0 ± 0.0	0.002 ± 0.0	0.001 ± 0.0	0.001 ± 0.0	11.4 (2) p < 0.05
	Nilgai	0.004 ± 0.0	0.002 ± 0.001	0.003 ± 0.0	0.003 ± 0.0	13.5 (2), p < 0.05
	Jackal	0.001 ± 0.003	0.001 ± 0.0	0.001 ± 0.0	0.001 ± 0.0	3.4 (2), p > 0.05
	Blackbuck	0.001 ± 0.0	0.001 ± 0.0	0.001 ± 0.0	0.001 ± 0.0	0.7 (2), p > 0.05
	Chinkara	0.002 ± 0.0	0.001 ± 0.0	0.001 ± 0.0	0.001 ± 0.0	2.7 (2), p > 0.05
	Wild boar	0.001 ± 0.0	0.001 ± 0.001	0.001 ± 0.0	0.0001 ± 0.0	0.2 (2), p > 0.05
	Hyena	0.001 ± 0.0	0	0	0	7.1 (2), p > 0.05
	Overall	0.008 ± 0.000	0.005 ± 0.001	0.008 ± 0.001	0.007 ± 0.0004	16.25(2), <0.01

highest average abundance of Chital and Wild Boar (0.004 ± 0.0 and 0.04 ± 0.0, respectively). However, the medium disturbed area had the highest mean abundance of Nilgai, Hare, and Elephant (0.003 ± 0.0, 0.002 ± 0.0, and 0.005 ± 0.0, respectively). Overall, Chital had the highest mean pellet abundance among the sanctuary's sampled plots. The Kruskal-Wallis test revealed a significant difference in mean pellet abundance of mammals among categorized sites based on the level of human disturbance within the sanctuary. The mean abundance of Sambar pellets in Kaimoor Wildlife Sanctuary was highest in the medium disturbed area (0.002 ± 0.0). The sanctuary's minimal disturbance area has the highest mean pellet group abundance of Nilgai (0.004 ± 0.0). The Kruskal-Wallis test demonstrated a significant difference in the average pellet group abundance of Chital, Sambar, and Nilgai across the sanctuary's high, medium, and low disturbed sections. However, there was no significant variation in mean pellet group abundance between the sanctuary's three categorized disturbed sites for Jackal, Blackbuck, Chinkara, Wild Boar, and Hyena.

4.3 Area Preference

Table 3 shows the area preferences of mammals in sampled plots inside the Katarniaghat Wildlife Sanctuary, which are equal to habitat preferences [21]. Chital liked the sanctuary's low disturbed areas (0.53 ± 0.1), followed by medium and high disturbed areas (0.24 ± 0.06 and 0.144 ± 0.07, respectively). Hare, elephant, and wild boar preferred low disturbed areas (0.84 ± 0.24, 0.52 ± 0.13, and 0.7 ± 0.12, respectively), but nilgai selected medium disturbed areas (0.08 ± 0.06). Barking deer showed a preference for low disturbance locations in the sanctuary (0.092 ± 0.3). The sanctuary's low disturbed area had the highest estimated area preference (0.52 ± 0.07). The Kruskal-Wallis test identified a substantial variation in area choice across recorded mammals in accordance with the pellet. In Kaimoor, Chital preferred low disturbed regions (0.75 ± 0.2), while Sambar preferred high disturbed areas (0.5 ± 0.13). Nilgai, jackal, blackbuck, wild boar, and hyena favored the most disturbed site (0.32 ± 0.04, 0.4 ± 0.12, 0.3 ± 0.1, 0.3 ± 0.16, and 0.63 ± 0.3, respectively). Overall, mammals preferred the sanctuary's minimal disturbed site the most.

The Kruskal-Wallis test results revealed a substantial difference in area preference among the pellet group of chital, sambar, and nilgai. However, no substantial variation was identified in area selection among jackal, blackbuck, chinkara, wild boar, and hyena.

4.4 Encounter Rate Estimation

Table 4 shows the mean encounter rates and chi-square values for mammalian species observed during transect monitoring at Katarniaghat Wildlife Sanctuary.

Table 3 Mean area preference of mammalian species [20]

Sanctuary name	Species	High	Medium	Low	Overall	Chi-square (degree of freedom), Significance level
Katarniaghat	Chital	0.144 ± 0.07	0.24 ± 0.06	0.53 ± 0.1	0.3 ± 0.04	16.6 (2), $p < 0.01$
	Barking deer	0	0	0.092 ± 0.3	0.3 ± 0.1	20.57 (2), $p < 0.01$
	Nilgai	0.17 ± 0.1	0.08 ± 0.06	0.6 ± 0.15	0.3 ± 0.06	14.9 (2), $p < 0.01$
	Hare	0	0.07 ± 0.07	0.84 ± 0.24	0.3 ± 0.08	19.1 (2), $p < 0.01$
	Elephant	0.28 ± 0.1	0.1 ± 0.06	0.52 ± 0.13	0.3 ± 0.06	7.1 (2), $p < 0.05$
	Wild boar	0.14 ± 0.06	0.05 ± 0.04	0.7 ± 0.12	0.3 ± 0.05	34.5 (2), $p < 0.01$
	Overall	0.2 ± 0.07	0.2 ± 0.04	0.52 ± 0.07	0.3 ± 0.03	18.1 (2), $p < 0.01$
Kaimoor	Jackal	0	0	0.75 ± 0.2	0.25 ± 0.1	24.7 (2), $p < 0.001$
	Chital	0.15 ± 0.01	0.07 ± 0.05	0.5 ± 0.13	0.25 ± 0.05	11.4 (2), $p < 0.05$
	Sambar	0.32 ± 0.04	0.2 ± 0.05	0.24 ± 0.04	0.25 ± 0.05	13.5 (2), $p < 0.05$
	Nilgai	0.4 ± 0.12	0.15 ± 0.1	0.2 ± 0.1	0.25 ± 0.05	3.4 (2), $p > 0.05$
	Blackbuck	0.3 ± 0.1	0.23 ± 0.1	0.2 ± 0.01	0.25 ± 0.05	0.67 (2), $p > 0.05$
	Chinkara	0.1 ± 0.07	0.4 ± 0.13	0.3 ± 0.1	0.25 ± 0.1	2.7 (2), $p > 0.05$
	Wild boar	0.3 ± 0.16	0.18 ± 0.13	0.3 ± 0.16	0.25 ± 0.1	0.26 (2), $p > 0.05$
	Hyena	0.63 ± 0.3	0.12 ± 0.1	0	0.25 ± 0.1	7.1 (2), $p > 0.05$
	Overall	0.28 ± 0.03	0.17 ± 0.03	0.3 ± 0.02	0.25 ± 0.01	16.25 (2), $p < 0.01$

Due to fewer sightings of mammals, encounter rates were estimated rather than density estimation. The sanctuary's least affected section had the highest chital contact rate (2.7 ± 0.8%).

The medium had the greatest encounter rate per kilometer with wild boar, while the low disturbed area had the lowest encounter rate (0.74 ± 0.45 and 0.22 ± 0.14, respectively). Similarly, barking deer had the highest contact rate in the medium, followed by the least disturbed area. Jackal contact rates were only recorded in the least affected area (0.08 ± 0.03). At 3.6 ± 0.4, the low disturbed region of the sanctuary had the greatest encounter rate per kilometer among all mammals. The Kruskal-Wallis test revealed a significant difference in the mean encounter rate per

Table 4 Mean encounter rate of mammalian species [20]

Sanctuary name	Species	High	Medium	Low	Overall	Chi-square (degree of freedom), Significance level
Katarniaghat	Chital	0.7 ± 0.28	1.08 ± 0.3	2.7 ± 0.8	1.6 ± 0.4	11.4 (2), $p < 0.05$
	Wild boar	0	0.74 ± 0.45	0.22 ± 0.14	0.33 ± 0.16	4.4 (2), $p > 0.05$
	Barking deer	0	0.04 ± 0.02	0.03 ± 0.02	0.02 ± 0.01	1.45 (2), $p > 0.05$
	Nilgai	0.1 ± 0.04	0.02 ± 0.02	0	0.02 ± 0.01	5.7 (2), $p > 0.05$
	Jackal	0	0	0.08 ± 0.03	0.03 ± 0.01	7.2 (2), $p < 0.05$
	Langur	0	0.5 ± 0.32	0.16 ± 0.1	0.22 ± 0.1	1.8 (2), $p > 0.05$
	Overall	0.91 ± 0.3	3.7 ± 0.7	3.6 ± 0.43	2.9 ± 0.43	25.4 (2), $p < 0.01$
Kaimoor	Nilgai	0.8 ± 0.28	0.64 ± 0.3	0.2 ± 0.1	0.6 ± 0.1	5.7 (2), $p > 0.05$
	Jackal	0	0.02 ± 0.01	0.13 ± 0.04	0.05 ± 0.01	15.1 (2), $p < 0.01$
	Langur	0.21 ± 0.14	0.21 ± 0.2	0.1 ± 0.07	0.2 ± 0.1	0.87 (2), $p > 0.05$
	Blackbuck	0.8 ± 0.4	0.25 ± 0.1	0	0.4 ± 0.1	3.7 (2), $p > 0.05$
	Overall	1.92 ± 0.5	1.16 ± 0.37	0.61 ± 0.12	1.26 ± 0.22	2.9 (2), $p > 0.05$

kilometer recorded by chital and jackal in the sanctuary's disturbed sections (high, medium, and low). In the sanctuary's categorized site, transect monitoring revealed no discernible variation in the mean encounter rate per kilometer for wild boar, barking deer, nilgai, and langur.

In Kaimoor Wildlife Sanctuary, the mean encounter rate of nilgai was highest in the high, followed by the medium and low disturbed areas (0.8 ± 0.28, 0.64 ± 0.3, and 0.2 ± 0.1). Jackal had the highest encounter rate in the low disturbed area (0.13 ± 0.04). When examining the total encounter rate per kilometer for various mammalian species, nilgai exhibited a dominant rate of 0.6 ± 0.1. In regions with significant human intervention, the mean encounter rate per kilometer for all mammals was highest (1.92 ± 0.5). The Kruskal-Wallis test results revealed an important difference in the mean encounter rate of jackals throughout the sanctuary's three designated locations.

4.5 Group Size Estimation

Table 5 shows the mean group size of recorded mammals along the sampled transect in high, medium, and low disturbed sections of the sanctuary. While the mean group size of langur, wild boar, and barking deer was largest in the medium disturbed area, the mean group size of chital was highest in the low disturbed region (2.7 ± 0.8).

In the sanctuary, the average group size of chital was the largest at 1.7 ± 0.4. The sanctuary's minimal disturbance area has the highest mean group size for all mammals (3.6 ± 0.4). The Kruskal-Wallis test revealed a significant difference in the mean group size of jackal, langur, and chital in the sanctuary's three categories. In Kaimoor Wildlife Sanctuary, jackal, langur, and blackbuck had the largest mean group sizes across the three disturbed sites (2.02 ± 0.7, 0.45 ± 0.03, and 1.7 ± 0.9, respectively). The mean group size was largest in locations with high human disturbance (4.05 ± 1.05). The Kruskal-Wallis test identified a significant difference in the mean group size of nilgai and langurs during transect monitoring at the sanctuary's three categorized sites.

Table 5 Mean group size of mammalian species [20]

Sanctuary name	Species	High	Medium	Low	Overall	Chi-square (degree of freedom), Significance level
Katarniaghat	Nilgai	0.1 ± 0.04	0.02 ± 0.02	0	0.02 ± 0.01	5.7(2), $p > 0.05$
	Jackal	0	0	0.08 ± 0.03	0.03 ± 0.01	7.2(2), $p < 0.01$
	Langur	0.13 ± 0.1	1.4 ± 0.6	0.3 ± 0.15	0.6 ± 0.2	8.5 (2), $p < 0.05$
	Chital	0.7 ± 0.3	1.1 ± 0.3	2.7 ± 0.8	1.7 ± 0.4	11.4 (2), $p < 0.05$
	Wild boar	0	0.74 ± 0.45	0.2 ± 0.14	0.03 ± 0.02	4.4(2), $p > 0.05$
	Barking deer	0	0.04 ± 0.02	0.03 ± 0.02	0.02 ± 0.01	1.4(2), $p > 0.055$
	Overall	0.91 ± 0.3	3.7 ± 0.7	3.6 ± 0.43	2.9 ± 0.43	25.4 (2), $p < 0.01$
Kaimoor	Nilgai	0.8 ± 0.28	0.64 ± 0.3	0.2 ± 0.1	0.6 ± 0.1	5.7 (2), $p > 0.05$
	Jackal	0	0.02 ± 0.01	0.13 ± 0.04	0.05 ± 0.01	15.1 (2), $p < 0.01$
	Langur	0.21 ± 0.14	0.21 ± 0.2	0.1 ± 0.07	0.2 ± 0.1	0.87 (2), $p > 0.05$
	Blackbuck	0.8 ± 0.4	0.25 ± 0.1	0	0.4 ± 0.1	3.7 (2), $p > 0.05$
	Overall	1.92 ± 0.5	1.16 ± 0.37	0.61 ± 0.12	1.26 ± 0.22	2.9 (2), $p > 0.05$

5 Discussions

Pellet group count and line transect methods were utilized in this study to measure animal status and distribution in relation to human-induced disturbance. Notwithstanding their respective shortcomings, both strategies were applied independently in this work to estimate group size, encounter rate, density, abundance, and preferred area. There seemed to be an association between human activity levels and the species densities in the sampled plots and transects in Katarniaghat. For example, barking deer were only found in regions with lower levels of human disturbance, suggesting that they preferred these settings. This result is in line with previous studies conducted by [22], who found that in order to reduce the risk of predation, barking deer select areas with a lot of canopy cover. Barking deer favor areas that are not used by people directly [23]. A similar pattern was found in the abundance estimation of numerous species in the sampled plots in Katarniaghat, with the lowest disturbed parts sustaining the highest abundance of pellet groups from different animal species. Similar conclusions regarding the majority of mammal density and its association with human disturbance were drawn from Kaimoor. A research conducted in Manas National Park after the current study in two protected areas shows the detrimental effects of anthropogenic disturbance on the density of ungulates [24]. Nonetheless, nilgai's mean pellet group density suggested a predilection for areas with significant disturbance. Since nilgai are known to prefer open spaces and are commonly observed close to human settlements, it is possible that their preference for severely damaged habitats stems from the availability of foraging opportunities in areas with changing grass cover and proximity to nutrient-rich crops. Unaffected by human activity, jackals and hyenas were seen in severely damaged regions, probably attracted by the presence of domestic animals.

The findings reveal how mammal species and human disturbance interact dynamically in the Kaimoor area. While certain species react reliably to human activities, others exhibit more nuanced activities driven by habitat preference, foraging possibilities, and predator–prey dynamics. Effective conservation and management processes aim at minimizing the adverse impacts of human activity on populations of wildlife in the Kaimoor and comparable landscapes depend on an understanding of these dynamics [25].

Blackbuck, being browsers, showed a preference for areas with medium intensity of human use, likely supporting areas with a higher population of seedlings. Chital, known to be favored by grass availability, exhibited a preference for grassy patches. In a previous research, same effects of human interference on populations of chital in terms of spatial segregation and population retrieval after a population decline of domestic animals [26]. Livestock grazing can have significant impacts on chital populations by altering habitat structure, resource availability, and interspecific competition [27, 28]. Understanding these effects is crucial for effective conservation and management of chital populations, particularly in human-subjugated landscapes where human–wildlife conflicts are prevalent. A study on patch-burn grazing practice and its impact on small mammals concluded it as an effective habitat restoring strategy [19]. Sambar, known to utilize open grasslands and forest patches, unveiled preference for areas not directly utilized by humans. In the present study, it showed an association toward low disturbed sites, supporting a heterogeneous vegetational stratum. In other studies, wild boar, known to tolerate logging effects, showed a preference for human interfered areas. In contrast to previous findings, the respective animal was highly associated with low disturbed sites.

Overall, the results regarding area preference by mammals indicate that sites with high levels of human activity are not generally preferred in both study areas. [29] found spatial and temporal exclusion of wildlife with human presence in Western Himalaya. [30–34] in their investigation found that anthropogenic disturbance in an area governs the species distribution. Similarly in the present study, the wildlife of Katarniaghat showed their preference toward site of minimal human interference, whereas in Kaimoor nilgai, jackal, blackbuck, and hyena preferred areas with high intensity of human use. These differences in habitat preferences/site selection among mammal species in the present study highlight the complexity of human–wildlife interactions and the importance of considering species-specific ecological requirements and behaviors in conservation planning and management. Because, extensive usage of an area by humans brings changes in the behavioral patterns of wildlife by switching to foraging in night to avoid human activities in the daytime, thus affecting diet overlap with other species [17, 35]. Similar outcome was documented from the Brazilian and Nepal forests [8, 36].

The estimation of encounter rate of recorded animal species is crucial for understanding their population status in relation to human disturbance and for healthier management and species conservation planning. It is a fundamental tool in wildlife ecology and conservation, providing essential data for evidence-based decision-making, adaptive management, and long-term sustainability of ecosystems and the species they support. [5] in their study documented about indicator species of Bornean forests in the context of anthropogenic disturbance. In Kaimoor, the highest encounters of nilgai were estimated in high disturbed areas of the sanctuary. Conversely, the low disturbed site of Katarniaghat supported the highest encounters of chital and jackal, with significant differences compared to sites with varying levels of human use. Overall to draw the attention, mammal encounters were highest in the low disturbed site of Katarniaghat and the high disturbed site of Kaimoor. This observation highlights the contrasting patterns of mammal distribution and habitat preference between the two study areas. These findings align with previous research, such as [9, 37], which suggested that canopy openings improve leaf quality and increase foliage biomass in disturbed forests, providing important resources for herbivores.

Group size is crucial for understanding individual's fitness activities such as foraging efficiency, reducing predation risks, social interactions, communication increasing vigilance and reproductive success [38, 39]. In Kaimoor, the occurrence of the highest mean group size of some species was supported by high disturbed areas. Moreover, the overall highest mean group size was estimated from high disturbed areas, indicating increased vigilance activities coupled with low predation risks. Conversely, most species with maximum mean group size were associated with areas of low human activity in Katarniaghat, with significant differences for jackal, langur, and chital. This could be attributed to larger group sizes facing increased competition for food, resulting in reduced foraging success. Therefore, presumably low disturbed sites support more food resources resulting in increased foraging success thus attracting larger groups of wild animals compared to highly disturbed sites. The findings of [40] came with a similar conclusion on ungulates that their group size is affected due to different ecological conditions driven by human interference.

6 Conclusions

Based on findings from both the sanctuaries, certain points need to be addressed. The local communities should be switched to alternate livelihood activities such as traditional handicraft production, eco-tourism ventures, non-timber forest products cultivation, small-scale enterprises by coordinating with their respective stakeholders. Implementation of educational programs and awareness campaigns regarding the importance of forests and its resources to inform local communities and engage community members in conservation activities, workshops, and awareness-raising events to foster a sense of ownership and stewardship toward the wildlife sanctuary. The prominence of community-based resource management through a participatory decision-making process should be actively done. Capacity-building programs should be encouraged by coordinating with government agencies, NGOs, and research institutions. By implementing these remedial measures in a holistic and participatory manner, it is possible to reduce forest resource dependency on mammalian communities in a wildlife sanctuary while promoting sustainable development and biodiversity conservation goals.

References

1. Allred BW, Fuhlendorf SD, Hamilton RG (2011) The role of herbivores in great plains conservation: comparative ecology of bison and cattle. Ecosphere 2(3):1–17
2. Blom A, Zalinge R, Heitkoing IMA, Prins HHT (2012) Factors influencing the distribution of large mammals within the protected African central forest. Oryx 39(4):381–388
3. Champion HG, Seth SK (1968) A revised survey of the forest types of India. Government of India Publications, Delhi, p 200
4. Chandra S (2010) Management plan for Kaimoor wildlife sanctuary. Wildlife Preservation Organization Forest Department, Lucknow, p 98
5. Cheyne SM, Sastramidjaja WJ, Ryadin MY, Macdonald DW (2016) Mammalian communities as indicators of disturbance across Indonesian Borneo. Global Ecol Conserv 7:157–173. https://doi.org/10.1016/j.gecco.2016.06.002
6. Dave C, Jhala Y (2011) Is competition with livestock detrimental for wild ungulates? A case study of chital (Axis axis) in Gir forest, India. J Trop Ecol 27(3):239–247. http://www.jstor.org/stable/23012360

7. Du-Plessis MA (1995) The effects of fuelwood removal on the diversity of some cavity using birds and mammals in South Africa. Biol Conserv 74(2):77–82
8. Ewart HE, Pasqualotto N, Paolino RM, Jensen K, Chiarello AG (2024) Effects of anthropogenic disturbance and land cover protection on the behavioral patterns and abundance of Brazilian mammals. Global Ecol Conserv 50:e02839. https://doi.org/10.1016/j.gecco.2024.e02839
9. Herrera J, Wright P (2011) The effects of habitat disturbance on Lemurs at Ranomafana National Park, Madagascar. Int J Primatol 32:1091–1108
10. Mammides C, Cords M, Peters MK (2008) Effects of habitat disturbance and food supply on population densities of three primate species in the Kakamega Forest, Kenya. Afr J Ecol 47(1): 87–96
11. Liedloff AC, Coughenour MB, Ludwig JA, Dyer R (2001) Modelling the trade-off between fire and grazing in a tropical savanna landscape, northern Australia. Environ Int 27:173–180. PMID:11697666
12. White LJT (1994) Biomass of rainforests mammals in the lope reserve, Gabon. J Ani Ecol 63: 499–512
13. Johns AD, Skorupa JP (1987) Responses of rainforest primates to habitat disturbance- a review. Int J Primatol 8:157–191
14. Stokes EJ, Strindberg S, Bakabana PC, Elkan PW, Iyenguet FC, Madzoke B, Malanda GAF, Mowawa BS, Moukoumbou C, Ouakabadio FK, Rainey HJ (2010) Monitoring great ape and elephant abundance at large spatial scale: making effectiveness of a conservation landscape. PLoS One 5(4):e10294. https://doi.org/10.1371/journal.pone.0010294
15. Strassmann BI (1987) Effects of cattle grazing and haying on wildlife conservation at National Wildlife Refuges in the United States. Environ Manag 11(1):35–44
16. Jha RN (2000) Management plan of Katerniaghat Wildlife Sanctuary. Forest Department, Uttar Pradesh, Lucknow, p 274
17. Lee E, Etten RCV (1956) Evaluation of the pellet group count as the deer census method. J Wildlife Manag 20(1):70–74
18. Neff L (1968) The pellet group count technique for big game trend, census and distribution: a review. J Wildlife Manag 32(3):597–614
19. Rogers G, Julander O, Robinette WL (1958) Pellet group count for deer census and range use index. J Wildlife Manag 22(2):193–199
20. Tahoor A (2017) An ecological study of protected area relationships and its implication for wildlife cosneravtion and mnagment in selcetd protected areas of Uttra Pradesh, India. Thesis. Aligarh Muslim University
21. Pokhrel CP (1996) Food habit and habitat utilization of swamp deer (Cervus duvauceli duvauceli) in the Royal Bardia National Park, Nepal. M.Sc. dissertation. Tribhuvan University, Kathmandu, p 38
22. Nagarkoti A, Thapa TB (2007) Distribution pattern and habitat preference of barking deer (Muntiacus muntjak) in Nagarjun forest, Kathmandu. Himalayan J Sci 4(6):70–74
23. Kushwaha SPS, Khan A, Habib B, Quadri A, Singh A (2004) Evaluation of sambar and Muntkak habitats using geospatial modeling. Curr Sci 86(10):1390–1400
24. Lahkar D, Ahmad MF, Begum RH, Das SK, Harihar A (2020) Responses of wild ungulate assemblage to anthropogenic influences in Manas National Park. Biol Conserv 243:108425. https://doi.org/10.1016/j.biocon.2020.108425
25. Laidlaw RK (2009) Effects of habitat disturbance on protected areas on mammals of peninsular Malaysia. Conserv Biol 14(6):1639–1648
26. Srivastava N, Krishnamurthy R, Sathyakumar S (2020) Avoidance or coexistence? The spatio-temporal patterns of wild animals in a human dominated landscape in the Western Hiamalaya. Mt Res Dev 40(2):R20–R31. https://doi.org/10.1659/MRD-JOURNAL-D-19-00046.1
27. Mathai M (1999) Habitat occupancy across anthropogenic disturbances by sympatric ungulate species in Panna Tiger Reserve. Dissertation. Saurashtra University, Rajkot, p 75

28. Mathur PK, Kumar H, Lehmkuhl JF, Tripathi A, Sawarkar VB, De R (2010) Mammal indicator species for protected areas and managed forests in a landscape conservation area of northern India. Biodivers Conserv 20:1–17. https://doi.org/10.1007/s10531-010-9851-8
29. Madhusudan M (2004) Recovery of wild herbivores following livestock decline in a tropical Indian wildlife reserve. J Appl Ecol 41:858–869. https://doi.org/10.1111/j.0021-8901.2004.00950.x
30. Lamichhane S, Khanal G, Karki JB, Aryal C, Acharya S (2020) Natural and anthropogenic correlates of habitat use by wild ungulates in Shuklaphanta National Park, Nepal. Global Ecol Conserv 24:e01338. https://doi.org/10.1016/j.gecco.2020.e01338
31. Radford IJ, Gibson LA, Corey B, Carnes K, Fairman R (2015) Influence of fire mosaics, habitat characteristics and cattle disturbance on mammals in fire-prone savanna landscapes of the northern Kimberley. PLoS One 10(6):e0130721. https://doi.org/10.1371/journal.pone.0130721
32. Ricketts AM, Sandercock BK (2016) Patch-burn grazing increases habitat heterogeneity and biodiversity of small mammals in managed rangelands. Ecosphere 7(8):e01431. https://doi.org/10.1002/ecs2.1431
33. Tudge SJS, Brittain F, Kentatchime T, C.T.K., Rowcliffe J.M. (2022) The impacts of human activity on mammals in a community forest near the Dja biosphere Reserve in Cameroon. Oryx 56(6):974–955. https://doi.org/10.1017/S0030605321000806
34. Valerie JH, Chambers CL, Dickson BG (2019) Grazing and wildfire effects on small mammals inhabiting montane meadows. J Wildlife Manag 83(3):534–543
35. Sulaksono N, Pudyatmoko S, Sumardi S, Wardhana W, Budiman A (2023) The effect of anthropogenic disturbances on the spatiotemporal patterns of meidium-large mammals in Tropical volcanic landscapes. Animals 13(20):3217. https://doi.org/10.3390/ani13203217
36. Kandel RC (2012) Wildlife use of Barandhabar forest corridor: between Chitwan National Park and Mahabharat foothills, central Terai, Nepal. J Ecol Nat Environ 4(5):119–125
37. Lawes MJ, Murphy BP, Fisher A, Woinarski JCZ, Edwards A, Russell-Smith J (2015) Small mammalsdecline with increasing fire extent in northern Australia: evidence from long-term monitoring in Kakadu National Park. Int J Wildland Fire 24:712–722
38. Treves A, Chapman CA (1996) Conspecific threat, predation avoidance, and resource defense: implications for grouping in langurs. Behav Ecol Sociobiol 39:43–53
39. Lee SXT, Amir Z, Moore JH, Gaynor KM, Luskin MS (2024) Effect of human distiurbances on wildlife behaviour and consequences of prey-predator overlap in Southeast Asia. Nat Commun 15:1521. https://doi.org/10.1038/s41467-024-45905-9
40. Rafael RH, Harald B, Altrichter M, Chapman CA, Bonnell TR, Desbeiz AAL, Moreira-Ramirez JF, O'Farrill G, Fragoso J, Naranjo EJ (2016) What ecological and anthropogenic factors affect group size in White-lipped peccaries. Biotropica 48(2):246–254

SPRINGER NATURE

GPSR Compliance

The European Union's (EU) General Product Safety Regulation (GPSR) is a set of rules that requires consumer products to be safe and our obligations to ensure this.

If you have any concerns about our products, you can contact us on ProductSafety@springernature.com

In case Publisher is established outside the EU, the EU authorized representative is:

Springer Nature Customer Service Center GmbH
Europaplatz 3
69115 Heidelberg, Germany

The manufacturer's authorised representative in the EU is Springer Nature Customer Service Centre GmbH, Europaplatz 3, 69115 Heidelberg, Germany. If you have any concerns regarding our products, please contact ProductSafety@springernature.com

Printed and bound by CPI Group (UK) Ltd, Croydon, CR0 4YY
01/10/2026
02225940-0004